Advances in
ENVIRONMENTAL SCIENCES

Volume 1

Advances in

ENVIRONMENTAL SCIENCES

Edited by

JAMES N. PITTS, JR.
University of California
Riverside, California

and

ROBERT L. METCALF
University of Illinois
Urbana, Illinois

Volume 1

WILEY-INTERSCIENCE
A DIVISION OF JOHN WILEY & SONS
NEW YORK / LONDON / SYDNEY / TORONTO

10 9 8 7 6 5 4 3 2

Library of Congress Catalog Card Number 69-18013
SBN 471 69080 5

Printed in the United States of America

INTRODUCTION TO THE SERIES

Advances in Environmental Sciences and Technology is a series of multiauthored books devoted to the study of the quality of the environment and to the technology of its conservation. Environmental sciences relate, therefore, to the chemical, physical, and biological changes in the environment through contamination or modification; to the physical nature and biological behavior of air, water, soil, food, and waste as they are affected by man's agricultural, industrial, and social activities; and to the application of science and technology to the control and improvement of environmental quality.

The deterioration of environmental quality, which began when man first assembled into villages and utilized fire, has existed as a serious problem since the industrial revolution. In the second half of the twentieth century, under the ever-increasing impacts of exponentially growing population and of industrializing society, environmental contamination of air, water, soil, and food has become a threat to the continued existence of many plant and animal communities of the ecosystem and may ultimately threaten the very survival of the human race.

It seems clear that if we are to preserve for future generations some semblance of the existing biological order and if we hope to improve on the deteriorating standards of urban public health, environmental sciences and technology must quickly come to play a dominant role in designing our social and industrial structure for tomorrow. Scientifically rigorous criteria of environmental quality must be developed and, based in part on these, realistic standards must be established, so that our technological progress can be tailored to meet such standards. Civilization will continue to require increasing amounts of fuels, transportation, industrial chemicals, fertilizers, pesticides, and countless other products, as well as to produce waste products of all descriptions. What is urgently needed is a total systems approach to modern civilization through which the pooled talents of scientists and engineers, in cooperation with social scientists and the medical profession, can be focused on the development of order and equilibrium among the presently disparate segments of the human environment. Most of the skills and tools that are needed already exist. Surely a technology that has created manifold environmental problems is also capable of solving them. It is our hope that the series in Environmental Sciences and Technology will not only serve to make this challenge more explicit to the established professional but will also help to stimulate the student toward the career opportunities in this vital area.

Finally, the chapters in this series of Advances are written by experts in their respective disciplines, who also are involved with the broad scope of environmental science. As editors, we asked the authors to give their "points of view" on key questions; we were not concerned simply with literature surveys. They have responded in a gratifying manner with thoughtful and challenging statements on critical environmental problems.

James N. Pitts, Jr.
Robert L. Metcalf

Contents

Outline of Environmental Sciences. By Robert L. Metcalf, *University of Illinois, Urbana, Illinois* and James N. Pitts, Jr., *University of California, Riverside, California* 1

The Federal Role in Pollution Abatement and Control. By John Tunney, *Congress of the United States* 25

Our Nation's Water: Its Pollution Control and Management. By Cornelius W. Kruse, *The Johns Hopkins University, School of Hygiene and Public Health, Baltimore, Maryland* 41

Oxides of Nitrogen. By Edward A. Shuck and Edgar R. Stephens, *Statewide Air Pollution Research Center, University of California, Riverside, California* 73

The Formation, Reactions, and Properties of Peroxyacyl Nitrates (PANS) in Photochemical Air Pollution. By Edgar R. Stephens, *Statewide Air Pollution Research Center, University of California, Riverside, California* 119

Biodegradable Detergents and Water Pollution. By Theodore E. Brenner, *Technical and Materials Division, The Soap and Detergent Association, New York, New York* 147

Aeroallergens and Public Health. By William R. Solomon, *University of Michigan, Ann Arbor, Michigan* 197

Catalytic Removal of Potential Air Pollutants from Auto Exhaust. By Robert H. Ebel 237

Photochemical Air Pollution: Singlet Molecular Oxygen as an Environmental Oxidant. By James N. Pitts, Jr. 289

Author Index 339

Subject Index 351

Outline of Environmental Sciences

ROBERT L. METCALF
Departments of Entomology and Zoology,
University of Illinois,
Urbana, Illinois
AND
JAMES N. PITTS, JR.
Department of Chemistry,
University of California,
Riverside, California

I. Introduction . . . 1
II. Man and Environmental Quality . . . 2
A. Urbanization . . . 3
B. Industrialization . . . 4
C. Chemical Contamination . . . 5
III. Air . . . 7
IV. Water . . . 12
A. Eutrophication . . . 12
B. Microchemical Pollution . . . 13
C. Thermal Pollution . . . 14
D. Sewage . . . 14
V. Soil . . . 14
A. Pesticides . . . 15
B. Fertilizers . . . 16
VI. Waste . . . 17
A. Animal Wastes . . . 18
B. Radionuclides . . . 18
VII. Ecology and the Environment . . . 19
VIII. Public Health and the Environment . . . 20
A. Food . . . 21
IX. A Global View of Pollution . . . 22
References . . . 23

I. INTRODUCTION

The human environment has been defined as the aggregate of all social, biological, and physical or chemical factors which comprise the surroundings of man. Environmental Science, as it applies to this Advances Series, is defined in more restrictive terms as basic and applied inquiry about changes in environmental quality resulting from the activities of man. Environmental Science therefore relates to (*a*) the chemical, physical, and biological changes in the environment through contamination or modification; (*b*) the chemical nature and biological behavior of air, water, soil, food, and waste as they are affected by man's agricultural, industrial, and social activities; and (*c*) the application of the natural sciences and technology as well as the social sciences, including political science and administration, to the control and improvement of environmental quality. From our anthropomorphic viewpoint there are three environments: (*1*) the internal environment of the body; (*2*) the immediate environment of air breathed, water drunk, and food eaten; and (*3*) the general environment of the earth as a whole.

Environmental contamination is the most drastic of man's alterations in environmental quality. It may be (*a*) *accidental* from conflagration, explosion, industrial effluent, auto exhaust, and from sewage and other wastes; (*b*) *deliberate* from trials of nuclear weapons, agricultural or industrial burning; or (*c*) *purposeful* from the application of fertilizers, pesticides, pharmaceuticals, and food preservatives.

In the following overview of environmental science, we have attempted to sketch some of the significant problems arising from the unprecedented rate of increase of the human species and from the impact of the agricultural and industrial revolutions which have brought it about. Obviously, we have touched upon but a few of the thousands of important questions for which early answers must be found if man is to be able to preserve a substantial amount of the quality of his natural environment. The endeavor is a worthy challenge to the resourcefulness of science and technology. The authors hope this Series will not only serve to make this challenge more explicit but also will help to stimulate the student to the career opportunities in this vital field as well as related areas in the social sciences and administration.

II. MAN AND ENVIRONMENTAL QUALITY

The deterioration of environmental quality, which started when man first began to collect into villages and to utilize fire, has existed as a serious problem only since the industrial revolution. The boundary conditions of this problem may be broadly defined as the confinement of a human population growing at a rate which will double its numbers within 30 years (Fig. 1) in a biosphere of fixed dimensions, encompassed in a layer about 10 miles thick over the 200 million square miles of earth surface. In the United States, generally considered to be bountifully

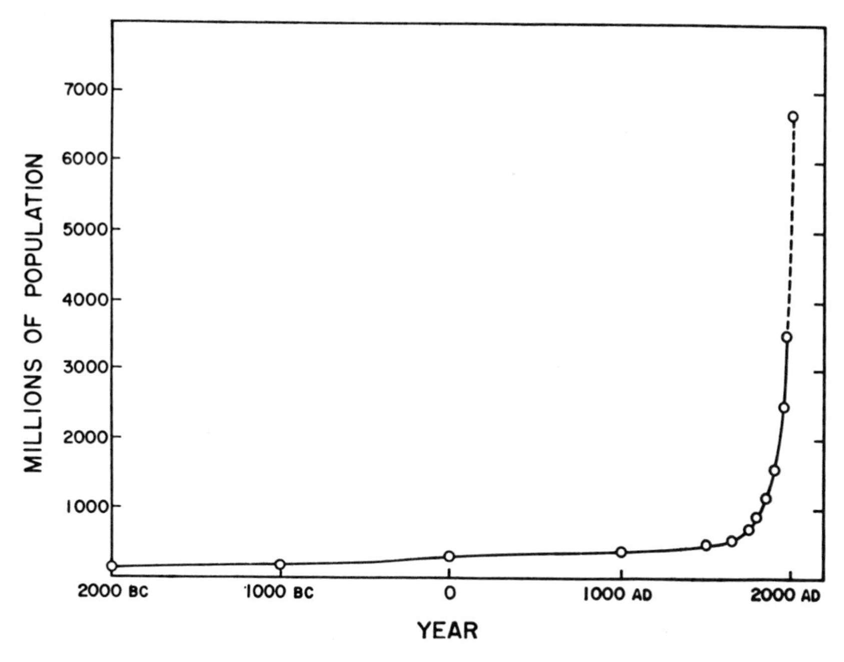

Fig. 1. Growth of world population over four thousand years (U.S. Dept. of Agriculture projections).

endowed with wide open spaces, the population has doubled in less than 50 years and is expected to double again by year 2030 (Table I). It is readily apparent that the materialistic progress of the rapidly expanding U.S. society has a progressively increasing impact on the quality of the environment and that we have entered into a period picturesquely described by *Time Magazine* as the "Age of Effluence" (May 10, 1968).

A. Urbanization

Many of the problems of environmental quality relate to modern man's relatively recent drive to abandon spacious rural living and to congregate into coalescing urban masses. The driving forces appear to be industrialization with advanced economic opportunity, suburban transportation, and the agricultural revolution through which the farmer who in 1850 could feed four is now able to feed thirty. The urban trend is relatively recent. Thus in 1920 only 14% of the world population was urban. This increased to 19% by 1940 and 25% by 1960. Recent estimates suggest that by 1980, 33%, and by 2000, 44% of the population will live in urban areas(8). In the U.S. the rush toward urbaniza-

TABLE I

Growth of the Population of the Continental United States

Year	Population[a]	Inhabitants per square mile
1800	5,308,483	6.1
1810	7,230,881	4.3
1820	9,638,453	5.5
1830	12,866,020	7.3
1840	17,069,453	9.7
1850	23,191,876	7.9
1860	31,443,321	10.6
1870	39,818,449	13.0
1880	50,155,783	16.9
1890	62,947,714	21.2
1900	75,994,575	25.6
1910	91,972,266	30.9
1920	105,710,620	35.5
1930	122,775,046	41.1
1940	131,669,275	44.2
1950	150,697,361	50.6
1960	178,464,236	60.0
1967	200,000,000	67.2
2000 estimate	300,000,000	99.0

[a] Data from U.S. Census.

tion has been much more dramatic, and whereas in 1790, 95% of the population was rural, by 2000 about 85% will be urban, despite "urban blight" and the rush to the suburbs (Fig. 2). The consequences of this profound sociological change are apparent in our urban air and water contamination problems, in public health problems arising from the noise and physical stresses of crowded living, and in the general degeneration of human values inherent in urban slums.

B. Industrialization

The rise of the modern industrial society is generally credited as beginning with inventions of textile machinery, such as the flying shuttle by Kay in 1733 and the spinning jenny by Hargreaves in 1770, together with the development of the steam engine by Watt in 1769. In the ensuing 200 years, the fossil fuels, coal, lignite, petroleum, and natural gas, have not only supplied the industrial age with fuel and energy and the consequent products of combustion, but they have also been the

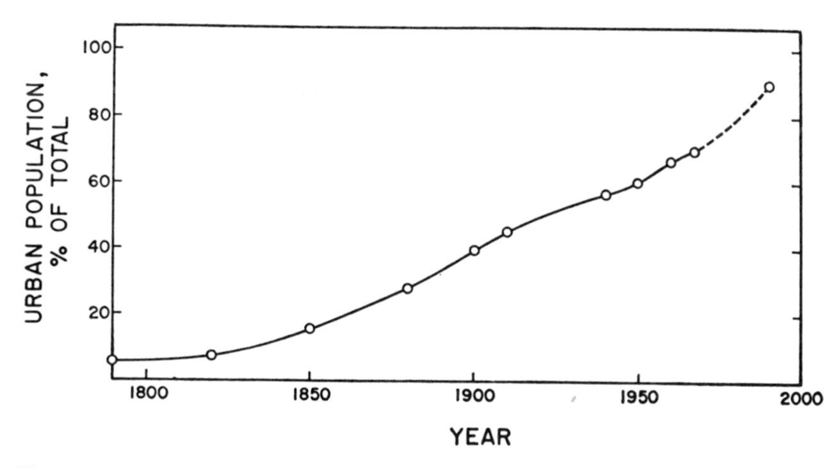

Fig. 2. Increase in percentage of population living in urban areas in the U.S. since 1790. (*Encyclopedia Brittanica;* (3)).

raw materials for the synthesis of the myriad of synthetic chemicals derived from coal tar and petroleum. Figure 3 shows the growth during the last century of coal production which has increased more than 12-fold and of petroleum production which has increased about 2000-fold. The President's Science Advisory Committee Report, *Restoring the Quality of Our Environment* (14), presents a remarkable discussion of the eventual environmental changes which may result from an annual 0.23% increase in the carbon dioxide content of the atmosphere resulting from combustion of these fuel sources. As shown in Figure 3, the rate of fossil fuel consumption is steadily increasing and the PSAC Report predicts that by the year 2000, a 14 to 30% increase in the present CO_2 content of the atmosphere may have occurred.

C. Chemical Contamination

Chemical contamination of the environment is the inevitable consequence of an expanding industrial society. The challenge of Environmental Science is to preserve maximum standards of environmental quality without unduly curtailing the materialistic approach to human welfare, which, if pressed, few of us would readily abandon. This challenge is immense, for human contributions to environmental contamination are already monumental. For example, the production of ferrous metals (pig iron) increased 100-fold from 1820 to 1947 (1 million to 98 million tons respectively; *Encyclopedia Britannica*). Its younger but environmentally longer lived rival, aluminum, has increased in availa-

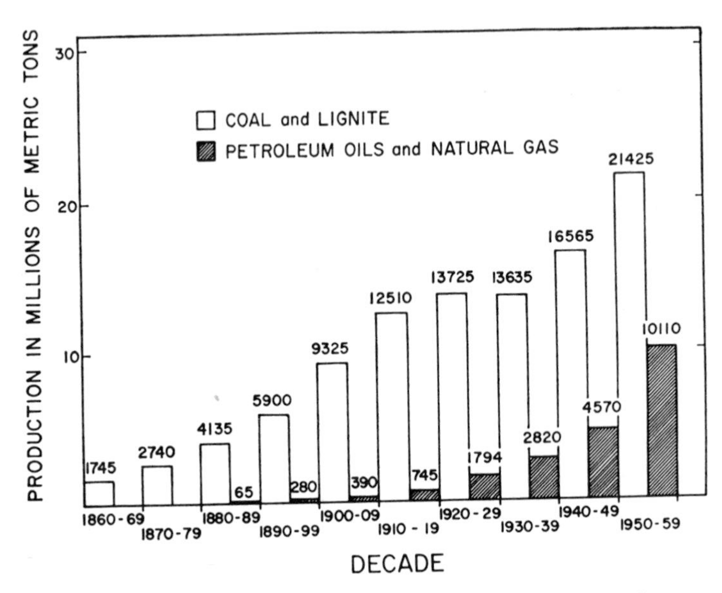

Fig. 3. World production of fossil fuels by decade since 1860; open bars, coal and lignite; shaded bars, petroleum and natural gas (14).

bility from 62,000 pounds in 1890 to 4.6 billion pounds in 1963. This has made possible the aluminum beer can which, unlike its tin-plated iron rival, does not eventually rust away but is virtually nondegradable. U.S. concrete production in 1963 was 407 billion pounds. This together with iron has erected the basic edifice of modern society and has made possible such convenient and life-saving developments as the 41,000 mile U.S. Interstate Highway system. However, ferro-concrete enters the environment much more easily than it is removed and it is worthy of reflection that this highway construction has permanently altered the character of more than 1.5 million acres of land! The age of plastics is at hand and while these may be less durable generally than iron and concrete, the eventual waste disposal problems from the products fabricated of the 15 billion pounds of plastic and resins produced in the U.S. during 1968, are challenging.

The production of synthetic organic chemicals in the U.S., as shown in Table II, has increased about 12-fold during the past 25 years, a rate which currently averages about 10% per year (*Chemical and Engineering News*, September 2, 1968). Included are fuels, plastics, plasticizers, fibers, elastomers, solvents, detergents, paints, pesticides, food additives, and pharmaceuticals. It has been estimated that more than 500 new

TABLE II

Production of Synthetic Organic Chemicals in the United States

Year	Billions of pounds[a]
1943	10
1953	24
1963	61
1964	77
1965	85
1966	100
1967	112
1968	122

[a] Approximations from U.S. Tariff Commission and *Chem. Eng. News*, Sept. 2, 1968.

and potentially toxic chemicals are produced each year on a scale large enough so that traces of them enter the environment through air, water, and indirectly, or directly into food (9). A WHO Report, *Microchemical Pollutants in the Environment* (18), points out that synthetic organic chemicals may be harmless, toxic, or carcinogenic. It is of great importance to determine the distribution, fate, and potentially hazardous effects of all of these synthetic organic chemicals in the environment, and to see that they are used properly and with due regard to environmental quality. Such studies necessitate development of suitable methods of microchemical analysis, determination of toxicological hazards from the ingestion of trace amounts, and assay of biodegradability and ecological distribution. The ultimate aim must be the establishment of criteria and standards for permissible levels of chemical contamination in water, air, and food.

III. AIR

Although often lyrically described as limitless, the air resource of the earth is the rather sharply defined region of the troposphere extending up from the earth's surface about 12 miles to the tropopause. The troposphere contains more than 95% of the gaseous constituents—nitrogen, oxygen, argon, carbon dioxide, neon, helium, krypton, xenon, ozone, radon, and water vapor. It is the region of weather, clouds, and rain which constantly dilute and scrub out natural and man-made pollution. Above the tropopause is the stratosphere which is clean and dry. With the absence of efficient scrubbing mechanisms, pollution in the stratosphere may remain for appreciable periods of time. One concern about future large-scale use of supersonic transport aircraft relates to possible long-term climatic effects of stratospheric pollution by particulate matter from jet engines.

On a more intimate level, our local air resources are restricted by meteorological and geographical effects which may cause surface stagnation and prevent the dilution of effluents arising from the earth's surface. The rate of the dilution in these local air sheds is affected by several factors including wind speeds and thermal convection currents which determine the mixing depth. Normally, daylight warming with rising air currents produces mixing depths of 700 to 2500 ft, but under conditions of high barometric pressure and cool surface air masses, basins of stagnant air characterized by strong daytime temperature inversions may form. These effectively trap air pollutants in zones of human activity; the classic Los Angeles type smog is one consequence of this phenomenon.

Historically, air pollution has been recognized as a major problem of urban man for more than three centuries! Thus in 1661 John Evelyn published *Fumifugium: or The Inconvenience of the Aer and Smoake of London Dissipated. Together with some Remedies humbly proposed.* Excerpts from the second printing of the book published in 1772 are shown in Figures 4–6. They point out qualitatively the nature of the problem (Fig. 4), illustrate an epidemiological approach (Fig. 5), and in Figure 6 show the technological ingenuity of the author whose suggestions for alleviating the situation are worthy of today's environmental scientists and engineers.

There are two general types of smog, the sooty, sulfurous "London smog" of John Evelyn and his ancestors, and a comparative newcomer (circa 1945) first identified in Los Angeles. It is a product of the age of transportation and special meteorological and geographical conditions, *photochemical air pollution*. Some areas of the world suffer from both types!

Table III, compiled from remarks in Professor Leighton's classic book, *The Photochemistry of Air Pollution* (10), compares the two types of smog. The typical London smog peaks early in the morning. It occurs at relatively low temperatures and at relatively high humidity. There is a radiation inversion and the atmosphere, which is polluted with particulates and oxides of sulfur, is chemically *reducing*. The most pronounced physiological manifestation of London smog is bronchial irritation.

—— piceaque gravatum
Fœdat nube diem[i];

It is this horrid Smoake which obſcures our Church and makes our Palaces look old, which fouls our Cloth and corrupts the Waters, ſo as the very Rain, and refre ing Dews which fall in the ſeveral Seaſons, precipitate t impure vapour, which, with its black and tenacious qu lity, ſpots and contaminates whatever is expoſed to it.

—— Calidoque involvitur undique fumo[k];

It is this which ſcatters and ſtrews about thoſe black a ſmutty *Atomes* upon all things where it comes, inſinuati itſelf into our very ſecret *Cabinets*, and moſt precio *Repoſitories:* Finally, it is this which diffuſes and ſpread Yellowneſſe upon our choyceſt Pictures and Hanging which does this miſchief at home, is [l] *Avernus* to *Fou* and kills our *Bees* and *Flowers* abroad, ſuffering nothing our Gardens to bud, diſplay themſelves, or ripen; ſo

[i] Claud. de rap. Proſ. l. i. [k] Ovid.

[l] A lake in Italy, which formerly emitted ſuch noxious fumes, that birds, wh attempted to fly over it, fell in and were ſuffocated; but it has loſt this bad quality many ages, and is at preſent well ſtocked with fiſh and fowl.

Fig. 4. Excerpt from *Fumifugium;* first published 1661 and reprinted 1772.

But, without the uſe of Calculations it is evident o every one who looks on the yearly Bill of Mor- ality, that near half the children that are born nd bred in *London* die under two years of age[e]. ome have attributed this amazing deſtruction to uxury and the abuſe of Spirituous Liquors: Theſe, o doubt, are powerful aſſiſtants; but the conſtant nd unremitting Poiſon is communicated by the oul Air, which, as the Town ſtill grows larger, has ade regular and ſteady advances in its fatal in- uence.

Corbyn Morris, Eſq;
A Child born in a Country Village has an even chance of living nea 40 years. Much has been ſaid againſt Mothers who put out their Chil- n to nurſe, and where they live in an healthy air, the practice is gene- y unjuſtifiable; but the chance for Life in infants, who are confined in preſent foul Air of *London*, is ſo ſmall, that it is highly prudent and mendable to remove them from it as early as poſſible.

Fig. 5. Excerpt from *Fumifugium.*

Till more effectual methods can take place, it would be of great ſervice, to oblige all thoſe Trades who make uſe of large Fires, to carry their Chimnies much higher into the air than they are at preſent; this expedient would frequently help to convey the Smoke away above the buildings, and in a great meaſure diſperſe it into diſtant parts without its falling on the houſes below.

Workmen ſhould be conſulted, and encouraged to make experiments, whether a particular conſtruction of the Chimnies would not aſſiſt in conveying off the Smoke, and in ſending it higher into the air before it is diſperſed.

A method of charring ſea-coal, ſo as to diveſt it of its Smoke, and yet leave it ſerviceable for many purpoſes, ſhould be made the object of a very ſtrict enquiry; and Premiums ſhould be given to thoſe that were ſucceſsful in it.

Fig. 6. Excerpt from *Fumifugium.*

TABLE III

Comparison of "London Type" and "Los Angeles Type" Smog

London	Los Angeles (photochemical air pollution)
Peaks early in A.M.	Peaks midday
Temperature 30–40°F	Temperature 75–90°F
Relative high humidity and fog	Low relative humidity and clear sky
Radiation or surface inversions	Subsidence or overhead inversion
Chemically reducing atmosphere	Chemically oxidizing atmosphere
Bronchial irritation	Eye irritation

Conversely, photochemical smog peaks in Los Angeles around noon to one o'clock in the afternoon and characteristically occurs on clear days with high temperatures, low relative humidities, and an overhead inversion or a subsidence. Another fundamental difference is that the atmosphere is chemically oxidizing due to the presence of such compounds as ozone and peroxyacetyl nitrate which are formed from the action of sunlight on the complex mixture of hydrocarbons and oxides of nitrogen which, in the Los Angeles Basin, are emitted primarily from auto exhausts.

Photochemical smog is additionally characterized by a variety of other effects. These include severe damage to many types of field crops, ornamental plants and trees, and to pine forests; pronounced eye irritation; reduced visibility; cracking of rubber exposed to the atmosphere; and finally, definite loss in vitality of an individual exercising on a smoggy day.

The condensation behavior and surface reactions of the spectrum of particulate matter (we include aerosols in this category) in the air, ranging from fogs through aeroallergens and into the submicron range of industrial smokes and fumes to particles of molecular dimensions such as proteins and viruses, are of great consequence in determining air quality. Of particular importance are the possible synergistic effects of atmospheric particulate matter acting together with pollutant gases such as sulfur dioxide. Thus toxicological and epidemiological data strongly suggest that it was the *combination* of particles and oxides of sulfur that was primarily responsible for most of the 4,000 excess deaths in the London killer smog of 1952.

Air contaminants, including the oxides of carbon, nitrogen, and sulfur and fluorides, smokes and dusts, arise in a large part from the combustion of fossil fuels, smelting, the manufacture of the tens of thousands of inorganic and organic chemical products, including paper, and incomplete combustion in automobile engines. Hundreds of chemical compounds have been identified as minor gaseous and particulate components of urban atmosphere.

The magnitude of the air contamination problem in the United States is illustrated in Table IV. In Los Angeles County alone 7.6 million gallons of gasoline are burned each day by 3.7 million automobiles, liberating in the process some 12 million grams of lead from the tetral-

TABLE IV

Emission of Atmospheric Contaminants in the United States

Contaminant	Millions of tons per year[a]
Carbon dioxide	5200
Carbon monoxide	105
Sulfur oxides	36
Hydrocarbons	23
Nitrogen oxides	4
Dusts	6
Smokes	5

[a] Estimates from PSAC Report (1965).

kyl lead antiknock compounds. The annual consumption of gasoline in the U.S. is 75 billion gallons each containing about 2 g of lead. Since 1923 more than 2.6 trillion grams of lead has been combusted in North America. If equally dispersed, this represents an average contamination by this persistent chemical element of >10 mg/m^2 (Chow and Johnstone 1965). The accumulation in metropolitan areas such as the Los Angeles Basin is several hundred times this rate. While the approximations in this estimate are admittedly large, so may be the problem.

Forest fires and deliberate slash and field burning are important local sources of air pollution; more than 100 organic compounds have been isolated from the combustion of wood. In 1962 approximately 1.3 billion board feet of lumber were accidentally burned in the United States.

Accurate values for the economic effects of air pollutants are difficult to obtain but they are known to be large. Certain classes of plants are particularly "smog sensitive." For example, tobacco plants are highly susceptible to ozone damage from photochemical smog. It is estimated that air pollution damage to agriculture approximates 500 million dollars per year in the U.S. alone (U.S. Department, of Health, Education, and Welfare, 1966, *The Effects of Air Pollution*, U.S.P.H.S. (17)).

IV. WATER

Man's earliest concern with water quality resulted from the establishment of the microbiological causation of disease and the role of waterborne pathogens in spreading epidemics of typhoid, cholera, salmonellosis, bacillary dysentery, and hepatitis. Most of our present highly developed water treatment practices are aimed at maintaining water purity in terms of bacterial contamination, usually as determined by the quantitative count of indicator organisms such as the coliform bacteria. A less imperative concern has been to preserve minimal standards of water quality in regard to taste and odors, salinity, and turbidity. Very recently with the development of supersensitive methods of detection such as gas chromatography, there is increasing concern with microchemical pollutants such as industrial effluents, detergents, and pesticides. Presently there is developing a total ecosystem approach to water quality with emphasis on the preservation of the normal flora and fauna of streams, rivers, lakes, and estuaries and long-term regard for the esthetic qualities of natural waters.

A. Eutrophication

Eutrophication of natural waters such as lakes and reservoirs is a process of ecological aging stimulated by increased levels of mineral nutrients especially orthophosphate and nitrate. These factors stimulate the growth of algae and other plants at very low levels (phosphorus at 0.05 ppm stimulates profuse growth). As a result of the decay of these plants and associated organisms, the oxygen utilization of the water, the Biochemical Oxygen Demand or B.O.D., is greatly increased and fish and other aquatic animals may find it impossible to exist. The serious consequences of this complex ecological process are illustrated by the eutrophication of Lake Erie. Here, wastes from large cities in Ohio, Indiana, Michigan, New York, and Pennsylvania (the effluents from the 300 miles of industrial complex along its southern shore) and the run-off from agricultural lands richly treated with fertilizers have combined to produce a lake that is rapidly dying in an ecological sense, with objectionable growths of algae and aquatic plants and a BOD so great that survival of many species is in doubt. As a result, many beaches have been closed and commercial fishing has been substantially changed. The most serious aspect of eutrophication is the great length of time required to reverse the process in large bodies of water. In Lake Erie this has been suggested to require 100 years or longer (5).

B. Microchemical Pollution

In theory it should be possible to detect in surface water a substantial number of molecules of any substance produced by man. In fact, as methods of chemical detection progress into the nanogram (10^{-9} g) and picogram (10^{-12} g) range, an astonishing array of substances can be identified as water pollutants in the ppb and ppt range. The waters of the Kanawha River in West Virginia contains as high as 457 parts of carbon-chloroform extractives per billion from which have been identified more the 100 synthetic organic chemicals including phenols, substituted benzenes, aldehydes, ketones, alcohols, chlorethyl ether, acetophenone, diphenyl ether, pyridine and other nitrogen bases, nitrites, acids, hydro-carbons including tetralin and naphthalene, detergents, and pesticides such as DDT and aldrin (12). Most of these contaminants have unknown environmental fates, their toxicity to aquatic organisms is poorly characterized, and they are often aesthetically undesirable.

Synthetic detergents used in households and industry are sold in the U.S. at a rate exceeding five billion pounds yearly. Their presence in surface and ground waters is evidenced by aesthetically undesirable foams in rivers, lakes, well waters, and sewage disposal plants. In addition, many of these compounds contain polyphosphates which provide essential nutrients for eutrophication. Replacement of alkylbenzene-sulfonates by the biodegradable linear alkylsulfonates is improving the situation.

The wide-scale agricultural use of the nonbiodegradable chlorinated hydrocarbon pesticides such as DDT, lindane, aldrin, dieldrin, chlordane, heptachlor, and endrin has resulted in their leaching from agricultural lands into surface waters where they are detected in the ppt to ppb range. The accumulation and biological magnification of DDT and other persistent pesticides in food chain organisms, where they may reach concentrations of several thousand ppm in carnivors, has occasioned widespread popular and scientific concern about the ultimate consequences on the reproduction of birds and fish (7).

Chemical carcinogens present in factory effluents such as 3,4-benzpyrene, 1,2-benzanthracene, β-naphthylamine, and benzidene have been detected in sewage sludge and industrial effluents, along with objectionable heavy metals such as copper, arsenic, lead, and mercury.

Many of these microchemical pollutants may be carried hundreds of miles down rivers into estuaries where they may be toxic to marine fishes and valuable invertebrates such as shrimps and oysters.

C. Thermal Pollution

Thermal pollution from industrial cooling operations, e.g., nuclear reactors, has become of increasing concern since upwards of 50% of the heat value of fuel must be dissipated in cooling water. A large operation may raise river temperatures by as much as 25°F and provide severe thermal effects on aquatic organisms especially by reducing the oxygen of water.

D. Sewage

Sewage is generally considered to aggregate 100 to 150 gallons per person per day with a daily U.S. output on the order of 20 billion gallons. Currently, sewage from only about 55% of the U.S. urban population is receiving adequate treatment. By 1973, 90% will require secondary treatment and 10% primary treatment to meet water quality standards. The total cost of the necessary programs between now and 1973 is estimated at 26 to 29 billion dollars (5). A growing source of pollution in the U.S. is the eight million pleasure boats whose occupants disperse over a huge area an estimated daily raw waste equivalent of a city of 500,000 inhabitants.

V. SOIL

Soils are among the most important of man's natural resources, and vast as is the land area of the world, its soils are yet finite enough to suffer deleterious modifications through environmental contamination. Thus, even the U.S. with an area of about 3.6 million square miles has begun to suffer serious soil contamination from pesticides, fertilizers, industrial wastes, saline waters, radionuclides, and from more permanent and less readily recognized pollutants such as concrete, asphalt, tin, iron, lead, aluminum, and polyethylene.

A. Pesticides

Pesticides applied to agricultural crops and to the soil itself are an important source of contamination. It is estimated (U.S.D.A., 1966) that of the approximately 350 million acres of cultivated land in the U.S., annual applications of herbicides are made to 95×10^6 A (27%), of insecticides to 42×10^6 A (12%), and of fungicides, nematocides, and crop control agents to 9×10^6 A (2.6%).

The persistence of the agents in the soil is extremely variable depending upon their intrinsic chemical reactivity, their water solubility, and their susceptibility to biochemical degradations by the thousands of species of soil microorganisms which have the capacity to ultimately metabolize every variety of organic substance. An indication of the relative soil persistence of various important types of pesticides is shown in Table V. It is apparent that the heavy metal pesticides containing arsenic, lead, copper, and mercury are deleterious contaminants which may affect the biological properties of the soil for many years. Fortunately, their use has been greatly curtailed. The persistent organochloride pesticides such as DDT, BHC, and the cyclodienes should be used with great care so that their rate of decomposition more than bal-

TABLE V

Persistence of Pesticides in Soils

Pesticide	Half-life ($T^{1/2}$) in years[a]
Lead, copper, arsenic	10–30
Dieldrin, BHC, DDT insecticides	2–4
Triazine Herbicides	1–2
Benzoic acid herbicides	0.2–1
Urea herbicides	0.3–0.8
2,4-D; 2,4,5-T Herbicides	0.1–0.4
Organophosphorus insecticides	0.02–0.2
Carbamate (carbaryl) insecticide	0.02

[a] Estimates from PSAC Report (1965), Lichtenstein (1966).

ances the annual increment of contamination. It is important to substitute readily biodegradable insecticides such as the organophosphates and carbamates for the more persistent compounds wherever possible and from the overall viewpoint to use selective furrow and spot applications and seed treatments for broadcast plantings. Photodecomposition of organic pesticides applied to plants is an important avenue in lessening soil contamination by agricultural residues.

Much more remains to be learned about the ecosystem effects of the more than 300 pesticidal compounds and their decomposition products which eventually find their way into the soil. Increased attention should be given to the movements of the more persistent pesticides in soils and in drainage waters and of their uptake by plants.

B. Fertilizers

Fertilizers have an almost ubiquitous and quantitatively expanding use in the agricultural lands of the world. For example, in 1968 the U.S. production of primary nutrients was 29 billion pounds. The major items are ammonium nitrate and phosphate, ammonia, potassium salts, and superphosphate but there is also significant application of minor elements such as iron, copper, manganese, molybdenum, and cobalt. Superphosphate fertilizers contain traces of undesirable elements such as arsenic, boron, fluorine, and uranium and its consequent radioactive decay products.

The principal deleterious aspect of the increasing use of fertilizers is their leaching into drainage water. Phosphorus is the principal element controlling the fertility of waters where it increases the growth of blue green algae, aquatic weeds, and other undesirable organisms. Phosphate fertilizers applied to soils are converted to water-insoluble phosphates within a few hours, so there is generally no important movement through ground water. However, leaching of accumulated phosphorus compounds through erosion of top soils is an important source of water contamination. It has been estimated that such losses may range from 10 to 80 pounds of phosphorus per acre per year of which up to 14 pounds per acre may be found in surface drainage water. As pointed out elsewhere, phosphorus at 0.05 ppm will stimulate algal growth so that good soil conservation practices are of far-reaching importance (16).

Nitrates are not only objectionable in trace amounts because of stimulation of eutrophication, but higher concentrations in drinking water, 8–9 ppm, have caused methemoglobinemia in infants and deleterious effects in livestock (14).

Salinity of soils is a major problem where irrigation is a major component of agriculture. The salt balance of soils is a function of soil types and balances between input and outgo of soluble salts such as Na^+, K^+, Mg^{++}, Ca^{++}, Cl^-, PO_4^{---}, SO_4^{--}, and CO_3^{--}. Salts will accumulate when the salt content of the irrigation water is excessive, when water supplies are limited and prevent proper leaching, and when there is inadequate drainage. Irreversible changes to the structural lattice of montmorillonite clays may result from infiltration with sodium hydrates when the concentrations of calcium and magnesium are low. These changes greatly reduce soil permeability and may ultimately make soils useless for agriculture.

Strip mining operations and acid mine drainage are said to have altered the top soil characteristics of as much as one million acres of land in the eastern U.S. (14).

VI. WASTE

A seemingly inexhaustible supply of gaseous, liquid, and solid waste materials are produced daily. They may be metabolic products of living organisms or by-products of living organisms or by-products of technology and obsolescence. In the U.S., waste management programs have been classified (14) as motivated by considerations of (*a*) public health and safety, (*b*) economic incentive, (*c*) conservation, and (*d*) esthetics. The magnitude of the total waste disposal problem in the U.S. may be gaged by Table VI. Among the fantastic diversity of human wastes from U.S. inhabitants are the 7.2 trillion gallons of liquid sewage and 320 billion pounds of municipal solid wastes per year. Included in this total are such significant items as 48 billion metal cans,

TABLE VI

Production of Solid Wastes in the United States

Contaminant	Millions of tons per year[a]
Mining debris	3300
Municipal	160
Paper	30
Industrial	15
Automobiles	5

[a] Estimates from PSAC Report (1965).

26 billion bottles and jars, 30 million tons of paper, and 5 million junked automobiles. Although much of this material is reclaimed and reprocessed, the collection of solid wastes costs the nation about 2.8 billion dollars annually. Furthermore, the magnitude of the problem is steadily increasing and presents a major source of esthetic pollution.

A. Animal Wastes

While human sewage is a widely recognized problem (*vide supra*), the excreta of farm animals is also a major source of water pollution as it seeps into streams, rivers, and lakes or enters them through runoff. The world population of domestic animals numbers about 1 billion cattle, 1 billion sheep, 350 million goats, 550 million pigs, 100 million buffalo, 64 million horses, 40 million asses, 15 million mules, and 3 billion fowls (1). The magnitude of the problem becomes apparent when it is realized that one cow generates about as much manure as 16 persons and a pig as much as two people. It has been stated that the farm animals of the United States produce ten times as much waste as the human population.

Such animal waste has disagreeable odors, serves as a breeding mixture for flies, and may contain undesirable pathogens. In addition, animal wastes contain substantial amounts of phosphorus. For example, the annual excretion of phosphorus per 1000 pounds of body weight is cow 17, poultry 30, and swine 45 pounds. This can contribute substantially to eutrophication of streams and lakes (16). The animal waste problem is compounded by modern practices in which poultry is often raised in establishments of 100,000 or more and in which beef cattle are confined in suburban feed lots with no grazing room. Major changes in animal management practices are necessary to solve these pollution problems.

B. Radionuclides

Radionuclides from fallout of atomic weapons eventually contaminate soil and may be taken up in plant products and accumulate in humans and animals. Of the approximately 200 isotopes of 35 elements that have been identified from fission products, the most important and their half-lives are shown in Table VII. Strontium 89 and 90 and ^{140}Ba resemble calcium in their biochemical behavior in soils, plants and animals. Strontium 90 is of particular importance because of its long half-life and its excretion in milk and deposition in bones and tissues. It is relatively immobile in soil.

Up to the time of the nuclear test suspension in 1966, approximately 174 megatons of activity had been released producing 4.5 tons of fission

TABLE VII

Persistence of Radionuclides

Element	Half-life ($T^{1/2}$)[a]
^{90}Sr	28 years
^{137}Cs	27 years
^{89}Sr	53 days
^{140}Ba	13 days
^{131}I	8 days
^{133}I	22 hours

[a] Data from PSAC Report (1965).

products. After 1966, the rate of deposition of ^{90}Sr fell below the amount lost by radioactive decay so that the total quantity in the soil will decrease unless the testing of nuclear weapons is resumed.

The potentially serious effects of radioactive fallout are illustrated by the effects of ^{137}Cs which accumulated on lichens in the Anaktuwuk Pass region of Alaska. The lichens are the principal dietary item of the arctic caribou which in turn comprises 90% of the diet of the Eskimos of the area, who thus ingest the skimmings from many acres of lichens. In 1963 it was found that Eskimos of this region contained a body burden averaging 790 nanocuries of ^{137}Cs. The maximum level of 1240 nanocuries was about 100 times that of the inhabitants of the lower U.S. (13).

The increasing industrial uses of nuclear reactors affords opportunities for the accidental contamination of the environment with radionuclides. In the Windscale accident of October 1957, an area of about 300 square miles in England was heavily contaminated with airborn ^{131}I released from fission of ^{235}U in ruptured reactor cores. Dairy cattle, feeding on contaminated graze, produced milk sufficiently contaminated with radioactive iodine to prevent its sale for several weeks and ^{131}I levels in human thyroids, especially of children, was substantially increased (2,15).

VII. ECOLOGY AND THE ENVIRONMENT

The science of ecology deals with the factors controlling the abundance and distribution of living organisms. It lies at the focus of environmental quality. Thus the environmental effects of our agricultural and industrial society are so profound and far reaching that throughout much of the world we are dealing not with the natural ecology which has arisen through millions of years of biological history but rather with the rapidly changing and disturbed ecology produced by the last 100 years of the industrial revolution. The concept of the "balance of nature" has little meaning under the pressure of man's activities in promoting the continuous growth of monocultures of corn, wheat, rice, and cotton; in damming streams and irrigating deserts; in the concrete and asphalt jungles of our megalopoli; in the almost global dispersion of living plants and animals; and in innumerable lesser alterations. Under these conditions it is fruitless to think of returning environmental quality to the pristine beauty of the age of our great grandfathers. We can only hope to reduce the intolerable to the tolerable through the accurate evaluation of *criteria* of environmental quality and the establishment and enforcement of *standards* of environmental quality which will permit the survival of man and the more adaptable of his associated plants and animals. Those wild creatures which cannot effectively meet the twin threat of industrialization and human population pressures, like the great auk and the passenger pigeon, seem almost certainly doomed to extinction or at best to survive in strictly controlled zoological havens.

VIII. PUBLIC HEALTH AND THE ENVIRONMENT

Although the conservation of wildlife and the esthetic aspects of our environment are of immense importance, the overriding concern about environmental quality relates to the effects of environmental pollution upon human health. It is from this source of concern that we must expect the maximum legislative pressures for standards of air and water quality.

The causal relationships between environmental contamination and human health are most clearly understood in the area of water quality. The spread of epidemics of typhoid and enteric diseases through contaminated drinking water has been repeatedly demonstrated, and in the western world the chlorination of city drinking water has become a standard public health practice. However, in contrast, infectious viral hepatitis is also transmitted by water supply and has not been controlled by conventional water treatment procedures. Fluoridation of drinking water for the prevention of dental cavities provides an interesting example of a violently controversial problem in environmental contamination which stands in contrast to the public acceptance of water chlorination. Almost nothing is known of the public health effects of the trace contamination of drinking water by the thousands of industrial effluents, by pesticides, fertilizers, detergents, and by heavy metals. Fortunately, nearly all of these can be removed by the existing technology of modern water treatment methods,

The public health problems of air pollution are much less clearly understood than those of water pollution. The poisonous fogs of Donora, Pennsylvania in 1948 and those of London in 1952 when 4,000 "excess deaths" were reported showed that air contaminants could accumulate to acutely hazardous levels. However, the chronic effects of daily exposure to photochemical smog, to carbon monoxide, or to sulfur oxides have not yet been clearly defined. The problem is complicated because of possible synergistic actions between various air contaminants and because of a wide range of effects upon the young and the aged and upon those suffering from chronic diseases (*vide supra*). While the health hazards of airborne allergens are fairly obvious, our knowledge of the effects of breathing air contaminated with particulate materials such as lead compounds, polynuclear hydrocarbons, and asbestos is meager. The answers to the public health consequences of the environmental contamination by such materials will require inspired and laborious toxicological and epidemiological studies extending perhaps over several human generations.

A. Food

Food has always contained a substantial burden of undesirable elements ranging from contaminating pathogens such as *Salmonella* and *Botulinus* to naturally occurring toxic substances such as oxalate, avidin, and cholesterol. During the past 50 years, public health emphasis on food quality has almost completely reversed from concern about suboptimum quantities of the vitamins and minerals to preoccupation with the presence of trace amounts of pesticides, fertilizers, preservatives, radioactive fission products of fallout, and carcinogens such as the polynuclear hydrocarbons and the aflotoxins. As with the other areas of environmental science, food quality will continue to reflect population pressures which give rise to new technologies of crop production, food processing, preservation, and storage, all of which introduce extraneous chemicals of chemical changes in food substances. It will also be modified by the expanding industrialization with its host of chemical effluents which inevitably introduce into food, traces of a vast variety of environmental contaminants. However, the production and quality of food is more carefully controlled than the water we drink, and far freer from contamination than our urban air.

Fortunately, the human organism is equipped with an efficient array of metabolizing enzymes which have arisen through countless generations of exposure to unwanted chemicals in natural foods. Thus, there seems little doubt but that trace amounts of any organic xenobiotic can be metabolized and eliminated from the body predominantly as water-soluble metabolites. The rate of disappearance of simple organic compounds, i.e., their biodegradability, ranges from aspirin with a biological half-life about 1 day through DDT with a half life of about 0.5 year. The slow rate of biodegradability of DDT has resulted in its presence, together with its primary degradation product DDE, in human fat to an average of about 10 ppm (6). Studies over the past 18 years have shown that this amount is in dynamic equilibrium. It is not increasing in the general population, nor is there any conclusive evidence that this level has an adverse effect on *human* health (6). Nevertheless, its ubiquitous presence indicates the complex problems of environmental quality and is disturbing to the biologist and nutritionist. Furthermore, it is a matter of major concern to the ecologist. The rapidly increasing dispersion of heavy metals such as lead which have much longer biological half-lives also poses serious problems to man's health.

IX. A GLOBAL VIEW OF POLLUTION

The international nature of many of the problems we have discussed above are now being officially recognized. Thus in the fall of 1968, Sweden formally asked the United Nations to convene a world conference on pollution in 1972. As the Swedish Ambassador Sverker Astrom pointed out, "In certain parts of the world the environment is, in fact, undergoing serious deterioration and in some cases destruction . . ." Processes are set in motion which if unchecked will drastically change and damage the conditions for human life.

"This phenomenon is clearly visible in countries in an advanced stage of industrialization and urbanization. As more countries approach and enter this phase of development . . . a world wide exchange of knowledge and experiences is required to prevent repetition of mistakes made." (*Los Angeles Times*, Oct. 28, 1968).

Another aspect of the international problem is illustrated by the conclusion recently reached by Swedish environmental scientists that air pollutants originating from industrial areas in northern England and carried across the North Sea have contaminated certain of Sweden's air and water resources. To date political difficulties, including national pride, have dulled virtually every attempt in North America and Europe to set up regional pollution control districts based on geographical and meteorological factors rather than county, state, or national boundaries.

However, times are changing. Thus, for example, the United States Clean Air Act of 1967 provides for the establishment of Air Quality Control Regions whose boundaries are based on the factors cited above, rather than arbitrary state lines.

Worldwide dispersion of trace amounts of environmental contaminants such as radionuclides, lead, and DDT occurs in air and ocean currents and these materials have been identified in the tissues of migratory birds and fish. Similar study will undoubtedly discover many more examples of the global distribution of relatively stable environmental contaminants. Truly, the day is at hand when one man's poison may become another man's meat.

Acknowledgment

One of us, JNP, acknowledges his indebtedness to the Research Grants Branch, National Air Pollution Control Administration, Consumer Protection and Environmental Health Service, U. S. Public Health Service which has generously supported his research through Grants AP 00109 and AP 00771. JNP also acknowledges with appreciation a sabbatical leave from the University of California, Riverside during which time his portion of the manuscript was completed.

References

1. T. C. Byerly, *Proc. 15th Ann. Meeting Agr. Res. Inst.*, Nat. Acad. Sci., 1966, p. 35.
2. A. C. Chamberlain and H. J. Dunstan, *Nature*, **182**, 629 (1958).
3. *Chem. Eng. News*, pp. 74A–130A, Sept. 2, 1968.
4. T. J. Chow and M. S. Johnstone, *Science*, **147**, 502 (1965).
5. *Environ. Sci. Tech.*, pp. 85, 88, Feb. 1968.
6. W. J. Hayes, Jr., *Sci. Aspects Pest Control*, Pub. 1402, Nat. Acad. Sci., 1966, p. 314.
7. E. Hunt, *Sci. Aspects Pest Control*, Pub. 1402, Nat Acad. Sci., 1966, p. 251.
8. D. Kiefer, *Chem. Eng. News*, p. 90, Oct. 7, p. 118, Oct. 18, 1968.
9. D. H. K. Lee, *Suppl. Amer. J. Pub. Health*, **54**, 7 (1964).
10. P. A. Leighton, *The Photochemistry of Air Pollution*, Academic Press, N. Y., 1961.
11. E. P. Lichtenstein, *Sci. Aspects Pest Control*, Pub. 1402, Nat. Acad. Sci. 1966, p. 221.
12. F. M. Middleton, *Proc. Conf. Physiol. Aspects Water Quality*, Washington, D. C., 1960.
13. H. E. Palmer, W. C. Hanson, B. I. Griffin, and D. M. Fleming, *Science*, **144**, 859 (1964).
14. President's Science Advisory Committee Report, *Restoring the Quality of Our Environment*, The White House, Washington, D. C., 1965.
15. N. O. Stewart and R. N. Crooks, *Nature*, **182**, 629 (1958).
16. A. W. Taylor, *J. Soil Water Conserv.*, p. 228, Nov.-Dec., (1967).
17. U.S. Department of Health, Education, and Welfare, *The Effects of Air Pollution*, Publication 1556, 1966.
18. World Health Organization, *Microchemical Pollution in the Environment*, MHO/PA/110.63, Geneva, Switzerland, March 18, 1963.

The Federal Role in Pollution Abatement and Control

JOHN TUNNEY
House of Representatives, Congress of the United States, Washington, D.C.

I. Public Health 25
II. Resource Conservation 26
III. Federal, State, and Local Roles 27
IV. The Sequence of Events Leading to Abatement 31
V. The Future 32

I. PUBLIC HEALTH

The Constitution of the United States established the responsibility of the federal government for public health in its clause "to promote the general welfare." This was many years, however, before contamination of the environment by the activities of human populations was recognized as a cause of public disease. Around the turn of the century waterborne organisms were shown to cause typhoid fever, and it was also shown that these organisms were spread by the contamination of municipal water supplies by sewage. Although chlorination could render the water safe to drink, by then the idea that water must be treated as a renewable resource had been established. Programs for pollution control in major streams were started.

Air pollution was originally viewed as a nuisance but a necessary evil; it was thought of as a sign of industrial activity and economic progress. As this idea is replaced by the concept that man must control his environment, the gross and obvious contaminants (particulate matter, eye irritants, and unpleasant odors) are beginning to be eliminated. The cities of Pittsburgh and St. Louis were moved by public indignation to attack the problem of smoke, and they succeeded, to a great extent, in cleaning their atmospheres of this contaminant. Los Angeles citizens were irritated into activity by the eye watering caused by photochemical smog. Once it was shown that this pollutant was a result of hydrocarbons in the air (from industry and automobiles), a vigorous abatement campaign was developed. In spite of the local successes in abating air pollution, populated areas have an air pollution problem once known in only a few cities.

More recently, subtle adverse effects on health from exposures to low concentrations of air pollutants over a long period of time have been recognized. Environmental epidemiologists have begun the difficult task of firmly establishing cause-and-effect relationships so that complex and often expensive abatement measures may be taken if necessary.

The definition of health has gradually broadened to include mental and social well-being, or even further, to represent the capacity of the individual to pursue happiness. Any impairment of a person's capabilities is considered a threat to health. On this basis, the interaction between human beings and their natural environment presents a host of problems for public health policy and practice.

II. RESOURCE CONSERVATION

An alternative and often complementary approach of the federal government to pollution is based on resource conservation. The early progress of the United States, in particular the winning of the West, required an attitude of struggle with nature. This policy was gradually replaced by one of harvesting the bounty which could be gained by intelligent exploitation of mining, forests, agriculture, and water power. Only recently has a more modern concept of renewal begun to influence conservation policy.

Only the energy from the sun is continuously supplied to the earth. Everything else which we use must be considered as a resource which must be renewed. The population explosion and the demands of technological affluence have brought all natural resources to the point of limited supply. A use-and-discard economy must be replaced by a recycle policy which includes, at the outset, provision for the renewal and reuse of the material things of life.

Pollution is often caused because recycle costs are more expensive than discard costs. In addition, contamination of air and water make it more costly for these natural resources to be reused as they must be. A polluted stream means excess costs to purify the water for a downstream drinking supply. Acid gases and mists in the air corrode metals and shorten machinery life so that special air conditioning which further pollutes the outside environment is often required.

Therefore, the concept of environmental quality goes beyond human health and includes all aspects of ecology. And, in terms of economics, a pollution-free environment is seen to be a part of wise usage of all our resources.

III. FEDERAL, STATE, AND LOCAL ROLES

Environmental pollution is not a single problem, and it is often not even a conglomeration of related ones. Pollution presents a new set of technical, institutional, and economic factors for different contaminants, local geographical and meteorological features, and specific industrial or commercial situations. Therefore, the essential federal role is to provide information on the causes of pollution and means of abatement. Research to identify pollutants and their effects on health or property is the necessary first step towards restoring and maintaining the quality of the environment. It would be wasteful if each state or region had to establish these facts independently. Relying on the normal course of scientific curiosity might be too slow. Development of practical abatement processes and devices would also provide widely applicable information. Thus the federal support of pollution-related research and development was called for in the early legislation concerning both air and water pollution. Table I summarizes the present magnitude of this funding—a total of $185 million in fiscal year 1968.

The second federal government responsibility is to exert leadership. The operation of federal installations and equipment should incorporate the very best pollution abatement practice. Executive orders have been issued to require secondary sewage treatment for most federal facilities and to require the use of low sulfur fuel in power plants and heating units in New York, Philadelphia, and Chicago. By specifying the performance with respect to exhaust emissions of motor vehicles, the federal government can encourage development of abatement devices.

At the present time, contractors or suppliers to the government do not have to comply with any pollution standards. It may be that procurement contracts could become a lever to enforce pollution control just as they have been used to promote other social objectives.

Although early legislation consistently recognized the limits of federal involvement and stated that enforcement must be a function of local government, a number of factors have tended to broaden the central government role. The most important barrier to state and local government control of environmental pollution is that surface waters and air masses do not correspond to arbitrary political jurisdictions. Air and water regions are determined by geography and weather. Pollutants move across state and county lines. Cleaning up one city will not help if another upstream or upwind refuses to control its effluents. Thus, the federal government is often the only government

TABLE I

Total Expenditures Reported by the Federal Government on Pollution Research Development and Demonstration (except pesticides) in Thousands of Dollars[a b c]

	Fiscal year 1967			Fiscal year 1968		
	Intra-mural	Extra-mural	Total	Intra-mural	Extra-mural	Total
Effects of pollution:	12,063.3	25,097.3	37,881.8	15,629.1	29,126.9	45,217.0
Directly on man	4,651.0	6,509.0	11,219.0	5,228.0	5,728.0	1,102.7
On crop plants and domestic animals	993.8	7,425.0	8,418.8	1,428.0	9,084.0	10,512.0
On nondomesticated plants and animals	273.0	4,391.0	5,203.0	2,429.3	3,710.0	7,529.3
On materials or structures	248.8	24.0	272.9	304.0	82.0	386.0
On environments:						
1. Air	322.0	2,641.3	2,963.3	283.0	2,006.0	2,289.0
2. Freshwater (eutrophication)	1,101.1	1,493.9	2,595.0	949.7	2,498.2	3,447.9
3. Marine	213.0	730.6	943.6	701.5	797.2	1,498.7
4. Urban		155.0	155.0	50.0	1,092.0	1,142.0
5. Rural	1,000.0	100.0	1,100.0	1,150.0	523.0	1,673.0
6. Wild		636.3	636.3		1,114.5	1,114.5
7. Soil	65.0		65.0	65.0		65.0
8. Mixed	30.0	294.1	324.1	19.5	655.0	674.5
Transport, distribution and fate of pollutants:	2,567.7	16,718.6	19,699.8	2,974.3	17,141.5	20,372.4
Movement	2,186.0	14,417.4	16,604.1	2,303.3	14,508.4	16,811.7
Degradation	381.0	1,470.2	1,851.2	671.0	2,099.1	2,770.1
Measurement and instrumentation	6,203.4	5,016.3	11,836.9	6,305.4	4,812.2	11,773.4
Exposure to and sources of pollution	2,017.3	653.2	2,670.5	2,563.0	792.2	3,355.2
Social, economic, and legal aspects of pollution	368.6	2,589.1	2,957.7	684.2	3,936.5	4,620.7
Prevention and control of pollution:	51,609.4	20,113.6	71,844.0	27,231.8	72,246.2	99,588.0
Research	17,535.2	14,220.0	32,876.2	21,248.8	16,338.7	37,698.0
Development	3,783.6	18,871.0	21,654.6	3,453.0	30,755.0	34,208.0
Demonstration	2,322.0	13,832.0	16,154.0	2,037.0	25,103.0	27,140.0
Total obligations	74,829.8	70,188.1	146,890.6	55,387.8	128,055.5	184,926.6

[a] Fiscal year totals and numbered category totals may exceed the sums of the respective rows and columns because some agencies did not provide detailed breakdowns. Total figures are more nearly correct.

to which the citizen whose environment is polluted can appeal for relief from his polluting neighbor.

Voluntary compacts among jurisdictions are certainly encouraged. The Ohio River Valley Water Sanitation Commission (ORSANCO) and the Delaware River Basin Commission are Congressionally approved interstate compacts which are proceeding to improve and maintain these streams. Present laws also provide for international agreements such as the Great Lakes abatement effort which will involve Canada.

Where jurisdictions cannot agree and where pollution originating in one area affects another, present laws allow the federal government to step in. Through a lengthy process of hearings and planning conferences, an abatement procedure is established. If progress is unsatisfactory, the federal government may obtain a court order to force abatement. So far no case has ever been brought to this point, but the threat of injunctive power is probably the reason behind compliance by some recalcitrant municipalities and industries.

The federal government has also assumed the role of giving financial and technical assistance in planning, establishing, and maintaining local air and water abatement programs. Grants for a portion of the costs are made to local government for surveys of pollution sources and measurement of current environmental quality. Training in the operation of monitoring devices is provided. The abatement plans resulting from these local programs must conform to federal standards in order to justify the partial funding grants-in-aid.

When a major abatement task is identified, such as sewage treatment, the federal government uses its tax-gathering power to provide substantial financing. Grants for sewage treatment plant construction are designed to speed the time when all municipal effluents will be purified to at least the secondary stage—that is, 70–90% of the organic matter removed. This incentive is necessary because of the inability of many local governments to raise the required funds by bond issue. Other areas simply are unwilling to place the long term gains of regional water quality ahead of short term educational, recreational, and other governmental services. Federal policy calls for stream improvement, and the construction grants make it possible.

---

[b] Funds transferred between agencies are shown only by the agency to which the funds are appropriated to avoid "double counting." A total of $9,683,200 was transferred in Fiscal Year 1967 and $9,574,600 in Fiscal Year 1968.

[c] $1,242,552 in Fiscal Year 1967 and $1,122,300 in Fiscal Year 1968 provided by State and local governments as part of cooperative research are not included in the total figures.

Other actions which the federal legislation has not, as yet, provided for, include economic incentives for industry and nationwide uniform emission standards. The economics of environmental quality are incomplete, so that no real cost–effectiveness accounting is possible. The costs incurred from a polluted environment are known only in fragments. It is impossible to place dollar values on esthetic factors or even on recreation, visibility, and irritation. Analyses have been made which place the damage or deterioration at hundreds of dollars per year per family. But when polls are taken to find out what citizens would pay for cleaner air or water, the average willingness (as a measure of damage) amounts to only a few dollars per year. Rationalizing abatement on the basis of savings is thus an unproductive approach.

Strict dependence on health relationships may also be impossible. The multiple factors in diseases which are caused or aggravated by environmental values preclude the usual dose–response sort of relationship. Long-term, low-level exposures may never be shown with precision to have a predictable health effect, even though there is every reason to conclude that they are not good for us. It is more realistic to set goals based on reasonably well established health effects and the principle that man simply should not degrade the environment of which he is a part.

The installation and operation of abatement processes and devices *is* subject to cost analysis. For example, the Department of the Interior estimates that $29 billion may be required in the years 1969–1973 to meet projected waste treatment, sanitary sewer, and water cooling requirements. Purchases of air pollution control equipment are estimated at $1 billion per year now. Alternative methods can be compared on the basis of cost and the improvements which they bring in effluent control. Improvements to the environment again suffer in measurement from the nonquantifiable aspect of subjective judgment.

The capital outlays for control equipment generally do not bring a direct return on investment (some offset may occur from waste recovery). These expenditures to industry are similar to those for safety and industrial hygiene. They are legitimate costs of doing business, but the speed with which installation occurs may be limited by capital availability. Further, the improvements in the environment accrue to the society as a whole. Therefore, it may be good public policy to provide economic incentives in the form of rapid amortization or income tax credit for pollution-control equipment.

The concept of uniform national emission restrictions is based on the belief that if one area requires more stringent controls than another, industry will move to the region with lax standards. Testimony before Congress has indicated that there are so many other unequal cost factors (labor, raw material supply, transportation, etc.) between locations that abatement costs would not sway the choice of new plant location and would certainly not cause existing investments to be moved. The 1967 Air Quality Act calls for a two-year study of this situation.

In the case of motor vehicles and watercraft, a policy other than that used for stationary pollution sources has been adopted. Because these mobile sources may move into an area where local conditions require very strict control, the federal government has preempted the standard setting authority. All automobiles conform to federal exhaust emission standards. Only California, which has the most severe smog problem, is exempted from preemption—and then only if its standards are as stringent or more so than the national values.

IV. THE SEQUENCE OF EVENTS LEADING TO ABATEMENT

The federal water and air quality laws follow the common theme that standards for the receiving environment are chosen locally, based on criteria or descriptive cause-and-effect relationships which are developed at the federal level. There is sufficient difference to warrant a separate description of the working of the legislation.

The Water Quality Act calls for states to submit standards for surface waters to the Department of the Interior Federal Water Pollution Control Administration. These standards prescribe measured values for stream quality such as dissolved oxygen, temperature, and salt content. The numerical values are chosen in consideration of local conditions, the use of the stream, and present quality of the stream. The standards are selected on the basis of descriptions such as the type of fish which will live and thrive in a certain water quality. These descriptions are the product of scientific research culminating in the deliberations of the National Technical Advisory Committee to the FWPCA on water quality criteria.

State standards were submitted by July 1, 1967. The Secretary of the Department of the Interior has approved a number of the plans, and others are being negotiated to assure that they result in enhancement or at least maintenance of present stream quality. If the states do not set satisfactory standards, the federal government may impose its own after a procedural interval of hearings and conferences. Enforcement implementation is initially up to the states. If they fail or reach an impasse in an interstate matter, the federal government can again step in.

The Air Quality Act of 1967 follows the general idea of the water law, but the sequence is just beginning. The Secretary of the Department of Health, Education, and Welfare is to promulgate descriptive criteria for each air pollutant along with recommended abatement technology which is economically acceptable. The states then set prescriptive standards for ambient air quality based on these criteria. Again local geographical, weather, and economic factors may mean different standards are chosen for different regions. If the state does not respond with suitable standards, the Secretary may issue his own and proceed to enforce them.

Both water and air standards apply to designated river basins and air regions—areas which share the same environment. The quality of the ambient environment is the governing factor, not the properties of any one emission or effluent. In water the most important criteria are those for aquatic life. Similarly, in air the determining criteria are those for breathing. The concept is that if the living creatures in air and water are protected, other desirable attributes will also be insured.

The ambient environment principle infers that different pollutant sources may be restricted to different degrees to give the best resulting quality at the least cost. For example, one sewage outfall into a stream may be much greater in volume than others. Equal efficiency of treatment would produce a certain water quality, but, if additional treatment could be accomplished in the large volume, then the stream's health could be increased at a lower overall cost. Similarly, it may be less costly to apply strict controls to major power plants than to bring uniform restrictions to all chimneys. Unfortunately, it is difficult to allocate costs back to all polluters if only a portion are chosen for control. Perhaps the continuing and evolving management of air and water resources on the basis of regions will bring the development of an efficient and equitable abatement plan. Until that time, the best policy is to control pollution at all sources to assure ambient environmental quality.

V. THE FUTURE

We have had a tendency to deal with problems in a piecemeal manner. We have not anticipated problems in a preventive fashion. Rather, we let problems reach a point where we must try to cope with them in order to keep them from getting even more out-of-hand. This certainly has been the case in the area of environmental quality control.

The Congress, state and local governments, and industries have begun to show their concern and awareness of the problems of air and water pollution in the face of the ever-increasing outcry of public dissatisfaction. We have, however, continued to act in our manner of responding to problems once they have become large enough to attract national attention. We have been caught ill-prepared to deal with these problems, and the indecisiveness of our legislation is indicative of our lack of foresight. While we are presently looking at the two giants in the area of environmental quality control, air and water pollution, we are not viewing them in a sophisticated enough manner, nor are we paying enough attention to their impact on each other and on the entire environment. We are not giving enough consideration to other problems which will increase with our national growth if they continue to go on unchecked.

We have a need to look after the entire environment and the ecology of that environment. We must know the relationship of air to water pollution, of each to solid waste matter, and of all to each other. We must understand the effects of radiation on all forms of pollution, the effects of pollution on the weather, the relationship of topography to pollution, and the beneficial combinations of urban planning to topography and the relationship of that combination to pollution abatement. We must understand the impact of environmental change on human beings, and how the changes in the quality of the atmosphere and the appearance of the environment affect man both physiologically and psychologically.

We have a need to understand our physical surroundings and the ecology of that environment better, so that we may direct our efforts at beneficially altering our environment, and so that we may be able to foresee future problems that could be averted at an early stage. We must have an overview of our surroundings so that we can understand our strengths, weaknesses, and needs, and act accordingly.

At present, we are dealing with many of the problems of our environment in many areas of the government. I do not question that each of these areas has a special and particular interest in its area of authority. The Public Health Service in the Department of Health, Education, and Welfare certainly has appropriate interest in air pollution, for air pollution has direct effects upon our health. The Department of Transportation also has an interest in air pollution as it relates to automobiles and their ability to create air pollution, and as it relates to decreased atmospheric visibility which affects air transportation. The Department of Commerce has obvious interests when one of the great sources of air pollution is industry, and The Department of Agriculture is keenly interested in the impact of polluted air on crops and vegetation. I do not deny that each of these departments has a specialized and necessary interest in air pollution.

The Department of the Interior has an important concern with water pollution, for it has jurisdiction over the billions of gallons of water which come from areas of its jurisdiction. The Department of Housing and Urban Development also has an intense interest in water use, for it must concern itself with the water needs, and water and sewage systems of the great cities. Agricultural need for high quality water speaks for itself. Industrial use of water again involves the Department of Commerce, and there are obvious health needs in water purification systems. Again, I maintain that these areas all have individual legitimate and necessary authority in these realms.

The Atomic Energy Commission has obvious authority in the area of radioactive materials. The Department of Health, Education, and Welfare has interests in the health aspects of these potentially harmful materials. The Department of Defense has needs for nuclear fuels, and every department involved with water which is interested in desalinization as a source of additional fresh water has considered atomic facilities for such processes. These interests each have certain special concerns, and I feel that they are rightfully exercizing authority in their own particular areas.

I could go on and on, but certainly it is not necessary. There is nothing wrong with this departmental specialization in related fields; it is advantageous for the government to look at the problems from a variety of specialized points of view. There are, however, some major needs which are not being met.

First and foremost is the need to view the entire environment and its total ecological interaction. It is essential to relate all of these areas of interest to each other. The environment is certainly composed of many more elements than have been mentioned here. The ecology of the environment—the interaction of all of those elements—is something that I could not explain here for we do not yet understand many aspects of it. That is my point. The understanding of our ecology is essential if we hope to successfully deal with the many problems of our environment. The understanding of our ecology is essential if we hope to create programs that will alleviate our environmental problems, both now and in the future. The understanding of our ecology is essential if we are to make the various individual programs in our government relate effectively to one another, and to advance our activities in the realm of improving the entire environment. An effective overall view of the environment and its ecology will enable us to evaluate the effectiveness of our present efforts throughout the government.

It is for this reason that I introduced the "Ecological Advisers Act of 1967." This bill proposes the creation of a Council of Ecological Advisers in the Executive Office of the President.

The purposes of this Council are manyfold. Primarily this branch of the Executive Office is to provide an overview of the problems of the ecology of the national environment, and to recommend and develop ideas and concepts for the implementation of programs designed to improve, protect, reclaim, restore, and conserve the various aspects of our environment. The Council is to establish devices for reviewing the effectiveness of, and the need for, programs throughout the federal government, or sponsored or supported by the federal government, in related areas of environmental or ecological quality.

The most important of the Council's tasks will be the relating of the various areas of environmental interest to each other, and the development of creative concepts and plans for the continual improvement of the ecological and environmental conditions of the nation.

The Council is also to direct the coordination of the efforts throughout the government by its appraisal of programs. Through its staff and research facilities, it is to streamline and coordinate the research activities of the various areas of Federal interest and involvement in ecological questions. The Council will also advise the President on the allocation of funds for the various federal areas involved with environmental questions.

The overview of the Council will be directed at the entire ecology of the environment—from the point of view of man and his needs. Ecology is not a concept which directs itself toward the effect of the interaction of the elements of the environment on one individual organism or element. However, in the case of the Council, we are interested in the environment and its ecology as it relates to man. The Council should not occupy itself with the narrow definition of each constituent element of the environment, but rather with the overall interaction of the constituent elements as they relate to man through their interaction with each other and with man. The Council must take a larger and not a smaller view of the picture of the environment. It must take a creative and comprehensive look at the ecology of our environment, concerning itself not only with the physical implications of the environment, but with the psychological and sociological implications of the conditions and interactions of the ecology of the environment on man. This will certainly include both the man-made and the natural elements of the environment.

The need for such a Council is clear. I have been in contact with representatives from industry, federal departments, and the Executive Offices and with scientific specialists; they all express the idea that, in one form or another, some type of overseeing body is necessary to deal with the ever-increasing and continually proliferating questions and areas of authority concerned with our environment. They all affirm that an ecological view is necessary. There is a need to develop a long-range view of the problem, and corresponding long-range plans. There is a need to see that those areas of the government dealing with various environmental problems are able to bring all resources to bear on those problems, are using all of the material available to the federal government, are not duplicating other efforts and programs, and are far-reaching and creative in their efforts—with an understanding of the relationship of their projects and work to other related undertakings in other places in the public and private realms.

It is only fair to ask questions concerning the placement of such a Council at the level of the Executive Office of the President. Once the need for such a body was determined, careful consideration was given for the placement of such an overseeing policy body. During the course of deliberations, the places considered for such a governmental function varied from a new Cabinet-level department to an Assistant Secretary of Health, Education, and Welfare.

It has become increasingly clear that a body created to deal with the entire environmental ecology must be in a commanding place in the government if it is to be in any position to get an effective overview of national efforts and is to be a far-reaching policy determiner. If it is to be such an overseeing body, it cannot be placed in the structure of any one department involved in any one area of ecological or environmental quality control. It could never oversee or direct the efforts of activities in other departments equal in stature to the department in which it was a subdivision. Therefore, creating an Assistant Secretary, or some such similar position in an existing agency would essentially be impractical and make such a body or position impotent.

On the other hand, there is no desire to create a "super-department" —a department to collect all of the various environmental quality research and control functions rooted throughout the various agencies and departments. As stated earlier, there are certain specific areas which are best kept where they now are, and such a massive reorganization would only postpone further the needed advances in this field.

The logical place for this Council, therefore, is at the level of the Executive. There, as a result of its position and its composition, and through its contact with the President, it is in a position to direct and enhance the activities in the federal interests in environmental quality control, and to exercise independent and creative judgment in a previously much neglected field.

The power of this council is derived from a number of areas. First and foremost is the position of the Council in the Executive Offices of the President. The ecological advisers should be the executive equivalent of the Council of Economic Advisers, and through their function of reporting directly to the Executive on a regular and frequent basis, and recommending policy, programs, and allocations, the advisers maintain a prominent and powerful position in the government.

The composition of the body also lends itself to authority both within and outside of the federal structure. The prestige value of a membership composed of experts and outstanding figures from a number of areas of private service should enable the Council to wield a great deal of influence in areas of environmental concern. The arrangement within the Council which enables the advisers to serve without leaving their important positions in public and private life, enables each individual member to maintain and enhance his own individual status in his area of specialty and influence.

The composition of the Council should be designed to include representatives of science, industry, and areas that are major concerns of environmental quality. The advisers themselves should be individuals who are capable of taking an effective overview of the situation and not become involved with the particulars of the various programs which come under the authority of the Council.

For this reason I feel that the larger part of the Council membership should be composed of social scientists, social and community planners, and public administrators. The great volume of the needed scientific expertise should come from the staff of the Council which will serve on a full-time basis. As previously mentioned, the Council members will retain their positions in public or private organizations in order to maintain positions of authority, and to help the member to maintain his specialized expertise and thus contribute more to the Council.

There is one additional underlying question which must be answered. This question deals with the concept of creating a new structure within the government each time a problem is newly recognized or appears to have grown or be growing. Should we create some new part of the bureaucracy every time we discover or redefine a problem? Of course we should not in every case, or even in most cases, for we surely have the facilities within our gigantic federal structure to handle most problems. This question can honestly be asked of any new proposal, and it certainly must be asked of a proposal of such far-reaching proportions.

In the case of this plan I have introduced, I believe wholeheartedly that the need for creation of the Council is clear. Our environment is our most immediate concern. It affects us every minute of every day, and the ecology of that environment can alter our lives. This is not a simple problem and therefore cannot be met by a simple solution. This is a problem that has roots throughout our country, and is dealt with in almost every area of the federal government. It is an area which includes environmental elements which must be actively related to each other if any valuable headway is to be made in the field of environmental quality control.

The possibilities of such a plan are very encouraging. The purview of the Council will deal not only with the interrelationship of the elements of the environment, but with the effects of those interrelationships on man himself. Only a Council in such a commanding position could be capable of collecting related information, coordinating efforts and projects, streamlining federal activities in this rapidly growing field of interest, and developing the long-range and creative plans involving all areas of the government which are necessarily involved in this realm.

Only a Council such as the one proposed could have the latitude to develop such new concepts as the psychological implications of life in an urban area in terms of total ecology, and only such a Council could be in a position to promote now unknown projects and concepts which are certain to develop in areas of urban and rural social ecology, and total concepts of waste disposal and related pollution abatement projects.

The need for such action is obvious. The environmental problems of this country are increasing daily and we must stop dealing with them only as they appear as blemishes on the national countenance. We must be far-reaching in our own efforts to establish a body with needed authority to view an immense problem from a comprehensive position and to develop effective solutions to complex and important problems.

The Subcommittee on Science, Research and Development to the Committee on Science and Astronautics of the U.S. House of Representatives held hearings during January, February, and March of 1968, as part of its continuing investigation of environmental quality and public policy. The Subcommittee considered my bill to create a Council of Ecological Advisers and other bills on federal government organization as well as the general subject of pollution control, ecology, and environmental quality.

A report was issued in June, 1968, entitled "Managing the Environment." The Subcommittee found that the Council of Ecological Advisers would be advantageous as "a focal point," and continued that:

> If given full backing by the President, the group could go far in setting national policy. It could also collect, generate, and publish the needed economic and ecologic baseline data which are now generally lacking. The subcommittee suggests that development of the Council concept be continued. In the event that the organizational changes recommended in this report do not achieve conformance to policy, the Congress should create a special Council to do so.

The Subcommittee made the following conclusion about the need for increasing our ecological knowledge:

> The past several years have demonstrated this need but there is today no federal government plan to satisfy it. The short-term, highly visible demands on scientific resources are a barrier to formulating this strategy for ecological research and environmental engineering. But the leadership of the nation, both public and private, must organize and carry out such a program. Otherwise, future Subcommittees will again study the problem of environmental management and come to the same conclusion as does this one:
>
> *A well intentioned, but poorly informed society is haphazardly deploying a powerful accelerating technology in a complex and somewhat fragile environment. The consequences are only vaguely discernible.*

Our Nation's Water: Its Pollution Control and Management

CORNELIUS W. KRUSÉ
Professor and Head, Department of Environmental Health, The Johns Hopkins University, School of Hygiene and Public Health, Baltimore, Maryland

I. The Sanitary Movement.......... 41
II. The New Sanitary Movement.......... 50
III. Status of Public Water Supply.......... 55
IV. Status of Waste Water Treatment.......... 59
V. Technological Issues.......... 62
1. Persistent Chemicals.......... 62
2. Transmission of Virus by the Water Route.......... 64
3. Eutrophication.......... 66
4. Adequacy of Water Supply.......... 67
References.......... 70

I. THE SANITARY MOVEMENT

Urbanization and industrialization began in the United States about 1820 without much understanding of the problems and hazards related to unfavorable environmental factors such as polluted water supplies, indescribable systems of feces disposal, and overcrowding with disreputable housing conditions. These conditions were compounded by filthy milk and food, myriads of flies and rats, poor nutrition, long hours of overwork, and gross ignorance and carelessness. The results were disastrous and a period of great urban epidemics followed. Today we may smile at the years of bitter quarreling and controversy between professional factions regarding the true causes of the epidemics. Although our forefathers were quite wrong in their theory that disease was transmitted by miasmata, it was wisely reasoned that if filth vitiated the air and caused disease, the obvious countermeasure was to remove the filth.

The sanitary reform movement was eventually launched soon after the Civil War, but not without many years of public debate in the newspapers, in periodicals, and by church and other opinion-leading groups. The cleanup of cities which took place resulted in outstanding reductions in illness and death, although organized public health was still unknown.

Engineers began serious study of the technology of water purification works and water carriage of sewage as developed abroad. The study of the relationship between bacteriology and epidemiology was about to dawn, and by 1880 the control of urban typhoid fever was no longer on an empirical basis. Nevertheless, existing water treatment practices still lacked a positive barrier against bacterial pathogens, and almost as many people died of typhoid then as are killed today in automobile accidents. Early in the 20th century the great effort to improve the water quality and treatment practices, including chlorination, resulted in the rapid reduction of typhoid fever death rates to about one-half the previous rates. Actually, another generation of progress in public health with extensions of sewers, treatment of sewage, pasteurization of milk, immunization, and general improvement in the standard of living was required to reduce typhoid and other filth-borne diseases to the present level of near eradication. In the conquest of typhoid, none of the other safeguards would have been adequate without the protection and improvement of our public water supplies.

The chronicle of the past seems banal in the light of the recent crises involving water supply and pollution control. In several respects, however, the recent furor over filth and unsightliness parallels the first sanitary reform. The movements were sparked primarily by small, but articulate, groups of laymen rather than by professionals. The public was aroused through propaganda techniques of fear and discontent. New experts and innovators directed criticism and ridicule at responsible authorities and prolonged the controversies regarding ways and means of attaining often poorly defined objectives. Both periods, fortunately, enjoyed the benefits of a growing body of knowledge and technology which could be applied upon the acceptance of the necessity of making more costly and complicated provisions required by urban life. There are, also, some significant differences between the old and new era. One difference is the shift away from state and local grassroot initiative and responsibility to one of greater intervention by the federal government. One hundred years ago environmental disease and death was a staggering reality, whereas today the apprehension for human health constitutes only a small part of the pressure for water management reform exerted by sportsmen, conservationists, and other citizen groups.

The orthodox definition of the police power of the state has long provided the power to regulate health, safety, morals, public order, protection of property, and welfare. The need for statutory control of water was widely recognized, and legislation designed to protect public supplies and to abate gross dangers to public health has been enacted in one form or another in every state. Purveyors of community water, whether a public- or private-owned utility, were required by law to furnish safe water to all consumers. However, in the absence of a documented threat to the safety of a water supply, it was difficult to prevent effectively the ever-increasing discharge of raw sewage and industrial wastes into natural water courses. In addition to remedial actions to abate major sources of sewage pollution for water supply protection, some control was accomplished solely on the basis of fish kill and general nuisance. Moreover, although thousands of public and private bodies were in operation for the control and distribution of water and for the collection and treatment of waste water, no national plan of comprehensive character existed. Thus each state assumed the responsibility for controlling pollution in its own way, some entering into various interstate compacts with their neighbors.

Since its origin with the old British Royal Commission on Sewage Disposal, a common practice among control agencies was to determine the required degree of sewage treatment on the basis of available dilution in the receiving waters. For example, it was empirically determined that the dry weather flow of a receiving stream must provide from 2.4 to 6 cu ft/sec of dilution water in order to handle without nuisance the raw, untreated sewage from a population equivalent of 1000 persons. Industrial wastes could be expressed in population equivalents and the dilution ratio chosen from the stream characteristic whether it was swift flowing or sluggish. The appropriate degree of treatment was specified to reduce the population equivalents of sewage when existing dilution water was insufficient for raw discharge. Even at that time it was known that, regardless of the degree of sewage treatment given, a minimum dilution of 1 cu ft/sec for a population of 1000 was required if algal growth problems (eutrophication) were to be avoided downstream. Dilution standards, now considered archaic, gave way to standards controlling the quality of waste water effluents and to the maintenance of a particular quality of water in the stream.

During the period 1930–1940 state and interstate agencies made some effort to classify public waters according to appropriate uses such as community water supply, fish or shellfish propagation, swimming, and recreation. These activities led to the establishment of criteria describing in general terms of coliform bacterial density,* dissolved oxygen, pH,

* The term "coliform density" refers to an average of the most probable number of organisms of the coli-aerogenes group per 100 ml in the water by the *APHA Standard Methods for the Examination of Water and Wastewater.*, 12th ed. N.Y., 1965, pp. 604–8.

and aesthetic quality the kinds of water desired for respective uses. In this regard the states followed the recommendations of the U.S. Public Health Service. Quality standards for drinking water were first promulgated by the U.S.P.H.S. in 1914 for water to be used on interstate carriers. Among other things, the standards called for a concentration of coliform organisms not to exceed 1/100 ml on the average and not to exceed 6 times the number over 5% of the total number of samples examined. Epidemiological evidence supported the use of the easily detected intestinal bacterial group for establishing the potential danger for water-borne transmission of cholera, typhoid, paratyphoid, and dysentery. The health-related sections of the drinking water standards, revised in 1925, 1943, 1946, and 1962 (1) have been widely accepted by all the states for public water supplies. It was, however, obvious that there was some practical limit to the extent of raw water pollution with which water treatment could deal. The 1942 revision, therefore, contained a manual of Recommended Water Sanitation Practice (2) which included classifications of raw waters with respect to the treatment required for producing a finished product meeting the drinking water standards. The maximum coliform permissible loading for waters to be given the most complete treatment was quite conservative in the light of actual performance at that time in works employing prechlorination, and the use of raw water coliform limits was discontinued. Many states continued to accept the upper limits given in the manual as the value which must not be exceeded for satisfactory drinking water in order to assure high quality water for public use.

The quality standards of water for shellfish growing also date back to the era of the conquest of typhoid. At the request of state health authorities and the shellfish industry, the Public Health Service Shellfish Certification program was inaugurated in 1925. The manual of recommended practice (3) was revised in 1937, 1946, and 1959 and gave to the Public Health Service the supervisory control under the interstate quarantine regulations and specified the unified procedures in and among the various states involved. Epidemiological evidence has not established the exact numerical relationship between bacterial water quality of growing areas and the degree of hazard to health. Shellfish certified as safe for human consumption can be harvested only from areas free from obvious or potential sewage pollution. The growing area having a satisfactory sanitary survey, furthermore, must have bacterially clean waters equivalent to a median coliform density of 70/100 ml as determined by routine examinations. There is no evidence that the shellfish water quality criteria are not equally effective against viral infectious hepatitis as well as against the common shellfish-borne enteric diseases.

Numerous epidemiological studies seeking the adverse health consequences of swimming in polluted waters have not produced sufficient evidence for establishing the upper coliform limits for such sports. It is not surprising, therefore, that a great divergence of opinion exists among states regarding the quality of natural water suitable for swimming. The states have tended to set bacterial classifications at levels that would provide adequate areas for these activities, putting strong emphasis on sanitary surveys. The task of reviewing from time to time the evidence leading to recommended practice pertaining to sanitation of bathing places has been undertaken, since 1925, by the Conference of State Sanitary Engineers and the Sanitary Engineering Section of the American Public Health Association. Some ten reports, the latest dated 1957 (4), have been accepted with minor variations by all of the states. The fact that water in swimming pools was treated and under control made it relatively easy to demand and obtain a high quality, which in most states is equivalent to drinking water.

The intent of the early dilution standards was to avoid anaerobic conditions which resulted in nuisance, such as the appearance and odors of putrification, blackening of waters, and destruction of fish and green plant life. It was long recognized that the depletion of dissolved oxygen in water resulted in fish kill. Fisheries experts studied the environmental habitats of various species of fish and recommended the minimum oxygen requirements for eggs, young fry, and adult fish. Around 1940 considerable information was being accumulated on specific effects of industrial wastes in addition to the effects of oxygen depletion and pH ranges (5,6). European experience had shown in pond fish farming enterprises that properly diluted domestic wastes were not harmful but could beneficially add a source of nutrient to the fish–food cycle. Unfortunately, this was not the case with many wastes of industrial character since fish, unable to adjust or adapt by metabolizing the foreign chemical molecules at the cellular level, were found to be quite sensitive to low levels of toxic substances. Thus, the practice of "minnow bioassay" became common for determining the toxicity of wastes. Furthermore, any pollutant which destroyed the fish food supply or mechanically covered the spawning areas, though not directly toxic, was equally evaluated in the life cycle of fishes.

The competence and capability of the federal government in giving the states assistance in these problems had long been recognized. The Public Health Service Stream Investigation Station in Cincinnati, Ohio, created in 1912, studied the pollution and self-purification of streams and the associated practices of water and sewage treatment. Under the direction of the epidemiologist, Wade Hampton Frost, a scientific team of engineers, chemists, biologists, and physicians uncovered many fundamental principles with outstanding thoroughness. Notable was the elucidation of the fate of bacteria in polluted waters which was vital to the protection of downstream sources of water supply. The formulation of the rate of deoxygenation in streams below sources of pollution and the rate of atmospheric reaeration in streams were studied. From these were developed the well known mathematical "dissolved oxygen sag curve" formulation of Phelps-Streeter. When this formulation was applied to field observations, reaeration constants could be calculated; however, the mathematical complexity of the sag formulation limited its usefulness. Imhoff and Fair (7) introduced the "self-purification constants" appropriate for typical streams in the United States and simplified the mathematics for solving practical stream sanitation problems. A common utilization involved the analysis of each stretch of river from point to point of pollution. In this manner the maximum permissible waste loading in BOD* or population equivalents could be forecast at points throughout the basin, indicating where and to what degree treatment must be ordered to maintain a desirable level of dissolved oxygen.

The states that undertook the task of classifying public waters as to usage discovered for the first time the detail location and extent of pollution. The scheme for classifying followed more or less the existing uses made of sections of streams, lakes, and estuaries, a process which did not materially upgrade the overall quality of the states' public waters. The criteria for the low quality uses were not clearly defined with respect to coliform density or dissolved oxygen and generally were described by the absence of nuisance odors and unsightly suspended and floating matter. Consequently, areas set aside for low quality uses were not conducive to desirable fish, shellfish, and water birds. There were many practical difficulties which prevented the establishment of a uniform policy regarding the minimum water quality or required degree of treatment because of the vast differences between geographical regions of the United States. In 1940, for example, almost 10,000 of the 13,000 community water systems utilized ground water sources which considerably reduced the urgency for maintaining high water quality in

* The BOD is an indirect measure of the amount of decomposable matter in waste determined by measuring the amount of oxygen demanded directly or indirectly by the living organisms responsible for decomposition.

local surface streams. Water for many of the large metropolitan communities, including Boston and New York, was unfiltered surface supplies collected from sparsely inhabited watersheds. Therefore there was no pressing need for high water quality in the lower reaches of the tidal streams. Coastal communities utilizing ocean outfalls had only to consider on-shore bacterial pollution and nuisance.

For much of the population served by community water and sewerage systems, the opposite situation prevailed. One city after another withdrew water and discharged wastes back into the same body of water. In such dual usage a much higher level of water quality control had to be provided because it was essential to consider taste-producing substances and the possibility of harmful concentrations of toxic substances in addition to bacterial loadings. It was, however, a gross exaggeration of fact to suggest that "one man's sewage is another man's water supply" since several protective devices were invariably provided. They included treatment of water and sewage in addition to the purification capacity and the available dilution of the streams. Reuse and multiple use of our water resources are inevitable and create the demand for better and more efficient methods of water renovation.

The sharing of the purification burden between water and sewage treatment and the natural self-purification assets of the stream has been scientifically and economically established as a principle of water resource management. The practical application of the principle was subject to abuse which led to the further deterioration of streams and lakes, particularly around large population and industrial centers. The fears of public health hazard, once the main target of pollution control, were being supplemented by concern for aquatic life and wildlife and despoilation of natural beauty. The main issue was the conflicting philosophies regarding the legitimate use of water resources which tended to polarize on one side the zealous conservation group and on the other those accepting some unfavorable ecology for the benefit of the majority of the people. Some states chose to compromise by requiring at least primary treatment (removal of settleable solids) of waste waters for all communities regardless of essential need. The advantage of such a policy was largely in the simplification of enforcement programs and was thought to result in less opposition from pollutors.

The years of the great depression were claimed to be a period of complacency, with the water-borne diseases laid to rest by public health controls and the reliable performance of the standard water filtration plant process regardless of the source of supply. Aside from improvements in mixing and flocculation, the only noteworthy advance in the process was in disinfection. Marginal chlorination, now known as "combined residual chlorination," was giving way to more efficient processes. The development proceeded from exploration in "pre- and postchlorination" to "superchlorination" and finally to "breakpoint chlorination," now called "free residual chlorination."

A review of the facts, however, shows that substantial progress was made in both water and waste water treatments during this period as shown in Figures 1 and 2. Despite the shortcomings of regulatory machineries, many states made advances in water pollution control and waste treatment practices. It was during the thirties that the fruits of basic research led for the first time in the United States to the construction of dozens of works for the oxidation of wastes by the activated sludge process. These compact units could give a high degree of removal of organic matter and bacteria surpassed only by the intermittent sand filter plant, a process well suited only for small communities. Many of

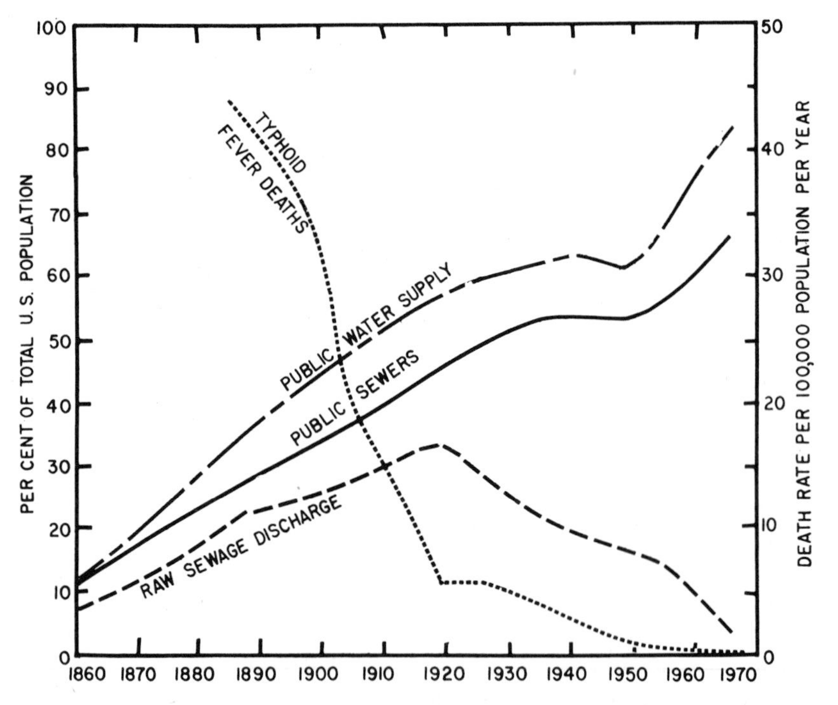

Fig. 1. Growth of public water supply and sewerage in the United States, showing the slow but steady decline of typhoid fever deaths during the first sanitary movement beginning in 1860, followed by the crisis period of World War II and the beginning of the second sanitary movement of 1950.

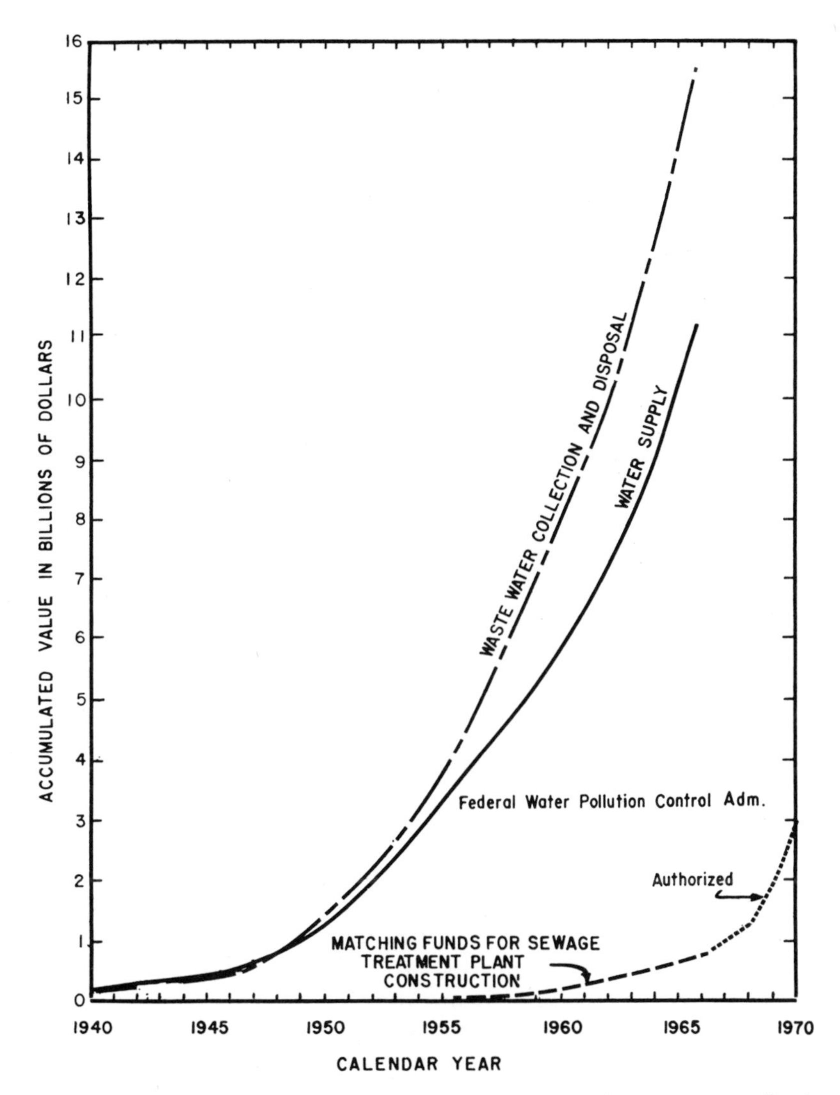

Fig. 2. The dollar value of new construction for water and waste water collection and disposal facilities following World War II in the United States. (Source: Department of Commerce, Bureau of the Census, Construction Reports Series C 30—1967.)

the plants were enormous in size and treated large volumes of sewage. Typical of these were the Calumet and Northside plants in Chicago, New York City's Ward Island and Coney Island plants, and plants in Milwaukee, Cleveland, and Columbus.

A new concept of biofiltration was introduced by the replacement of the spray nozzle on trickling filters with rotary distributors and the provision of single or stage filtration with recirculation and high hydraulic dosing rates. Mechanization in grit and sludge removal, vacuum filtration for sludge dewatering, and manufacture of heat-dried sludge fertilizers were all developments of the prewar period.

By 1940 the wastes of some 30 million American and even greater millions of population equivalents of wastes from industries continued to be discharged without treatment into oceans, large rivers, and lakes, but a definite trend toward treatment was apparent (Fig. 1). The admission that water pollution problems cut across municipal, county, state, and international boundaries was exemplified by numerous instruments and institutions for better management. The Tri-State Treaty (1932), concluded by New York, New Jersey, and Connecticut, set treatment standards for effluents affecting the waters of the metropolitan area of New York City. An international treaty with Canada ordered all cities on the Great Lakes to stop discharging untreated sewage into boundary waters. The Interstate Commission of the Delaware River and the Potomac River Interstate Commission searched for the best ways and means of setting quality standards for the improvement and maximum beneficial use of these rivers. But even the regional and basin approach could not overcome the apparent need for adequate and comprehensive state law enforcement. Metropolitan district solutions were being actively pursued for both water and sewerage and were being carried out as rapidly as funds permitted in such large scale undertakings as the city–county of San Francisco, Minneapolis–St. Paul, and in the District of Columbia. Much of the construction of water and sewage treatment was made with the assistance of the Federal Public Works Administration, which purchased the construction bonds of local government bodies at par with interest at 4% per annum.

The momentum was not carried far into the next decade because the nation was engaged in a global war, and priorities were reordered toward more critical national objectives.

II. THE NEW SANITARY MOVEMENT

Even before the end of the second world war there were indications that water and sewerage works were rapidly approaching design capacities and suffering from inadequate maintenance. Except for emergencies almost all additions and extensions to systems had been postponed for the duration of the war. In an effort to provide for a less chaotic transition from war to peace time, municipalities encouraged various planning bodies to bring proposals of public works up to date for immediate construction as soon as conditions permitted. Major public works in the average municipal practice traditionally require considerable time from the planning to the actual initiation of construction. Consequently, the blueprints ready for immediate construction involved a little more than long delayed minor projects meeting obvious and immediate need. The U.S.P.H.S. had already revealed the magnitude of the immediate water and sewerage needs based on a state by state inventory made in 1947–1948. The estimated cost of the backlog of construction made during a period of steadily increasing prices was about $8 billion. When the whole of the United States was considered, the estimates were unbelievably low and clearly demonstrated the lack of awareness by municipal governments for impending urban crises for which solutions could not be found at the customary leisurely pace of operations.

The lures of the suburbs began when the expanded use of the automobile facilitated the exodus of, primarily, the well-to-do from the city to outlying low density areas. Large, planned suburban areas were believed to be unnecessary and it was generally assumed that all cities had unused subdivision land which could accommodate the demand for new housing. The cities were properly directing plans toward rehabilitation of blighted areas and slum clearance in an effort to provide a satisfactory city environment which might hold and attract an economically viable population. The chronic housing shortage, intensified by the depression, became acute during the war years and after the release of restrictions of building materials could no longer be contained within the overcrowded, substandard existing structures. From a wartime building level of some 300 thousand new housing starts per year, the building rate jumped to nearly 1 million units in 1946 and steadily increased each year to a peak of almost 2 million new houses in 1950. In the metropolitan areas the construction of new homes has been at a fantastic average of almost 1 million per year over the two postwar decades. The cities had their share of the housing boom, but many sites with existing water and sewer services were by-passed to satisfy the great demand for the low tax–low density areas outside the city limits. The unexpected and unplanned phenomenal development of raw land in the metropolitan areas caught the rural political subdivision totally unprepared to give essential community services. One of the greatest gaps in public administration has been the failure to provide early and promptly the machinery for engineering, administrating, and financing of water and sewerage services to meet the requirements of the rapid outward movement of population. The precedents for such arrangements were already extensive, but the technical, economic, and social intricacies involved rested upon local leadership and responsibility and upon the will of the people who benefited from and paid for the services. Where sanitary districts or metropolitan water districts had already existed, the response to need was relatively rapid and public water, at least, could be extended into the new subdivisions. More often, however, neither public water nor sewers could be supplied, and home builders resorted to the temporary expedient of household wells and septic tank systems which were never intended for other than sparsely populated areas. Relatively few local health departments were staffed for regulating such installations and at that time one-half of the counties in the United States had no sanitation services for maintaining adequate control on well protection and on lot sewerage disposal schemes. The exact number is unknown, but it is estimated that between 4 and 5 million wells and septic tanks were provided for new homes in the period 1946–1950. The former city resident soon became disenchanted with the responsibility of operating his own water and sewer systems and voiced strong dissatisfaction with "septic tank suburbia."

While the flood of suburban home building continued unabated, the problems of sewage nuisance, health hazard, water shortage, and economic hardship intensified. A crisis was developing which the state and local governments could no longer avoid. Although very little notice was given by the news media, health departments feared the resurgence of diseases of insanitation. Miraculously, the protective barriers held and no increases in reported water-borne disease outbreaks or poisonings were attributed to public, semiprivate, or private water supplies as shown in Table I.

The great effort of the decade 1950–1960 in the area of comprehensive and master planning, establishing new service districts, and passing legislation governing new subdivisions began to produce some effect. Federal legislation designed to assist in bringing desperately needed water and sewerage to the population spill into undeveloped areas was slow in coming. The Water Pollution Control Act of 1948, in some respect, exacerbated the situation by diverting attention and limited resources to stream pollution problems. States were encouraged to forbid extensions of or connections to existing public sewers until certain

TABLE I

Average Annual Number of Reported Water-Borne Disease Outbreaks in the United States, 1938–1960[a]

Period years	Outbreaks	Cases per outbreak
1938–1940	45	583
1941–1945	39	201
1946–1950	23	121
1951–1950	8	139
1956–1960	7	121

[a] S. R. Weible, F. R. Dixon, R. B. Weidner, and L. S. McCabe, "Waterborne Disease Outbreaks, 1946–60," in *J. Am. Water Works, Assoc.*, **56,** 948 (1964).

stream pollution abatement requirements had been met. Permission for subdivision sewers and temporary sewage treatment plants were often refused for fear of contributing to the pollution of streams.

The drought of 1953–1954 clearly disclosed serious inadequacies for the provision of public water brought about by the increased population and water use, supply failure, or inadequate distribution system capacity. A high priority program amounting to $4.7 billion was carried out during the period by the water works industry. The massive expenditures would not have been possible without strong public approval and support, although the primary beneficiaries lived outside the central cities. An estimated additional 42 million population was connected to public water supplies and in so doing displaced some 500,000 home well–pressure tank systems. For the first time the number of individual home well systems that were retired exceeded the number of new wells installed.

The growth of public sewerage has always fallen behind public water supply because of the difficulties of establishing an equitable source of revenue to meet the bond obligations and the inability to function with dignity and effectiveness of a public utility. The revenue for sewerage construction, operation, maintenance, and replacement reserve has been best found in a sewer service charge system formulated to assure a self-supporting enterprise. The philosophy of the sewer service charge has been the equitable allocation of costs to those who benefit and most commonly has been based on water consumption. During the decade $6.6 billion was expended for public sewerage systems. Of this amount approximately one-half was applied to stream pollution abatement through interceptor construction and additions to treatment. The effort was impressive, but very little headway was made in the struggle to eliminate the septic tank and cesspools from congested areas. Perhaps 10 million on-lot disposal units were taken out of operation by the availability of new public sewers and this represents a number greater than all such systems in existence in 1940. Nevertheless, the proliferation of new on-lot sewage disposal systems was such that by 1960 some 21 million units were being relied upon in areas served both by public water and home well systems. The latter situation constituted the more serious threat to health in that water well and sewage disposal were too intimately associated as evidenced by the occasional off-tastes and appearance of detergent foam in home water taps.

The costs of extending water and sewerage services into the suburban areas were found to be quite high because of generally larger lot sizes and sparse population densities located along sections of the proposed mains. The experiences indicated that reliance upon water rates alone or in conjunction with front foot assessments would not be adequate in producing the necessary capital and that all property which would benefit from the facilities in the future should share a reasonable part of the cost. The state and local subdivision regulations attempted with varying success to limit individual home systems to slowly developing subdivisions in predominantly rural areas and to exercise control over the design, construction, and maintenance of such systems. The codes were written with minimum lot size restrictions for on-site systems which would encourage housing development on land where public water or both public water and sewerage were planned for construction.

During the period there were some state grants and loans for planning, design, construction, operation, and maintenance of public sanitary works, but for the most part the accomplishments made during the fifties were paid for with community funds through the traditional manner of assuming local responsibility. Conservation interests, unhappy with the progress in stream pollution abatement, fought hard for federal legislation to relieve the states from the responsibility of enforcing pollution regulations. The Federal Water Pollution Control Act (P.L. 660 amended) did not succeed in this endeavor but established the pattern of appropriating annual sums, beginning in 1956, for furnishing matching grants-in-aid to communities for stimulating the construction of waste treatment works. The formula for allocating federal aid was based not so much on pollution abatement need, but rather on the economic rating of the various states and territories. The total U.S. investment in water and sewerage facilities in the postwar years was quite substantial as can be seen in Figure 2. A portion of the capital improvements under sewerage works includes some storm sewer construction, but approximately one-half represents investment in sewage treatment facilities.

The improvement in the sanitary standard of living continued to increase and can be seen by the figures in Table II. The rise in the number of homes served by some type of piped water and water carriage of wastes was phenomenal and reflects the growth of urbanization from 52% of the population in 1940 to 70% in 1960. Therefore, it is not surprising that the averages for the entire United States, rural and urban, reveals that the great majority of the population utilized modern sanitary conveniences. The table also indicates the unfinished business of extending public services wherever feasible and the sizable population of the country that exists with relatively primitive arrangements. The takeoff point for such corrections began in 1950. Any major diversion of effort or resources from these goals would not be in the best interest of public health.

TABLE II

Growth of Water and Sewerage Systems in the U.S.
Estimated Population (millions) [a]

	1940		1950		1960	
	pop.	%	pop.	%	pop.	%
Piped water supplies	98.3	74.5	125	82.8	165.4	92.9
Public and private utility	85.0	64.4	92	61.0	134.0	75.3
Home and farm systems	13.3	10.1	33	21.8	31.4	17.6
Nonpiped water supplies	33.7	25.5	26	17.2	12.6	7.1
Total	132.0	100.0	151	100.0	178.0	100.0
Water carriage sewerage	84.8	64.2	113	74.5	160	89.8
Public sewers	72.0	54.5	80	53.0	111	62.3
Septic tank and cesspool	12.8	9.7	33	21.5	49	27.5
Nonwater carriage, privy	47.2	35.8	38.5	25.5	18	10.2
Total	132.0	100.0	151.0	100.0	178.0	100.0

[a] Based on Dept. of Commerce, Bureau of the Census, 16th Census Reports 1940 Housing, Vol. 2, and U.S. Census of Housing 1950, Vol. 1 and 1960, Vol. 1, and unpublished data.

III. STATUS OF PUBLIC WATER SUPPLY

In 1967 it was estimated that in the United States 157,000,000 persons in 20,000 communities enjoyed public water service under the regulation of state agencies. Without the wide acceptance of standardization among these utilities—large and small, public and privately owned—it is doubtful whether they would have been able to economically sustain the growth and development made necessary by expanding population and industry. Much of the credit for standards must be given to the American Water Works Association, a professional society of more than 18,000 members, who have worked to improve the industry through the formulation of standards for basic products, materials, and procedures. Since its founding in 1881, AWWA has remained alert to new methods, equipment, and materials and provided the bridge between basic research and practice.

Despite the remarkable postwar progress made, the industry cannot afford to coast because recurring water shortages continue year by year even in normally humid regions of the country. In each instance the provision of public water is not due to natural scarcity, but due to deficiency in facilities for transmission, proper storage, treatment, and distribution or reuse of water.

The U.S. Department of Commerce in its periodic check of adequacy of water supply capacity came to the conclusion that in 1963 32% of the population of the United States was served by public systems that were showing hazard of deficiencies or were definitely deficient in meeting current demand. Furthermore, it was surprising to the state agencies concerned to learn from a Health, Education and Welfare task force report (8) that despite the dramatic control and decline of water-borne disease 6,000 public water systems are not able to meet federal standards for the quality of drinking water. The conclusion was based on inferences derived from the results of the Public Health Service survey of 742 interstate carrier water supplies, of which 21% did not completely meet the federal standard. To many readers this statement signified that 6,000 public water supplies served unsafe water and, of course, this is not true. The federal standard imposes two types of limits, those which when exceeded shall be grounds for rejection due to an adverse effect on public health and those, not mandatory, which shall not be exceeded whenever a more suitable source may be developed at a reasonable cost. The non-health related aspects of water quality include limits pertaining to clarity, color, taste, odor, hardness, staining, corrosivity, and temperature. These total quality goals are attainable by known treatment procedures, but because of highly mineralized water in some communities, especially from ground water sources, the expense of completely conforming to the federal standard may be very high. The management and the public customer have to weigh the value of the benefits to be derived against the cost of producing the finer quality water.

The task force report could have mentioned that 3000 systems serving more than 50,000,000 people practice the prophylactic addition of fluoride to drinking water. Controlled fluoridation has reduced the dental caries incidence in children approximately 60–65%. An additional benefit of controlled fluoridation is the maintenance of normal skeleton and the prevention of osteoporosis (9). Also, evidence is accumulating which suggests that fluoride and other mineral constituents of drinking water may significantly lower death incidence by heart attack.

The infusion of federal funds for research has significantly extended the fundamental understanding of the unit processes employed in water treatment. Today it is common practice to utilize long contact time free residual chlorination ahead of and through filters, so that it is most unusual to recover bacteria in the water applied to the filter. Consequently, modern chlorination practice has largely eliminated the function of filters for bacterial removal and permits modification in filter design and operation. Improved water coagulation with coagulation aids including the polyelectrolytes* permits high water filtration rates with relatively coarse filter media for greater efficiency and reduction in cost. It has long been known that in the backwashing process of sand filters, hydraulic classification places all the fine grains at the top and all of the coarse material at the bottom and results in the top few inches of the filter doing most of the work with the remaining 23 in. of the filter media contributing very little to filtration. This problem has recently been overcome by the development of new multilayered filters. Essentially the filter media is composed of three or more layers of materials of widely different densities. For example, the bottom of the bed will have a layer of small grain size high density garnet sand, over which is a layer of larger size less dense silica sand and a top layer composed of still larger size, low density anthracite coal. With improved coagulation and settling the new filters may double the capacity of existing plants with finished water of as good or better quality than that obtained with conventional filter units. Plants are available with considerable automation and control systems monitoring of all factors related to treatment including raw water turbidity, coagulant and coagulant aid dosages, flow rate, backwashing, and final effluent turbidity.

The greatest challenge of the future is for communities to finance the

* Specifically polyacrylamides and polysaccharides.

enormous construction expenditures required to overcome current deficiencies, obsolescence, and demand by future population growth. The amount of federal participation in water supply construction for communities seems arbitrarily fixed, and the distribution of appropriation takes into account such things as the state's population and its personal income. The Housing and Home Financing Agency and the Land and Facilities Development Administration of the U.S. Department of Housing and Urban Development have offered aid in public works undertakings if communities are unable to sell their bond offerings at reasonable rates. In 1966 the requests to LAFDA for bond underwriting for water and sewage construction totalled $2.5 billion, yet only $100 million was available for funding. The expenditure for water utility construction alone in 1966 was $1.3 billion, which is about one-half the average annual needs of $2.1 billion through 1980 as shown in Table III.

TABLE III

Average Annual Construction Requirements for Water Utilities, 1967–1980 [a]

	Millions of 1966 dollars			
Region	Deficiencies	Obsolescence, depreciation	Population growth	Total
Northeast	103	156	269	528
North central	112	171	293	576
South	109	166	286	561
West	77	118	203	398
Total	401	611	1,051	2,063

[a] Source: BDSA, U.S. Department of Commerce, Business and Defense Services Administration.

Historically, local entities have served the population with public water supplies efficiently and economically, and these agencies must continue their primary responsibility under state control. Agencies, public and private, such as water districts, cities, towns, villages, investor-owned water companies, commissions, and authorities, are responsible for planning, financing, constructing, and operating public water supply systems as self-sustaining utility-type enterprises. A sound water resources policy must have as its primary objective the provision of an adequate supply of water for our people, carefully planned and properly managed. It is obvious that in the race for new sources of supply of good quality raw water the public water industry must operate within the framework of national needs for wise managements of water resources. If the water works industry is to continue to support and finance itself with a minimum of public assistance as it has done in the past, it must stop being proud of slogans like "water cheaper than dirt" and begin to be ashamed of poor service caused by low water rates. The cost of public water service in competition with backyard wells and cisterns of the first sanitary movement is now history. New standards of quality and service and the greater employment of inferior raw supplies and inconvenient sites will require a more rational study of costs for high water quality and service to meet new market demands.

IV. STATUS OF WASTE WATER TREATMENT

Obviously, municipal and industrial wastes result in organic and chemical pollutants and a temperature increase. However, almost any use that man makes of our water resources results in some degree of impairment of quality which may affect reuse. Irrigation represents the largest withdrawal of water from ground and surface supply and results in silting and significant increase in dissolved solids in the drainage water. The modern trend toward massive 1000 MW steam–electric generation stations make considerable demand for single pass cooling water ranging in volume per kilowatt per year from 500,000 gallons for fossil fuel to 800,000 gallons for nuclear fuel. Aside from thermal pollution, the cooling water from nuclear plants will contain low levels of radioactivity which must be considered in terms of reconcentration in fish or shellfish. Impounding reservoirs for hydroelectric power, flood control, and multiple uses including low-flow augmentation for pollution abatement is generally beneficial to water quality; however during thermal stratification periods, deep reservoirs may release poor quality waters downstream. Many land uses and practices of man contribute to the degradation of water through drainage and runoff representing an infinite variety of water pollutants detectable in storm flushing of urban and rural land, soil erosion from farms and from construction projects, and the acid drainage from mining operations.

The responsibility for pollution control and abatement rests with those who return undesirable waste products to our streams, lakes, and underground sources of water. That existing organizations have been successful as much as they have in this regard is a tribute to the professionals who have been quietly working with communities and industry on a day-to-day basis to obtain compliance without prolonged and costly legal procedures. It is a relatively simple matter compared to the financial aspects to identify pollutants and degree of treatment required in the controllable discharges of municipal and industrial wastes. Waste water treatment is a nonrevenue-producing enterprise which traditionally lacks appeal or strong financial support. Progress in terms of number of treatment plants built is meaningless if support for skilled operation and maintenance is not provided. The fixing of responsibility for the prevention and control of water pollution originating from land use practices is not so simple.

Many state regulatory agencies welcomed the greater participation of the federal government in the water pollution field. They anticipated financial support to municipalities for construction of necessary sewage treatment works, grants-in-aid to states for upgrading administration of pollution control agencies, research and training activities, and the development of comprehensive river basin studies.

While the federal laws recognize the states' primary responsibility, conservation interests have worked hard to expand the enforcement powers of the federal government. Consequently, several amendments have been made to the act which in each case substantially provided a more militant abatement program at the federal level. The Public Health Service was relieved of responsibility and the Federal Water Pollution Control Administration transferred from the Department of Health, Education, and Welfare to the Department of the Interior (1966). The Department of the Interior was chosen since it was acknowledged that pollution of water goes beyond public health considerations and that the Geological Survey, Fish and Wildlife Service, Bureau of Mines, and Reclamation Services appeared best qualified to get the stream pollution abatement job done.

The Water Quality Act of 1965 established stronger enforcement through the promulgation of water quality standards for interstate water and satisfactory plans for their implementation. The states with only limited criteria and guidelines of what was required had until June 30, 1967, to submit standards and plans for review and approval by the Federal Water Pollution Control Administration. If the states were unable to submit mutually acceptable programs, the Administration would supply them.

The muscle of the Water Quality Program is claimed to be the enforcement power of the Secretary of the Interior who can proceed with suits against violators without the usual conferences and hearings. Attitudes for compliance by communities are sweetened with subsidy given under the Clear Rivers Restoration Act of 1966, which authorized $3.4 billion for the building secondary sewage treatment in the next six years. Nothing was provided for industrial wastes, but there are currently over 100 bills before the congress proposing tax incentives for industrial pollution abatement. It is not clear whether these vast federal sums will be expended wisely, since the current state and federal philosophy calls for "complete" treatment and year round chlorination regardless of whether or not it is actually needed. Money will obviously be wasted if such requirements are imposed on communities and industries discharging to the ocean or large river systems. Until the Water Quality Standard program has been established in all of the states with federally approved plans, no one can estimate the cost of water pollution control. As of December 1967 only 10 states have approved plans.

It is becoming evident that corrective federal legislation has outrun the logical and practical objectives for regulation and intelligent action. State and federal health agencies have the legal and moral obligation to prevent hazards to health, and this responsibility is being discharged. Potentially toxic or dangerous wastes, whether radioactivity or viruses, are constantly under surveillance. The federal infusion of construction money, large as it appears to be, has been actually seed money which encourages communities to wait for grants rather than to assume financial responsibility for upgrading their waste water treatment. Construction funds authorized under the Clean Rivers Act were to escalate from $150 million in 1967 to $1.25 billion in 1971. However, there is reason to suspect that great gaps occur between the amount of money authorized, appropriated, and actually expended for water pollution control—despite the proclamation of the need for more action now. So long as a population of 6.8 million continues to discharge raw sewage, a population of 24.6 million discharges inadequately treated sewage and a population of 6.2 million desperately needs public sewers, the emphasis should be on the obvious solutions to the pollution problem.

The estimated cost of this backlog in 1966 was $2.7 billion (10). The expenditure in 1966 for municipal waste water construction came to about $1 billion. Treatment accounted for about 35% of the total or $400 million. Not included in the estimated expenditure is $200 million made by industry. At this time large diversions of resources and effort will not be realistic in such areas as large base computer studies, automation of round-the-clock river-monitoring stations, storm water treatment schemes, or premature application of advanced waste water technology such as osmosis, filtration, and deionization.

The public is more concerned with the obvious needs for water pollution control through the construction of necessary treatment facilities and proper operation through frequent inspections by regulatory agencies and improved training for competent operation. In regard to the latter it is discouraging to note that only 15 states have mandatory certification of operation of water, sewage, and industrial waste treatment works.

It is acknowledged that many problems of water pollution extend beyond the boundary of governmental subdivision of city, county, and state. Solving problems by the regional and basin approach is encouraged by a 10% bonus in addition to customary federal matching funds for planning and construction. Eight of such interstate agencies are now recognized for federal aid, all of which were compacts initiated by the interested states. The largest, the Ohio River Valley Water Sanitation Commission, involves eight states. The achievements of ORSANCO have been most successful (11).

Over the past several years the advances in waste water treatment and water pollution control have been published in detail in many excellent books and documents. In addition current research in improved biological, chemical and physiochemical processes in waste water treatment are found in proceedings of numerous conferences and technical journals such as the *Journal of the Water Pollution Control Federation*. Municipal and most industrial wastes can be treated by known process to a high degree of purification. The cost of higher treatment may be expected to double and the conditioning and disposal of the solids and sludges may develop to be the most difficult and costly aspect of pollution control for the large community. The apparent conservation aspect of returning sludge as a safe inoffensive organic humus to agricultural land has been extensively undertaken and decisively found to be an economic failure. Ultimate sludge disposal in this manner, if truly valuable for conservation, should receive subsidy to overcome the excessive cost of preparation and transport now borne by city taxpayers.

V. TECHNOLOGICAL ISSUES

1. Persistent Chemicals

Paralleling the increase in water pollution and water requirements as a result of the great postwar urban expansion is the problem of industrial wastes and, in particular, the problem of the number and complexity of persistent, refractory organic and inorganic chemicals. These products and wastes of our industrial technology are discharged in industrial and domestic wastes but also enter water in subsoil seepage and runoff from rural and urban areas. These persistent chemicals resist removal or destruction by standard water and waste water treatment processes. Included are such widely used compounds as synthetic detergents, complex organic pesticides, and waste products of chemical and pharmaceutical manufacture. The number of the exotic chemical organics is growing so rapidly that the composition and toxicity of many are largely unknown.

The gross contamination of these materials in water supply is determined by the chloroform-soluble carbon filter extract method referred to as the CCE concentration. Waters derived from sources remote from industrial activities or human population usually show CCE concentrations less than 0.04 mg/liter. Where concentrations of CCE of 0.2 mg/liter are found, the taste and odor of the water are always poor.

Of principal concern is the lack of knowledge regarding the health effects of these low concentrations in drinking water. While no adverse health effects are experienced at existing levels, there is evidence of the toxicity to fish, the unpleasant taste and odor in water supplies, and the interference with some treatment processes such as ion exchange and softening.

In order to maintain the low concentration levels in water, industry must assume the major responsibility for analyzing these complex organic compounds and for determining their safe concentrations for use or discharge to receiving water. The registration and labeling of economic poison under federal and state laws and regulation is an example controlling exotic chemicals which might be detrimental to the public health. Much can and has been done by industry in the substitution of less stable or more biodegradable products. The synthetic detergent content of lakes, streams, and groundwater has decreased markedly since the industry has replaced "hard detergents," the branched chain type, with the "soft" biodegradable type. Substitution of organophosphorus for organochlorine pesticides is another example of reducing the persistence and accumulation of complex organic chemicals in aquatic environment and related food-chain processes.

Should the CCE concentration approach or exceed 0.2 mg/liter, the limit set in the Federal Drinking Water Standard, treatment can be provided for the removal of organic compounds. Powdered activated carbon applied prior to filtration is an old established water treatment practice. However, if the taste- and odor-producing substances require high dosages of activated carbon, the treatment cost will be prohibitively high. While some small filters of granular carbon have long been in use, only recently has the development of granular carbon filters for large scale water treatment become feasible. The economic application of carbon for large scale treatment is contingent upon on-site reactivation of spent carbon, for which practical techniques for granular carbons are available. The principal features of the process involve replacement of the customary filter sand in the conventional rapid sand filtration plant with a hard coal based granular activated carbon. Filtration rate and bed depth are adjusted to provide the required water–carbon contact time. The carbon bed effectively removes both turbidity and CCE concentrations from chemically coagulated water. The filter is routinely backwashed in the regular way for cleaning until laboratory tests reveal the need for reactivation. The novel feature of the system is the removal of spent carbon, thermal reactivation, and return of activated carbon to the filter.

Carbon adsorption by multiple granular beds or the use of powdered activated carbon in upward flow towers is an essential unit process in complete waste water renovation schemes. Aside from the waste from specific manufacturing processes, the persistent exotics are rarely discharged in municipal wastes, and treatment cannot remedy the pollution of waterways by these materials. Complete waste water renovation, though technically feasible, will not be urgently needed in the immediate future.

2. Transmission of Virus by the Water Route

In a public water supply it is desirable to maintain a free chlorine residual throughout the distribution system since the safety of the water must be determined at the tap of the consumer. Coliform organisms are often absent in properly protected ground water supplies; nevertheless, the application of a free chlorine residual will provide some protection against possible contamination through faulty plumbing or defects in the water distribution system. In surface water treatment plants, application of a free chlorine residual is commonly practiced, which provides for a long contact period in relatively low pH waters throughout coagulation, sedimentation, and filtration processes.

In terms of magnitude, gastroenteritis and diarrhea are still the most important water-borne diseases. The potential of the common source outbreak in a public system is apparent and utilities cannot afford to become complacent and careless in disinfection practices. Untreated ground water, along with distribution system difficulties, has been the most frequent cause of recent outbreaks. The water-borne epidemic of *Salmonella typhimurium* involving 18,000 persons at Riverside, California, in May and June of 1965 is such an example.

Except for the single notorious Delhi, India, epidemic, less than 5000 cases of viral infectious hepatitis in some 49 outbreaks have been ascribed with certainty to water-borne transmission. Water-borne viral transmission can occur, but all evidence points to the fact that, like polio, water-borne infectious hepatitis is rare and has never been associated with a properly treated public water supply.

It has been repeatedly shown that free residual chlorination effectively inactivates many enteric viruses in waters of varying temperature and pH. However, it is difficult to maintain free chlorine in heavily polluted water or sewage due to the presence of many interfering substances which combine with free chlorine and reduce its disinfection properties. Nitrogenous substances including ammonia, albumose, proteins, and free amino acids combine with chlorine to give inorganic and organic chlorine species. Against bacteria the application of combined chlorine residuals can at proper levels and contact time be an effective disinfectant. For virus inactivity, however, it is known that the inorganic chloramines are very poor and that the organic combined chlorine compounds have no virucidal properties at all. Therefore, it is possible to have virus passing through a complete water filtration plant when disinfection is by combined rather than free chlorination and despite the fact that the water quality as determined by the coliform test indicates satisfactory disinfection. The knowledge explains the New Delhi epidemic of infectious hepatitis where 28,745 consumers of the public water supply were infected. The episode was associated with heavy sewage pollution of the raw water for which combined residual chlorination was adequate only for bacterial disinfection.

Terminal disinfection of sewage effluents is a common procedure for protecting recreational waters and shellfish growing areas. The dosage of chlorine applied is determined by the desired most probable number of coliforms in the effluent. The combined chlorine residual normally maintained at the end of a 15–30 min contact period will not inactivate enteric viruses. Combined chlorine residuals, much higher than normally employed, equivalent to reducing the coliform MPN approximately to that of the drinking water standard, will achieve about 99% viral inactivation in treated sewage effluents. It is believed that the viral inactivation is due to the uncombined free chlorine in the system which combines with viral nucleic acids and interfering nitrogenous substances at a rate governed by the pH and temperature of sewage effluent. The applicaton of chlorine to produce a free residual chlorine in treated sewage effluent would be prohibitively expensive. Where a high degree of a virus inactivation is demanded, as for the protection of shellfish-growing areas, iodine may be an economic alternative since this halogen, unlike chlorine or bromine, does not readily combine with interfering substances.

3. Eutrophication

Until very recently, the major consideration in water pollution abatement had been directed to reasonable water quality standards as measured by dissolved oxygen and coliform organisms. Of great concern today is the increasing amount of plant nutrients which reach the surface waters from various activities of man. The over fertilization of waters accelerates the natural aging process of primarily lakes and reservoirs, but also rivers, harbors, and estuaries. The overabundance of algae and aquatic plants causes problems of color and odor and increases the cost of water treatment. The lush algae and plant growths greatly reduce the value of water for recreation purposes and alter the composition of fish population to less desirable species.

The plant nutrients such as phosphorus and nitrogen enter the surface waters from many sources including runoff from rural and urban land and from waste water effluents. The greatest single contribution of nitrogen and phosphorus in natural water originates in runoff from farming land. The quantities of nitrogen and phosphorus contained in the drainage waters from agricultural lands is equal to or greater than the amount of chemical fertilizers applied. Farm animals in the United States generate 10 times as much waste as the whole human population, but it is difficult to determine the concentrations of nutrient pollution reaching ground and surface waters from these sources.

The distribution of nutrient sources in the environment makes control difficult if not impossible. Domestic sewage and perhaps some industrial waste appear to be the only sources in which phosphate concentration is sufficiently high to make removal feasible. Perhaps one-third of the estimated one billion pounds of phosphorus reaching our lakes and streams each year will come from domestic wastes. The general impression is that the removal of phosphorus from domestic wastes should reduce the rate of eutrophication downstream from points of discharge. The limnologist report that the ability of the blue-green algae or their associated bacteria to fix nitrogen from the atmosphere makes phosphorus the ultimate limiting factor in algae blooms. However, without a better understanding of the nature of eutrophication and the role of trace elements and bottom deposits in recycling essential nutrients, the benefits of such a costly operation are difficult to forecast.

Lake eutrophication may be controlled in some instances by diverting waste water effluents as was done at Lake Zurich, Switzerland. Where effluent diversion is not feasible, tertiary treatment is indicated. The typical aerobic biological processes of sewage treatment take up nutrients in the production of microbial protoplasm. In waste oxidation ponds both bacteria and algae utilize the nutrients in the sewage, but in each biosystem the nitrogen and phosphorus available is in excess of microbial requirements for the stabilization of wastes. The unit processes for the separation of excess microbes from the stabilized waste water permit the leaking of captured nutrients from the cells under anaerobic conditions. Conventional complete sewage treatment may be expected to remove 50% of the total nitrogen, but only 10% of the total phosphorus based on raw waste concentrations. The practice of discharging the liquid fractions of anaerobic sludge digestion and sludge elutriation water at the incoming end of the treatment process adds to the enrichment of the treated sewage effluent.

The conventional complete treatment followed by a tertiary process of lime–alum coagulation–sedimentation can provide 95% removal of phosphorus, but adds nothing to the removal of nitrogen. The added step will almost double the cost of conventional treatment. Research is needed for developing cheaper nutrient-removal processes, especially since evidence is available that some activated sludge plants are operated in a manner which efficiently removes phosphorus (San Antonio, Texas, and Baltimore, Maryland).

4. Adequacy of Water Supply

The hydrological cycle ordains that water, unlike natural resources such as fossil fuels, is an inexhaustible resource on earth. The distribution of fresh water resources is not uniform but quite variable with the capriciousness of nature establishing local, regional, and continental patterns ranging from too much to too little rainfall. In dry climates man suffers a double penalty in that precipitation (rain and snow) is unpredictable and meagre in quantity, and a relatively greater portion of the water that does fall is quickly evaporated into the dry atmosphere.

Over the United States the average precipitation yields 4300 billion gallons of fresh water each day. Seventy-one per cent of the water returns to the atmosphere through evaporation from the soil, water surface, and through transpiration from terrestrial vegetation of both beneficial and nonbeneficial varieties. An estimated 1250 billion gallons per day run off into streams and lakes and percolate into the soil to recharge the level of ground water storage and are available for development by man. Man-made regulation and storage, including land management for improved ground water utilization, is an expensive undertaking. It has been estimated by water resource experts (12) that the ultimate economic development of water supply in the United States might be a sustained flow of 650 billion gallons per day. Channel uses, such as navigation, hydroelectric power, fish, and wildlife, may require about one-half, leaving an estimated dependable supply of 315 billion gallons per day for irrigation, industry, steam–electric cooling water, and domestic uses in urban and rural areas.

It has been the popular thesis of the "prophets of doom" that current demand for water has already exhausted our fresh water supply, and they stress the urgency in developing more water from the clouds or from the sea. The fallacy in this reasoning stems from the misinterpretation of the terms "total use or withdrawal" and "consumptive use," the amount of water used that returns to the atmosphere and is unavailable for reuse. Table IV gives a current estimate of the major water uses in the United States in terms of total water withdrawn, consumed, and returned for possible reuse. It is acknowledged that the total water withdrawn is in excess of dependable supply, but more important is that the consumptive use is considerably less than the available supply. It is evident that there should be no natural water shortage in the foreseeable future and with wise management the future water problem in the United States will not be quantity but rather quality for multiple reuse.

TABLE IV

Estimated Water Use in 1965
(billion gallons per day)

Category use	Withdrawal	Consumed	Returned
Irrigation	142	116	26
Steam–electric power	111	1	110
Industrial	74	12	62
Municipal (domestic–commercial)	23	5	18
Rural (domestic and farm)	6	4	2
U.S. total	356	138	218
Per cent	(100)	(38.5)	(61.5)
Available dependable supply:	315 billion gal/day		

The allocation of water resources development shows that past undertakings were largely justified for their immediate contribution to economic growth. Ninety-two per cent of water withdrawn in the West, principally in five western states, was for the high consumptive use of irrigation while 81% of water in the East was withdrawn for industrial purposes. The range of dollar values for the leading categories of beneficial water use have been estimated by Renshaw (13) as follows:

TABLE V

Value of Water Use in Cents/1000 Gallons

Use	Mean	Maximum
Domestic	32	72
Industrial	12	50
Irrigation	0.5	8

The highest dollar value of water is from domestic purposes and will always command the highest priority for quality and use allocation. Below the categories listed above will be many other important uses of water such as hydroelectric power, navigation, waste disposal, fisheries, and a host of intangibles for which true dollar values are most difficult to estimate.

Actual withdrawal of water for irrigation varies widely with a number of factors including rate of evapo-transpiration, type of crop, methods of applying the water to the land, and, most importantly, the real cost of water. Whereas on the average 80% of the water withdrawn for irrigation is consumed, all other categories of use return 70% or more for possible multiple reuse. Consequently, future shortages will be apparent for wasteful or uneconomic uses and emphasize the need for comprehensive natural water resources planning. The nation's broad interest and objectives in water resource management are made most difficult to carry out through complex laws, policies, procedures, and traditions, and the sole objective of economic improvement of communities, states, and regions. A new emphasis in planning and decision making is emerging which, hopefully, will consider all interests and alternative solutions to water problems. In many quarters there is a strong pressure exerted for greater consideration for intangible and nonmarket values of water, and compromises must be made to accommodate these values; however, economically inefficient water supply projects are not in the best interest of the public.

Among technical alternatives for water supply is the possibility of sea water conversion. The larger, dual purpose desalting nuclear power plants advocated by the Office of Saline Water or the Atomic Energy

Commission might demineralize water competitive in price with some natural water supplies for domestic use (30¢/1000 gal), but it is evident that the cost would be three times the maximum value of water for irrigation. More promising approaches in solving the "economic" shortage of water is in the reclamation of waste waters and water conservation techniques to reduce evaporation and transpiration from nonbeneficial vegetation. Practically all major water resources development concentrates on the visible surface waters, whereas most of the total fresh water resources are in the ground. The effective and economical use of ground water reservoirs in the total unified water management approach must receive greater support. Research and adequate technology in artificial recharge for solution to problems of salt water intrusion, contamination, and depletion are among the prerequisites to the exploitation of the ground water resources.

A significant step toward comprehensive national water resources planning was made with the passage of the Water Resources Planning Act of 1965. The act provides for the Water Resources Council composed of the Departments of Interior, Agriculture, Defense, and Health, Education and Welfare, and the chairman of the Federal Power Commission. The Council, cooperating with state and local groups, will formulate policies and review plans developed regionally regarding water and related land resources. Financial assistance will be provided to improve state water planning and for river basin planning commissions composed of state and federal representatives. By sharing the responsibilities among federal, state, local, and private interests, larger input of diverse views may be integrated into the management and conservation of the nation's water resources.

Undoubtedly some reorganization of public agencies may be necessary, and improvement of water laws and their administration are strongly indicated in order to plan for and operate interstate river basin developments. No one should doubt that the problem-solving genius of modern science can meet the present and foreseeable requirements of the nation's demand for water for the public good. Science can solve the technical problems, but management will largely depend upon statesmanship.

References

1. *U.S. Public Health Serv. Publ.* No. **956,** 1962.
2. *U.S. Public Health Serv. Bull.* No. **296,** 1942, reissued. Publ. No. **525.**
3. "A Manual of Recommended Practice for the Sanitary Control of the Shellfish Industry," *U.S. Public Health Serv. Publ.* No. **33,** 1959.

4. *Recommended Practice for Design, Equipment, and Operation of Swimming Pools and Other Bathing Places,* 10th ed., American Public Health Association, Inc., New York, 1957.
5. M. M. Ellis, *Bureau of Fisheries Bull.,* U.S. Dept. of Commerce, **22,** 1937, p. 365.
6. Arch E. Cole, *The Effect of Polluted Wastes on Fish Life, A Symposium on Hydrobiology,* The University of Wisconsin Press, 1941, p. 241.
7. Karl Imhoff and G. M. Fair, *Sewage Treatment,* 1st ed., Wiley, New York, 1940, Chapter XVII.
8. "The Strategy for a Livable Environment"—A report to the Secretary, Health, Education, and Welfare Task Force on Environmental Health and Related Problems, July 1967.
9. D. M. Hegsted, "The Beneficial and Detrimental Effects of Fluoride in the Environment—Proceedings University of Missouri's 1st Annual Conference on Trace Substances in Environmental Health, July 1967, p. 105.
10. Summary Report, 6th Annual Survey of Municipal Waste Treatment Needs by Conference of State Sanitary Engineers, January 1966.
11. Edward J. Cleary, *The ORSANCO Story,* Johns Hopkins Press, 1967.
12. Estimated for various sources from the Select Committee on Natural Water Resources, U.S. Senate, 1961.
13. E. F. Renshaw, "Value of an Acre-Foot of Water," *J. Am. Water Works Assoc.,* **50,** 303 (1958).

Oxides of Nitrogen

EDWARD A. SCHUCK AND EDGAR R. STEPHENS
Statewide Air Pollution Research Center, University of California, Riverside, California

I. Oxides of Nitrogen and Air Pollution . . . 73
II. Mechanism of Nitrogen Dioxide Photolysis . . . 75
III. Rate Constants . . . 81
A. Constants Required . . . 81
B. Reaction of Nitric Oxide with Oxygen . . . 81
C. Reaction of Oxygen Atoms with Nitric Oxide . . . 83
D. Reaction of Oxygen Atoms with Oxygen . . . 83
E. Reaction of Oxygen Atoms with Nitrogen Dioxide . . . 84
F. Summary . . . 84
IV. Determination of Nitrogen Dioxide Rate Constant Ratios . . . 84
V. Tests of Mechanism and Rate Constant Ratios . . . 88
A. Rate Constant Ratios and Oxygen Atom Concentration . . . 88
B. Rate Constant Ratios and Overall Quantum Yield . . . 89
C. Rate Constant Ratios and Rate Constants . . . 96
VI. Nitrogen Dioxide and Actinometry . . . 98
VII. Nitrogen Dioxide Photolysis in Presence of Other Species . . . 102
A. Interaction at Dilute Concentrations in Air . . . 102
B. Interaction of Oxygen Atoms with Hydrocarbons . . . 104
C. Projected Studies of Interaction with Hydrocarbons . . . 106
VIII. Summary and Remarks . . . 109
IX. Recent Developments . . . 110
A. Atmospheric Conversion of NO to NO_2 . . . 111
B. Leighton's Review . . . 112
C. Modification in NO_2 Kinetic Scheme . . . 113
D. Role of Singlet Oxygen . . . 115
E. Reactivity of NO . . . 116
References . . . 117

I. OXIDES OF NITROGEN AND AIR POLLUTION

In the following discussion it may appear that nitrogen dioxide (NO_2) is given more attention than other oxides of nitrogen. In reality this is only superficially true, since examination of the known or postulated details of oxides of nitrogen chemistry indicates that discussion of a single oxide is not possible because of the complex relationships which exist. Thus our discussion actually involves many other oxides such as nitric oxide (NO), nitrogen trioxide (NO_3), nitrogen pentoxide (N_2O_5), and nitrogen tetroxide (N_2O_4). Focusing our attention on NO_2 serves to give the discussion greater continuity. This is true largely because NO_2 is one of the more stable oxides and thus is better understood.

The importance of oxides of nitrogen, and in particular of NO_2, to air pollution problems can hardly be overemphasized. The unique importance of NO_2 is related to its many roles in polluted air masses. Thus NO_2 exists in the atmosphere of many urbanized areas at concentrations which are suspected of adversely affecting human health and which suppress growth of vegetative matter. Being a highly colored gas, NO_2 reduces atmospheric visibility and is responsible for most of the brownish discoloration of air masses near urbanized areas. Of equal importance is the role of NO_2 in the formation of photochemical air pollution products. Thus, in addition to absorbing visible light, NO_2 is an efficient absorber of the ultraviolet light which reaches the earth's surface in the 3000–4000 Å region (1). These latter wavelengths contain sufficient energy to dissociate NO_2 into NO and O. The dissociation energy required for production of NO and O from NO_2 is about 72 kcal/mole which is the amount of energy available from 4000 Å light. Thus, while absorption and excited molecule formation occur at wavelengths longer than 4000 Å, dissociation does not (2,3). When this dissociation reaction takes place in an atmosphere which is also polluted with certain hydrocarbons, additional reactions occur which involve the hydrocarbons, NO_2, NO, O_2, and oxygen atoms.

Some of the products resulting from these reactions are powerful eye irritants (4) which annoy the populace, and phytotoxicants which cause substantial economic losses to crops. It has been estimated (5) that in California alone crop losses resulting from air pollutants are approaching $100 million per year. If, as usually is the case, sulfur dioxide is also present in these polluted atmospheres, the otherwise slow conversion of sulfur dioxide to sulfuric acid becomes quite rapid and results in formation of a submicron aerosol which further drastically reduces the visibility. It is clear that in the absence of NO_2 there would be no absorption of sunlight energy and therefore virtually no formation of aerosols, eye irritants, or new phytotoxicants.

Yet exclusion of NO_2 from our atmosphere is impossible because of our dependence on high temperature combustion processes for power production, in spite of the fact that the emissions from these combustion processes contain mainly NO, which is a colorless, relatively nontoxic gas.

One of the reasons why the NO_2/NO ratio in the atmosphere is substantially higher than in combustion gases is associated with a reaction which takes place during atmospheric dilution of these exhaust emissions. The particular reaction responsible is the associative reaction which occurs between NO and O_2:

$$2NO + O_2 \rightarrow 2NO_2 \qquad R_{1a} = k_{1a}[NO]^2[O_2] \qquad (1a)$$

Knowledge of k_{1a} (see Sec. III) in addition to the dilution rate and concentration of NO in a given exhaust gas will permit calculation of the NO_2/NO ratio expected to be found in the atmosphere. The squared NO concentration function in eq. 1a indicates that the rate of conversion at the 10^3 parts per million (0.76 torr) NO level, characteristic of cruising auto exhaust, will be up to a million times greater than the conversion rate at fractional ppm NO levels characteristic of observed atmospheric concentrations. The numerical value of k_{1a} assures us that this conversion is negligible at ppm NO levels, therefore leading to an almost constant NO_2/NO ratio once dilution is complete. If we know the atmospheric dilution rate, we can calculate what this NO_2/NO ratio will be. Unfortunately we do not know this dilution rate; however, we can predict from eq. 1a the approximate per cent of conversion, assuming dilution occurs in a regular manner on some finite time scale.

The instantaneous per minute rate at which NO, at an auto exhaust concentration of 1000 ppm, is converted to NO_2 during atmospheric dilution is shown in Figure 1 as a function of per cent dilution. During the 20–60% dilution phase this represents a 5–10% conversion on a per minute basis. The rate represented in Figure 1 is based on the value of k_{1a} at 25°C. As will be discussed in Section III, k_{1a} is somewhat temperature dependent so that this rate, and therefore the fraction of NO converted to NO_2, will be a function of atmospheric temperature as well as exhaust gas temperature. For example, in the Los Angeles area, where the deviation of average atmospheric temperature is ±10°C, we can expect a 16% variation in the fraction of NO converted to NO_2 during exhaust dilution. The resulting variation in the NO_2/NO ratio should be detectable in air-monitoring data and probably accounts for some of the observed variations. However, other variations, such as the effect of temperature on rate of dilution, may also be contributing to changes in this ratio. It should also be noted from eq. 1a that changes in the NO exhaust concentration will drastically alter the rate shown in Figure 1. As noted we do not know the atmospheric dilution rate and thus we are not able to accurately generate the expected NO_2/NO ratio. However, the fact that midnight to 4 A.M. NO_2 levels in the Los Angeles area are as much as 25% as great at the NO levels is probably an indication of the importance of reaction 1a during exhaust dilution. We are fortunate, indeed, that NO auto exhaust concentrations are relatively

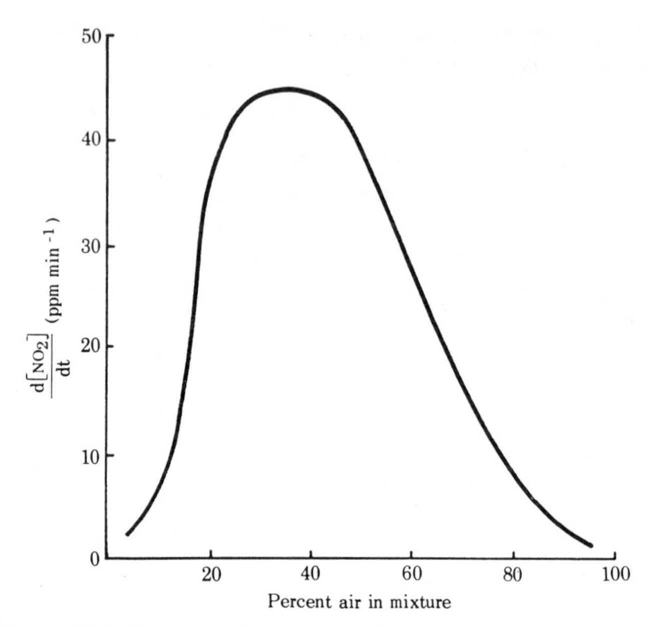

Fig. 1. Rate of NO_2 formation during atmospheric dilution of an exhaust gas containing 1000 ppm NO. $d[NO_2]/dt = 2k_{1a}[NO]^2[O_2]$.

small, since a more rapid conversion of NO to NO_2 during dilution could easily lead to a highway condition dangerous to human life. Even so, the information derivable from eq. 1a should serve as a guideline for the design of future highways. Unlimited horizontal and vertical expansion of our present highway system can also lead to dangerous localized NO_2 levels due to interference with the rate of exhaust dilution. Also, automotive schemes which result in an appreciable increase in NO exhaust levels should be approached with caution because of the squared relationship in eq. 1a of the NO concentration to the rate of NO_2 production during exhaust dilution.

Returning our attention to currently existing atmospheric conditions we become aware of other reactions which influence the ratio NO_2 to NO. During daytime hours observation of the atmosphere shows a rapid conversion of NO to NO_2 even though eq. 1a assures a reasonably stable ratio once exhaust dilution is complete. Thus, some facet of the photochemical reactions involving NO_2 and hydrocarbons promotes conversion of NO to NO_2 in the fractional ppm concentration range. This strange phenomenon prompted Philip A. Leighton to remark that this is the only photochemical system with which he is familiar where the concentration of the light absorber (NO_2) increases with time.

One of the products of photochemical air pollution reactions which is partially responsible for this rapid and unexpected conversion of NO to NO_2 is ozone (O_3). The reaction (see Sec. III) between NO and O_3 to form NO_2 and O_2 is so rapid that NO and O_3 in the fractional ppm range can coexist for only a few minutes. Since some excess ozone persists after sunset in certain polluted atmospheres, we would expect continued conversion of NO to NO_2 during the early evening hours. This excess ozone is rapidly exhausted; therefore, this reaction probably does not contribute to NO_2 levels, particularly after midnight. If conversion of NO to NO_2 during exhaust dilution were unimportant, we would thus expect to observe a decrease in the NO_2/NO ratio after midnight. In reality, in the Los Angeles area an increase in this ratio is observed between midnight and 4 A.M. This consideration, however, does not prove the importance of the conversion reaction. Obviously additional data concerning the mechanism of exhaust gas dilution is required in order to assess the contribution of reaction 1a to NO_2 atmospheric levels.

II. MECHANISM OF NITROGEN DIOXIDE PHOTOLYSIS

The foregoing discussion clearly indicates the importance of NO_2 as a direct air pollutant and of its involvement in the formation of photochemical air pollution products. Thus, it is not surprising that considerable interest has been generated in the photolytic dissociation of NO_2. In many ways the study of this dissociation and its relationship to photochemical air pollution products has made possible a revolution in gas phase photochemistry. Prior to 1950 the importance of gas phase reactions occurring in the ppm range (7.6×10^{-4} torr) was not generally recognized. Once recognized, scientists interested in these reactions were faced with the task of devising instruments and techniques which were a million times as sensitive as those previously available. In 1956 (6) the first long-path (100 m) infrared spectrophotometer made it possible to study reactions in the ppm range. Subsequent development and application of chromatographic detectors such as flame ionization, electron capture, and cross section detectors have steadily decreased the operational range until we can now study with confidence reactions in the fractional pphm range ($\sim 10^{-6}$ torr). The biggest advantage of these analytical methods in gas phase photochemistry is the ability to study reactions in a concentration range where side reactions such as polymerization are eliminated and where the effects of wall reactions can be minimized. Thus, more meaningful probes of gas phase photochemical reactions can be carried out with less confusion. Photochemists have, however, been rather slow in taking advantage of this ability to operate at these low concentration ranges. To some extent this is due to the "applied" connotation of photochemical air pollution research and to its confusing beginnings. Photochemists have, however, been quick to use the available increased sensitivity for investigations in which the reaction occurs to a very small extent.

Studies of the photolytic dissociation of NO_2 in the 0.1–1 ppm range were first carried out by Ford and Endow (7) in 1957. In spite of the obvious interest in this subject very little progress has been reported until quite recently (8). In fact, examination of the literature shows that in 1963 (9) data were reported which strongly suggested we knew little, if anything, about photolytic dissociation of NO_2 in the ppm concentration range (see Sec. V-B). This indeed presented a distressing situation, since a lack of knowledge of this dissociation meant that elucidation of the subsequent reactions with hydrocarbons to form products could not proceed in a logical manner. The fact that only limited data concerning the specific details of NO_2 dissociation can be found in the literature tends to gloss over the tremendous number of investigations carried out since 1952 which involved photolysis of NO_2 in the presence of hydrocarbons (10) from the standpoint of photochemical air pollution problems. However, to a large extent these studies were of a more practical nature with many diverse goals and therefore did not quantitatively contribute to a knowledge of the NO_2 dissociation reaction.

The reactions and attending rate equations which Ford and Endow (7) considered to be important during NO_2 photolysis are:

$$NO_2 \xrightarrow{h\nu} NO + O \qquad R_a = \phi k_a[NO_2] \qquad (a)$$

$$O + O_2 + M \rightarrow O_3 + M \qquad R_3 = k_3[M]\,[O]\,[O_2] \qquad (3)$$

$$O_3 + NO \rightarrow NO_2 + O_2 \qquad R_{31} = k_{31}[O_3]\,[NO] \qquad (31)$$

$$O + NO_2 \rightarrow NO + O_2 \qquad R_{28} = k_{28}[O]\,[NO_2] \qquad (28)$$

$$O + NO_2 + M \rightarrow NO_3 + M \qquad R_{29} = k_{29}[M]\,[O]\,[NO_2] \qquad (29)$$

$$NO_3 + NO \rightarrow 2NO_2 \qquad R_9 = k_9[NO_3]\,[NO] \qquad (9)$$

$$O + NO + M \rightarrow NO_2 + M \qquad R_{27} = k_{27}[M]\,[O]\,[NO] \qquad (27)$$

$$2NO + O_2 \rightarrow 2NO_2 \qquad R_{1a} = k_{1a}[NO]^2[O_2] \qquad (1a)$$

$$NO_3 + NO_2 \rightarrow N_2O_5 \qquad R_{10a} = k_{10a}[NO_3]\,[NO_2] \qquad (10a)$$

$$N_2O_5 \rightarrow NO_3 + NO_2 \qquad R_{10b} = k_{10b}[N_2O_5] \qquad (10b)$$

It will be noted that the rate constant numbering system used here is identical to that employed by Leighton (10). This attempt to standardize the nomenclature of the rate constants should eliminate much of the confusion inherent in assigning a new number each time a specific reaction is discussed.

Ford and Endow found that reaction 1a was negligible at the ppm concentrations employed and that reactions 10a and 10b were not required to explain their results. Thus, on the basis of the reaction scheme they derived the overall quantum yield (Φ_{NO2}) of NO_2 as:

$$\frac{2}{\Phi_{NO2}} = 1 + \frac{k_{29}[M]}{k_{28}} + \frac{k_{27}[M]}{k_{28}} \frac{[NO]}{[NO_2]} + \frac{k_3[M]}{k_{28}} \frac{[O_2]}{[NO_2]} \tag{32}$$

As will become evident later, eq. 32 expresses a series of very useful and fascinating relationships. It leads one to contemplate the possible reasons why this mechanism has remained relatively unexploited in the literature for the past nine years. Part of this explanation undoubtedly is related to the difficulties of operating in the ppm concentration range. The fact that the overall quantum yields determined by Ford and Endow (7) were in disagreement with previous data (11) generated in the 200-torr range may have also contributed to a prejudicial attitude toward eq. 32. Thus, why explore the use of an expression when the experimental data raises doubts concerning knowledge of the mechanism. A third contributing factor is the relative complexity of the quantum yield expression. The usefulness and implications of eq. 32 can be appreciated only by a thorough examination. This in turn demands a detailed knowledge of its derivation which, due to space limitations, is usually prohibitive in journal publication. Space in this discussion is no problem so that we can closely examine these details. In the derivation of eq. 32 we will, as did Ford and Endow, ignore reactions 10a and 10b since they are not necessary to explain the data and thus we cannot generate knowledge of them. We will, however, include reaction 1a since we intend to apply our derived expression to certain data where contribution from this reaction can be appreciable.

The rate of change of oxygen atom concentration from the reaction scheme is given by:

$$d[O]/dt = R_a - R_3 - R_{28} - R_{29} - R_{27} \tag{33}$$

If approximate numerical values are inserted into eq. 33, it is found that $d(O)/dt$ is small compared to the rates of oxygen atom formation and destruction reactions (12). Therefore, eq. 33 becomes:

$$R_a \cong R_3 + R_{28} + R_{29} + R_{27} \tag{34}$$

By inserting the required rates in eq. 34, rearranging, and dividing by the concentration of NO_2, we find that the oxygen atom concentration is:

$$[O] = \frac{\phi k_a}{k_{28} + k_{29}[M] + \frac{k_{27}[M]\,[NO]}{[NO_2]} + \frac{k_3[M]\,[O_2]}{[NO_2]}} \tag{35}$$

If the mechanism is correct, eq. 35 provides us with a method for calculating the oxygen atom concentration during photolysis of NO_2. This, of course, assumes that we know the value of various rate constants as well as the concentrations of certain molecular entities. Unfortunately, our knowledge of the rate constants involved is far from complete. In a general sense we know k_{28} and k_{29} within an order of magnitude, and as shown later (see Sect. III), we know k_{27} and k_3 within 22%. Within such a framework the use of eq. 35 can lead to a large variation in the predicted oxygen atom concentration. Even more important than this lack of precise rate constant data is the relationship of eq. 35 to the kinetic scheme. If eq. 35 is to be useful, the kinetic mechanism from which it is derived must be shown to be valid. If we could actually measure the oxygen atom concentration, we might evaluate eq. 35 directly. However, such measurements in the ppm range where our interest lies are at the moment beyond our experimental ability. The technique of investigating NO_2 concentrations within the capabilities of a mass spectrometer is of doubtful advantage since these higher NO_2 concentrations lead to new reaction paths which are not applicable in the ppm range.

Returning our attention to the derivation of the NO_2 quantum yield expression we note from the mechanism that the overall rate of NO_2 disappearance is given by:

$$d[NO_2]/dt = R_{31} + 2R_9 + R_{27} + 2R_{1a} - R_a - R_{28} - R_{29} \tag{36}$$

Again using the steady state approximation for rate of change of O_3 and NO_3 we can assume that:

$$R_3 \cong R_{31} \quad \text{and} \quad R_{29} \cong R_9$$

Substituting these latter approximations into eq. 36 along with the R_a approximation from eq. 34 we obtain:

$$d[NO_2]/dt = 2R_{1a} - 2R_{28} = 2k_{1a}[NO]^2[O_2] - 2k_{28}[O]\,[NO_2] \tag{37}$$

By inserting the oxygen atom concentration of eq. 35 into eq. 37, rearranging and dividing by k_{28}, we obtain:

$$\frac{-2\theta k_a[NO_2]}{d[NO_2]/dt - 2k_{1a}[NO]^2[O_2]} = 1 + \frac{k_{29}[M]}{k_{28}} + \frac{k_{27}[M]}{k_{28}}\frac{[NO]}{[NO_2]} + \frac{k_3[M]}{k_{28}}\frac{[O_2]}{[NO_2]} \quad (38)$$

Examination of eq. 38 indicates the generation of a rate expression in which all quantities except the rate constant ratios are known or can be measured. Thus, as with eq. 35 we need to know the values of certain rate constants in order to use or evaluate eq. 38.

III. RATE CONSTANTS

A. Constants Required

It is evident from the preceding section that the usefulness of eq. 38 to describe NO_2 photolysis and of eq. 35 to provide calculated oxygen atom concentrations is dependent on a knowledge of certain rate constants.

Ideally, we want to apply rate constants to eqs. 35 and 38 which have been determined by some method other than photolysis of NO_2. Fortunately some of these rate constants have been measured by independent methods.

B. Reaction of Nitric Oxide with Oxygen

The importance of this associative reaction to NO_2 atmospheric levels was discussed in Section I. Usually the contribution of this reaction to eq. 38 is negligible; however, there are concentration conditions which we will explore where reaction 1a contributes and therefore we require knowledge of k_{1a}.

Various reported values of k_{1a} which have been corrected for temperature effects are shown in Table I. The temperature at which the various determinations were made varied from 20 to 45°C. Bodenstein's data (13,14) indicate an 8% decrease in the value of k_{1a} for a 10°C increase in this temperature range. Thus the k_{1a} values in Table I were corrected to 25°C in order to provide a more exact comparison basis. The large number of determinations in Table I encourages statistical treatment of the k_{1a} values. The deliberate ordered variation of the k_{1a} values in Table I does not appear to be related to the obvious random variation in the O_2 and NO concentrations. This fact adds credence to the assumption that the relationship of reaction 1a closely approximates the experimental data. Analysis of the mean and standard deviation of the 15 rate constants in Table I indicates that the last two studies listed (15,16) differ from the mean by more than 2 standard deviations. When these two studies are eliminated from the data the remaining 13 values yield a k_{1a} of $4.5 \pm 0.6 \times 10^{-12}$ pphm^{-2} hr^{-1} ($7.5 \pm 1.0 \times 10^3$ liter^2 mole^{-2} sec^{-1}). This 13% standard deviation in k_{1a} is probably a measure of the precision attainable from the various analytical procedures employed. The result indicates that k_{1a} is one of our better known rate constants and in addition causes some doubt concerning the accuracy of the last two studies in Table I. Actually such doubts relative to the Treacy and Daniels (15) results have already been raised by Glasson and Tuesday (21). Thus Treacy and Daniels on the basis of their results proposed a seven-step mechanism involving NO_3 and N_2O_5 as well as NO and O_2. However, Glasson and Tuesday, working at concentrations more characteristic of observed atmospheric levels, obtained results which were consistent with the mechanism suggested by Trautz (30):

$$NO + O_2 \rightleftharpoons NO_3 \tag{39}$$

$$NO_3 + NO \rightarrow 2NO_2 \tag{40}$$

TABLE I

Values at 25°C of the Rate Constant for the Reaction, $2NO + O_2 \xrightarrow{k_{1a}} 2NO_2$

$$k_{1a} = \frac{d[NO_2]/dt}{2[NO]^2[O_2]}$$

k_{1a},pphm^{-2}hr^{-1} [a]	$[O_2]$, torr	[NO], torr	Investigator
5.85×10^{-12}	135	$3.8–68 \times 10^{-3}$	Greig and Hall (17)
5.25	152	$2.3–15 \times 10^{-3}$	Bufalini and Stephens (18)
5.00	16–25	19–23	Hasche and Patrick (19)
4.52	8–22	0.02–0.13	Brown and Crist (20)
4.46	3–13	7–15	Bodenstein (13,14)
4.46	15.2–152	$2.6–65 \times 10^{-3}$	Glasson and Tuesday (21)
4.46	—	—	Šolc (22)
4.28	1–430	8.1–340	Johnston and Slentz (23)
4.16	—	—	Lunge and Berl (24,25)
4.16	—	—	Briner et al. (26)
4.10	0.7–2	1.3–3.6	Kornfeld and Klingler (27)
3.92	23–190	$1.5–56 \times 10^{-3}$	Morrison et al. (28)
3.62	1–25	1–50	Smith (29)
2.29	5–20	1–15	Treacy and Daniels (15)
1.81	152	2.3×10^{-3}	Altshuller et al. (16)

[a] pphm^{-2} hr^{-1} $\times 1.66 \times 10^{15}$ = liter^2 mole^{-2} sec^{-1}.

Glasson and Tuesday further found that there is no effect on this thermal oxidation, described overall by reaction 1a, due to presence of olefins or due to photolysis of the NO_2 formed in the reaction.

C. Reaction of Oxygen Atom with Nitric Oxide

The only oxygen atom–nitric oxide reaction in the mechanism under discussion is reaction 27. Again we become aware, as pointed out by Leighton (10), that this reaction does not proceed directly as shown. The evidence indicates formation of an excited NO_2 molecule which is then stabilized by collision. In the absence of sufficient concentrations of the third body the excited NO_2 molecule dissociates to NO and O.

The values of k_{27} from seven investigations are shown in Table II. Calculation shows that only the Kistiakowsky and Volpi value differs from the average by about 2 standard deviations. After eliminating this value the remaining six studies yield a k_{27} of $1.6 \pm .036 \times 10^{-5}$ $pphm^{-2}\ hr^{-1}$($2.7 \pm 0.6 \times 10^{10}$ $liter^2\ mole^{-2}\ sec^{-1}$). Thus we apparently know k_{27} to within $\pm 22\%$.

TABLE II

Rate Constant for: $O + NO + M \rightarrow NO_2 + M$, at $25 \pm 4°C$

k_{27}, $pphm^{-2}hr^{-1}$	M	Method	Reference
2.8×10^{-5}	N_2	Mass spec.	Kistiakowsky and Volpi (31)
2.2×10^{-5}	N_2	Mass spec.	Klein and Herron (32)
1.7×10^{-5}	O_2	EPR	Westenberg and de Hass (33)
1.6×10^{-5}	O_2	Air afterglow	Clyne and Thrush (34)
1.6×10^{-5}	O_2, Ar	Air afterglow	Harteck and Dondes (35)
1.5×10^{-5}	O_2, N_2, Ar	Air afterglow	Kaufman (36)
1.1×10^{-5}	O_2, Ar, He	Catalytic probe	Ogryzlo and Schiff (37)

D. Reaction of Oxygen Atoms with Oxygen

In terms of the proposed NO_2 mechanism the only reaction involved in this category is the termolecular reaction leading to ozone formation (reaction 3). As with the previously discussed reactions, a stepwise mechanism (10) involving third-body stabilization of an excited O_3 molecule is indicated.

A value for k_3 first became available in 1957 (38). Quite recently several new investigations (39–42) have provided new data as well as a reinterpretation of the 1957 study. These more recent studies suggest a value for k_3 of $8.9 \pm 1.9 \times 10^{-8}$ $pphm^{-2}\ hr^{-1}$ ($1.5 \pm 0.3 \times 10^8$ $liter^2$

mole^{-2} sec^{-1}). Thus we also apparently know k_3 within a standard deviation of $\pm 22\%$.

E. Reaction of Oxygen Atoms with Nitrogen Dioxide

In this category we are concerned with reactions 28 and 29 of the mechanism and, unfortunately, it is in relation to these rate constants that our knowledge becomes very limited. For k_{29} the only available value other than estimates is that generated during photolysis of NO_2 by Ford and Endow (7) in 1957. Obviously we cannot use this k_{29} value to check on the validity of the mechanism from which it was derived. It should be noted that the value reported (7) for k_{29} of 6×10^{-5} pphm^{-2} hr^{-1} was based on the then accepted value of k_3 as applied to the Ford-Endow rate constant ratios. As we have just discussed, the value of k_3 has recently been reevaluated thus leading to a value of 1.3×10^{-4} pphm^{-2} hr^{-1} for k_{29} from the Ford-Endow data. This value is a factor of 20 greater than the theoretical termolecular collision factor (10) of 6×10^{-6} pphm^{-2} hr^{-1}.

Values for k_{28} are also few in number. The two independent values (43,44) of 2.2×10^3 pphm^{-1} hr^{-1} and 4.8×10^3 pphm^{-1} hr^{-1} are different from each other by a factor of 2 and also quite different from the corrected Ford-Endow value of 6.7×10^3 pphm^{-1} hr^{-1}.

F. Summary

It is obvious that many other possible reactions which may be important during NO_2 photolysis have not been considered here. However, these other reactions have been covered in detail by Leighton (10) and are not directly involved in our test of the Ford-Endow mechanism. We have succeeded in establishing the k_{1a}, k_{27}, and k_3 rate constants within an acceptable limit of error. The data concerning k_{28} and k_{29} permit no definite decision as to their reliability, although we might guess that we know their values within one order of magnitude. The two remaining reactions in the mechanism, reactions 31 and 9, do not appear in eqs. 35 and 38 and therefore cannot be evaluated by use of these equations. Thus from the standpoint of our knowledge of rate constants alone we do not have sufficient information for testing the Ford-Endow mechanism.

IV. DETERMINATION OF RATE CONSTANT RATIOS

Two methods have been used to generate values for the rate constant ratios of eq. 38. The method employed by Ford and Endow (7) (see eq. 32) takes advantage of the negligible contribution of reaction 1a in the ppm NO_2 concentration range. Under these conditions the rate-containing term in eq. 38 reduces to 2 over the NO_2 overall quantum yield. This follows from the fact that in weakly absorbing systems (10), k_a, the specific absorption coefficient, is a product of the NO_2 absorption cross section (K) and the incident light intensity (I_i). Ford and Endow assumed a value of unity for θ, the primary quantum yield; therefore the rate-containing expression of eq. 38 reduces to:

$$\frac{2KI_i[NO_2]}{d[NO_2]dt} = 2/\Phi_{NO2} \qquad (41)$$

By investigating photolysis of ppm concentrations of NO_2 in an atmosphere of nitrogen, Ford and Endow were able to generate a linear expression from eq. 32 when the inverse function of Φ_{NO2} is plotted against the NO/NO_2 ratio. Under this condition the intercept of the linear function has a value of $1 + k_{29}[M]/k_{28}$ and the slope of the line is given by $k_{27}[M]/k_{28}$. In a similar manner the data from NO_2 photolysis in nitrogen containing various amounts of O_2 yield values for $k_3[M]/k_{28}$.

The second method (8) for evaluating the rate constant ratios in eq. 38 takes advantage of the effects caused by changing the concentration of the third body (M). Again as in the Ford-Endow studies the contribution of $k_{1a}[NO]^2[O_2]$ is negligible when investigating ppm NO_2 concentrations in the absence of appreciable concentrations of NO and O_2. The ability to vary the concentration of M leads to expressions for evaluating the rate constant ratios which are more direct than the method employed by Ford and Endow. Thus in the absence of large concentrations of M, eq. 38 reduces to:

$$(d[NO_2]/dt)_{M \to O} = -2\phi k_a[NO_2] \qquad (42)$$

Confidence in the reliability of eq. 42 is warranted by reflection on the order of magnitude values of the rate constant ratios generated by Ford and Endow (see Table III). Thus it would appear that in the absence of M or large concentrations of NO or O_2 the rate constant ratio terms in eq. 38 have values which are very small compared to unity.

Equation 42 predicts the existence of linear relationship between the logarithm of the NO_2 concentration and irradiation time. Furthermore, the slope of the resulting straight line will be $2\phi k_a$. It is important to note that this value is obtained from the rate of NO_2 photolysis without prior knowledge of primary quantum yield or light intensity. As is discussed in Section VI, this has important implications for actinometry.

Reevaluating eq. 38 in the presence of an atmosphere of nitrogen ($M = 10^8$pphm), but in the absence of appreciable quantities of NO and O_2 yields:

$$\frac{-2\phi k_a[NO_2]}{(d[NO_2]/dt)_{M=1\text{ atm}}} = 1 + k_{29}[M]/k_{28} \tag{43}$$

In reality the concentration of M in eq. 43 need not be 1 atm. The criteria is that the $k_{29}[M]/k_{28}$ term be appreciable in relation to unity.

Again in eq. 43 as in eq. 42 a linear relationship is implied. Inserting the value of $2\theta k_a$ from eq. 42 into eq. 43 gives:

$$\frac{\{d(\ln [NO_2])/dt\}_{M=0}}{\{d(\ln [NO_2])/dt\}_{M=1\text{ atm}}} = 1 + k_{29}[M]/k_{28} \tag{44}$$

Thus the ratio of the slope of ln $[NO_2]$ versus time in the absence of M to that in the presence of 1 atm of the third body will be a simple function of $k_{29}[M]/k_{28}$.

There are, of course, several implied conditions which must be true before the relationships in eqs. 42 and 44 are correct. Thus it is recognized that M in these equations can be any molecule in the system. This includes NO_2 itself and its photolytic products. Therefore, the rate constant ratios (k_{29}/k_{28}, k_{27}/k_{28}, and k_3/k_{28}) must be less than 1 by several orders of magnitude when using concentration units of pphm and time in hours.

This assumption is most apparent when deriving eq. 43, which concerns NO_2 photolysis in an atmosphere of M. Knowing that the end products of NO_2 photolysis in an inert atmosphere are NO and O_2, we note that the NO/NO_2 and O_2/NO_2 ratios of the last two terms in eq. 38 are zero only at zero time. Thus we have in reality assumed that $k_{27}[M]/k_{28}$ and $k_3[M]/k_{28}$ are small compared to $1 + k_{29}[M]/k_{28}$. The extent to which experimental data fit these assumptions is illustrated in Figure 2, where the logarithm of NO_2 concentration is plotted as a function of irradiation time for a system in which 13 ppm of NO_2 was photolyzed in an atmosphere of nitrogen. As is typical for kinetic tests of this type, other possible mechanisms are not excluded. However, as will be seen, the Ford-Endow mechanism provides a consistent interpretation of a variety of experimental evidence. An additional check is provided by independent evaluation of the $2\phi k_a$ term in eq. 42. Recalling that in weakly absorbing systems k_a is the product of the known absorption cross section of NO_2 and the measured incident light intensity, we can calculate the rate of NO_2 photolysis from the measured

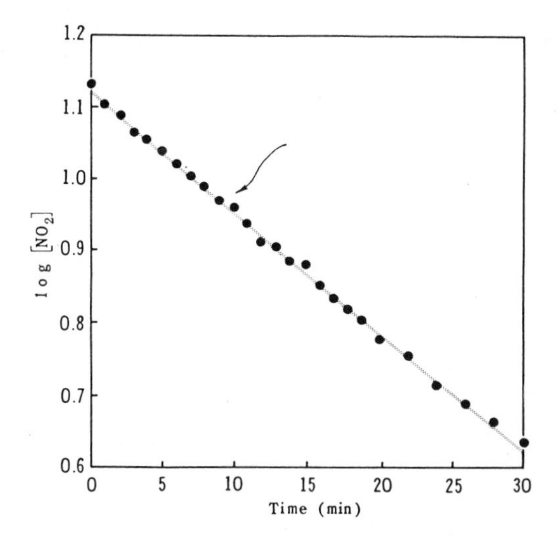

Fig. 2. Results of irradiating 13 ppm NO_2 in an atmosphere of nitrogen. Curve determined by least squares application of log $[NO_2] = A - Bt + Ct^2$.

light intensity plus the published primary quantum yield data and measured NO_2 concentrations. The value of $2\phi k_a$ generated by this method is statistically indistinguishable from the $2\phi k_a$ obtained from the slope of the ln NO_2 versus time plot whem ppm concentrations of NO_2 are photolyzed in the absence of added gases (8). Thus again the values of the last two terms in eq. 38 must be negligible under these conditions.

Returning our attention to eq. 38 we evaluate the system where photolysis of ppm concentrations of NO_2 is conducted in the presence of an atmosphere of nitrogen which also contains added amounts of NO. In this case eq. 38 becomes:

$$\frac{\{d(\ln [NO_2])/dt\}_{M=0}}{\{d(\ln [NO_2])/dt\}_{M=1 \text{ atm } N_2;\ NO=0.3 \text{ torr}}} = 1 + k_{29}[M]/k_{28} + (k_{27}[M]/k_{28})([NO]/[NO_2]) \quad (45)$$

Equation 45 predicts a direct relationship between the ratio of photolytic rates and the NO/NO_2 ratio under certain conditions. The most stringent of these conditions demands that the NO/NO_2 ratio be large enough to make the term containing this ratio of an equal order of magnitude as $1 + k_{29}[M]/k_{28}$. At the very least it must furnish an appreciable contribution, otherwise the results from eq. 45 would be indistinguishable from those obtained with eq. 44. Experimentation indicates that an initial NO/NO_2 ratio of 25 or more produces the required effect. This, of course, is further indication that the assumed order of magnitude for $k_{27}[M]/k_{28}$ is correct. It is also apparent that the value of the slope used in the denominator of eq. 45 for determination of $k_{27}[M]/k_{28}$ must be the extrapolated value at zero irradiation time since the NO/NO_2 ratio which determines the per cent contribution to the rate expression in eq. 45 will be changing appreciably with irradiation time due to the rapid disappearance of NO_2. Fortunately, concentration conditions can be chosen which allow a significant contribution from $(k_{27}[M]/k_{28})([NO]/[NO_2])$ without destroying the accuracy of this extrapolation. Thus photolysis of 20 ppm NO_2 in an atmosphere of nitrogen containing approximately 4×10^2 ppm NO generates NO_2 photolytic curves which can by the least squares treatment be accurately extrapolated to zero irradiation time (8).

When ppm concentrations of NO_2 are photolyzed in nitrogen with added O_2, eq. 38 reduces to:

$$\frac{\{d(\ln [NO_2])/dt\}_{M=0}}{\{d(\ln [NO_2])/dt\}_{M=1 \text{ atm } N_2;\ O_2=10 \text{ torr}}} = 1 + k_{29}[M]/k_{28} + (k_3[M]/k_{28})([O_2]/[NO_2]) \quad (46)$$

Again, in a manner similar to the previous case, the O_2/NO_2 ratio must result in a substantial contribution from the $(k_3[M]/k_{28})([O_2]/[NO_2])$ term. The $k_3[M]/k_{28}$ ratio turns out to be $\sim 10^2$ smaller than the $k_{27}[M]/k_{28}$ ratio. Thus, accurate values of $k_3[M]/k_{28}$ can be obtained by using O_2/NO_2 ratios of about 10^3.

The single most important feature of eqs. 44–46 is that they provide values for the rate constant ratios of eq. 38 which are independent of light intensity, absorption cross sections, and primary quantum yields. This independence becomes a distinct advantage (see Sec. V) when attempting to establish the validity of the numerical values of the rate constant ratios.

V. TESTS OF MECHANISM AND RATE CONSTANT RATIOS

A. Rate Constant Ratios and Oxygen Atom Concentration

The rate constant ratios generated in this laboratory by application of eqs. 44–46 are shown in Table III compared to those generated by Ford and Endow by application of eq. 32. The question to decide is which set of values in Table III is more nearly correct and what are the consequences of the numerical differences between the two sets of values.

TABLE III

Rate Constant Ratios

Ratio	Ford-Endow (7) (25°C)	Schuck et al. (8) (24°C)
$k_{29}[M]/k_{28}$	1.9	0.33 ± 0.08
$k_{27}[M]/k_{28}$	0.36	0.18 ± 0.04
$k_3[M]/k_{28}$	1.33×10^{-3}	$1.15 \pm 0.03 \times 10^{-3}$

This latter point can readily be evaluated by rewriting eq. 35 in terms of rate constant ratios rather than rate constants. By dividing eq. 35 by k_{28} we obtain:

$$[O] = \frac{1}{k_{28}} \frac{\phi k_a}{1 + \dfrac{k_{29}[M]}{k_{28}} + \dfrac{k_{27}[M]}{k_{28}} \dfrac{[NO]}{[NO_2]} + \dfrac{k_3[M]}{k_{28}} \dfrac{[O_2]}{[NO_2]}} \tag{47}$$

Clearly in certain cases a factor of 2 in calculated oxygen atom concentration will result depending on which set of ratios from Table III is applied to eq. 47. This is most obvious in systems where the $k_{29}[M]/k_{28}$ ratio predominates. A particular experimental system where this can be a serious effect is the determination of rate constants of olefin–oxygen atom reactions in which oxygen atoms are generated by photolysis of NO_2 in nitrogen. Obviously there are also systems which produce very similar values for calculated oxygen atom concentrations regardless of which set of data from Table III is employed. Thus the differences in calculated oxygen atom concentrations would decrease in systems where the $k_{29}[M]/k_{28}$ ratio is small compared to the other rate constant ratios in eq. 47.

B. Rate Constant Ratios and Overall Quantum Yields

The question concerning the validity of the data in Table III as well as the applicability of eq. 38 is of considerable importance to the field of photochemical air pollution since reaction of oxygen atoms with certain hydrocarbons is considered to be one of the major reactions occurring (10). Fortunately, there exists a considerable body of data pertaining to NO_2 photolysis which can be used to test the Table III data. Most of this data is expressed in terms of overall quantum yields. As pointed out in relation to the Ford-Endow study the definition of Φ_{NO_2} is contained in eq. 38. Thus, in the absence of large concentrations of a third body, and at NO_2 concentrations less than 10 torr, eq. 38 becomes:

$$(d[NO_2]/dt)_{M=0}/k_a[NO_2] = \Phi_{NO_2} = 2\theta \qquad (48)$$

This relationship as such is not too informative since it merely points out the method used to generate primary quantum yield values from the very studies we intend to examine. A more revealing and useful expression is obtained if we evaluate eq. 38 in the presence of an atmosphere of M. Thus eq. 38 becomes:

$$\Phi_{NO_2} = \frac{2\phi}{1 + k_{29}[M]/k_{28}} \qquad (49)$$

Predicting Φ_{NO_2} values from eq. 49 should allow us to assess the validity of the $k_{29}[M]/k_{28}$ ratios in Table III. This is particularly true for the ratio in the Schuck et al. study (8), since this ratio was determined by a method which was independent of the primary quantum yield whereas the Ford-Endow ratio (7) depends on an assumed quantum yield. Further consideration of eq. 49 shows that NO_2 itself should act as a third body above a concentration of 10 torr. Rewriting eq. 49 using the 0.33 ratio value of study 8 gives:

$$\Phi_{NO_2} = \frac{2\phi}{1 + 4.34 \times 10^{-4}[NO_2]} \qquad (50)$$

where the NO_2 concentration is expressed in torr. This, of course, assumes NO_2 has the same third-body efficiency as nitrogen. In order to calculate Φ_{NO_2} values from eqs. 48–50 we need to know the value of θ as a function of wavelength (10,45). This relationship is shown graphically in Figure 3. Published experimental Φ_{NO_2} values arranged in terms of decreasing NO_2 concentration investigated are shown in Table IV. Also shown are the Φ_{NO_2} values calculated by appropriate application of eqs. 48–50, and the $k_{29}[M]/k_{28}$ value from reference 8. With the exception of the reference 7 data and a portion of the reference 9 data, the comparison between experimental and calculated values in Table IV is in fair agreement. This agreement lends credence to the applicability of the mechanism as expressed in eq. 38 over a 10^4 NO_2 concentration range. It also indicates that the 0.33 value for $k_{29}[M]/k_{28}$ from reference 8 is more nearly correct. However, it is the deviations from this agreement which need to be explored.

Our attention is first drawn to reference 9 in Table IV which indicated that Φ_{NO_2} approached zero as the NO_2 concentration approached zero.

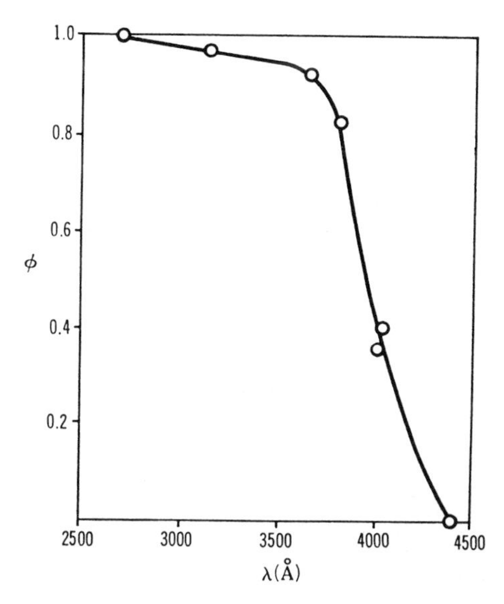

Fig. 3. Primary quantum yield of NO_2 photolysis (1,11,45,46,47).

TABLE IV

Experimental and Calculated[a] Quantum Yields

Study	$[NO_2]$, torr	Wave-length, Å	Φ, NO_2 alone Exptl.	Φ, NO_2 alone Calc.	Φ, $NO_2 + 1$ atm N_2 Exptl.	Φ, $NO_2 + 1$ atm N_2 Calc.
Norrish, 1929 (46)	22–268	3660	2.10	1.73		
	22–213	2900	2.07	1.90		
Holmes and Daniels, 1934 (11)	30–200	3660	1.83	1.75	1.47	1.38
	30–200	3130	1.93	1.85	1.27	1.46
Dickinson and Baxter, 1928 (47)	3.5–39.7	3660	1.54	1.82		
Pitts et al., 1964 (45)	10	3660	1.84	1.83		
	10	3130	1.94	1.93		
Blacet et al., 1962 (48)	6	3130	1.92	1.93		
	1.7	3130	1.44	1.94		
Ford and Jaffe, 1963 (9)	0.25–5.8	3660	0.6–2.00	1.84	0.4–1.40	1.38
Ford and Endow, 1957 (7)	2×10^{-3}	3660			0.8	1.38
Schuck et al., 1966 (8)	1×10^{-2}	3500	1.90	1.88	1.41	1.41
King, 1965 (49)	1×10^{-2}	3500			1.39	1.41
Schuck and Stephens, 1966	0.5–5.0	3500	1.87	1.88	1.43	1.41

[a] Calculated on basis of Ford-Endow mechanism and reference 8 value for $k_{29}[M]/k_{28}$.

The last study listed in Table IV shows the results of experiments in this laboratory which attempted to duplicate the reference 9 data. The results do not show this Φ_{NO_2} dependence on NO_2 concentration even though the equipment and techniques used were similar to those used in reference 9. Ford and Jaffe concluded (9), as was confirmed in the last study in Table IV, that the experimental system used in these particular studies can lead to large errors when the NO_2 concentration is less than 1 torr. Ford and Jaffe also pointed out the discrepancy between the data in references 9 and 7. Reference 9 (data in Table IV) demands a Φ_{NO_2} which approaches zero. Yet the data of reference 7 at much lower NO_2 concentrations while exhibiting a low Φ_{NO_2}, nevertheless, was not close to zero. Further evidence that this quantum yield is not zero at ppm NO_2 concentrations is shown in reference 49 (data in Table IV) where King gives NO_2 quantum yield data generated by application of an azo-methane actinometer. It is worthy of note that experimental and calculated Φ_{NO_2} values for the reference 9 data are in agreement at the higher, and therefore more accurately determinable, NO_2 concentrations.

The data from reference 7 (Table IV) in which the mechanism we are now examining was proposed, is also in disagreement with the calculated Φ_{NO_2} value. Several discussions with Hadley W. Ford have failed to shed light on this disagreement. However, during the course of these discussions an interesting, but perhaps fortuitous, relationship was discovered. Figure 4 shows a comparison between the data of references 7 and 8 when the inverse Φ_{NO_2} function is plotted against the NO/NO_2 ratio in the manner discussed in relation to eq. 32 (see Sect. IV). The surprising fact noted in comparing these figures is that the numerical values of the slopes and intercepts are very nearly the same even though the vertical axes differ by a factor of 2. Thus, but for this factor of 2, the rate ratios generated from the two studies are in close agreement. Needless to say, many frustrating hours were spent in an unsuccessful attempt to explain this observation. It will be noted that the slope and intercept from Figure 4-B of reference 8 result in values of the rate ratios which agree closely with the more independently determined reference 8 ratios in Table III.

As noted in Table IV, the calculated Φ_{NO_2} values at a given wavelength decrease appreciably at high NO_2 concentrations due to NO_2 itself acting as a third body. This third body effect of NO_2, evident in eq. 50, should also be apparent within those studies in Table IV which cover a wide NO_2 concentration range. Thus in relation to reference 46 data, eq. 50 predicts that Norrish should have detected a 10% drop in Φ_{NO_2} over the

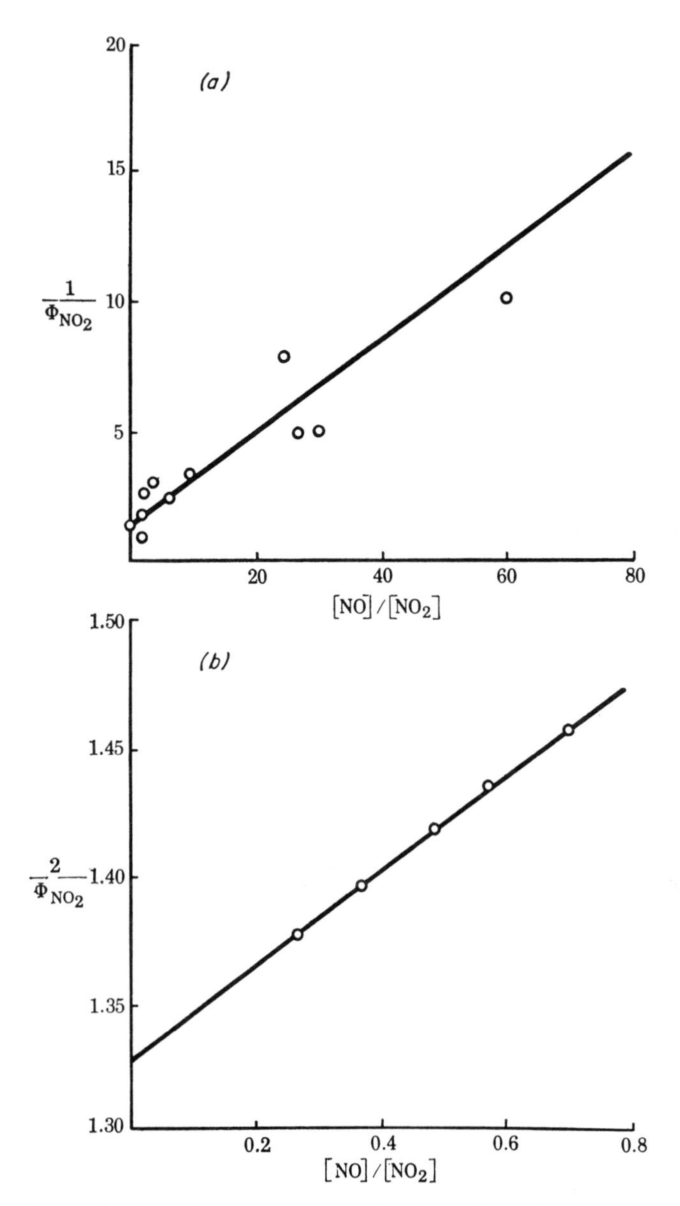

Fig. 4. Determination of rate constant ratios from data of references 7 and 8. (*a*) Reference 7 data: slope = 0.178, intercept = 1.45. (*b*) Reference 8 data: irradiation = 20 ppm NO_2 in 1 atm N_2, slope = 0.173 = $k_{27}[M]/k_{28}$, intercept = 1.33 = $1 + k_{29}[M]/k_{28}$.

NO_2 range investigated. When the individual experiments in reference 46 are examined, this predicted effect becomes apparent, as shown in Figure 5. This finding is indeed a tribute to the careful manner in which this particular study was conducted. The 20% difference between experimental and calculated Φ_{NO_2} values, as illustrated in Figure 5, is less easily explained. Although a similar difference is not in evidence in reference 11 (Table IV) thus providing a case for attributing the difference to accuracy, it is of interest to attempt determination of the nature of the observed variation.

The difference cannot be attributed to a difference in third-body efficiency between NO_2 and N_2 since this would prevent the slopes in Figure 5 from being equal, and in any event, the upper limit of Φ_{NO_2} at 3600 Å is 1.84 according to eq. 50 regardless of the third-body efficiency.

Another factor which might influence the results is the amount of light absorbed in the reaction cell. In the experiments of reference 46 this varied between 69 and 100%. It has been shown (50) theoretically that the measured rate of a reaction can be seriously affected if more than 50% of the light is absorbed. However, as pointed out by Calvert and Pitts (51), this theory fails to take diffusion into account. This diffusion in the gas phase, particularly, at the reduced pressures studied, can minimize or even eliminate this absorbed-light effect. What makes this an unlikely effect in this case is the negative slope observed in

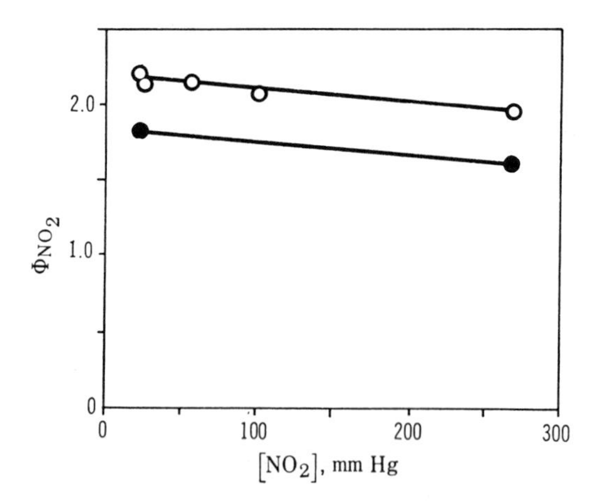

Fig. 5. Observed and calculated Φ_{NO_2} values for reference 46 data. Theoretical $\Phi_{NO_2} = 2\Phi/(1 + k_{29}[M]/k_{28}) = 1.84/(1 + 4.34 \times 10^4[NO_2])$. (○) Norrish, 1929; (●) theory, 1966.

Figure 5, whereas application of the absorbed-light theory predicts a positive slope. One might also question the fate of NO_2 when it acts as a third body in reaction 29. However, here again we have no way of ascertaining the fate of an excited NO_2 molecule, and again we fail to see a similar effect in the data of reference 11 (Table IV).

Norrish (46) recognized that above an NO_2 concentration of a few torr the recombination reaction 1a became important. Thus, he irradiated NO_2 systems until equilibrium was established—that is, until $d[NO_2]/dt$ became zero as evidenced by a stabilization of pressure in the reaction cell. Under this condition the rate of NO_2 destruction is equal to the rate of its formation:

$$(d[NO_2]/dt)_{(R_a+R_{28})} = k_{1a}[NO]^2[O_2] \quad (51)$$

By relating pressure change to oxygen concentration, the concentrations of NO could be calculated, and since k_{1a1} was known, the rate of NO_2 destruction could be determined by use of eq. 51. A portion of the results from reference 11 (Table IV) were also generated by application of eq. 51. The authors of this latter study point out that the experimental details do not indicate that the results in reference 46 (Table IV) were corrected for the fraction of light absorbed by N_2O_4. That this is a serious effect is guaranteed by the appreciable concentrations of N_2O_4 which are present at NO_2 concentrations above 10 torr (52) and by the fact that the absorption coefficient (1,10) of N_2O_4 is of the same order of magnitude as that of NO_2 in the 3100–4600 Å region. The effect of this absorption by N_2O_4 is not photolytic (2) but merely serves to dissociate the dimer back to NO_2. If this N_2O_4 absorption were not taken into account, it could result in the higher-than-calculated Φ_{NO_2} values of the reference 46 data. The fact is that the first reference to N_2O_4 absorption was not published until one year after completion of the reference 46 study. However, the application of knowledge relative to N_2O_4 absorption to the data of reference 46 has dubious value at this time and will not be attempted here. It is important to note, however, that this discussion of references 46 and 11 clearly indicates many of the added complexities which are associated with NO_2 photolysis above a 10-torr concentration.

In view of the interesting application of reaction 1a to the data of references 46 and 11, we should reconsider the potential significance of this recombination reaction to eq. 38. This reaction 1a was found to have a negligible effect in the NO_2 concentration range used to evaluate the rate constant ratios in Table III. However, further examination of eq. 38 reveals that continued irradiation of any NO_2-containing system will, as indicated by Leighton (10), produce an equilibrium where $d[NO_2]/dt$ is zero at some finite NO_2 concentration. At this equilibrium point eq. 38 becomes:

$$\frac{\phi k_a[NO_2]}{1 + \frac{k_{29}[M]}{k_{28}} + \frac{k_{27}[M]}{k_{28}}\frac{[NO]}{[NO_2]} + \frac{k_3[M]}{k_{28}}\frac{[O_2]}{[NO_2]}} = k_{1a}[NO]^2[O_2] \quad (52)$$

In the ppm NO_2 concentration range and in the absence of M, eq. 52 reduces to:

$$\phi k_a[NO_2] = k_{1a}[NO]^2[O_2] \quad (53)$$

Equation 53 is interesting in that it provides a method for evaluating k_{1a} during irradiation of certain systems. Thus in our laboratory irradiation of a mixture of 500 ppm each of NO and O_2 equilibrated at 3.5 ppm NO_2. Inserting these values into eq. 53 along with measured light intensity, absorption cross section, and primary quantum yield data resulted in a k_{1a} of 4.9×10^{-12} pphm^{-2} hr^{-1}. This is well within the 13% standard deviation of the average k_{1a} previously discussed (see Sect. III-B). This result also confirms the Glasson and Tuesday (21) result in that photolysis of the product of reaction 1a (NO_2) does not affect the thermal reaction between NO and O_2. The fact is that we were able to generate a reliable k_{1a} value at a total pressure which was 10^3 times as low as (0.76 torr) that employed by Glasson and Tuesday. This virtually eliminates the possibility of a collisional deactivation reaction and adds further credence to the mechanism of Trautz (see Sec. III).

C. Rate Constant Ratios and Rate Constants

Thus far in this discussion we have been able to support the NO_2 mechanism as described by eq. 38 by its obvious applicability to Φ_{NO_2} values. These same studies support the validity of the $k_{29}[M]/k_{28}$ value from reference 8. However, we have not yet developed information concerning the $k_{27}[M]/k_{28}$ and $k_3[M]/k_{28}$ ratios. Yet, if we are to use eq. 47 for calculating oxygen atom concentrations, we must prove the validity of these ratios. We must therefore find a method for relating these ratios to other published data. Reflection shows these latter ratios each contain a specific rate constant shown in Section III to have well established independent values. Thus applying k_3 or k_{27} from Section III to the rate ratios in Table III should result in consistent data since these ratios are linked by the common factor of k_{28}. The results of applying the independent k_3 and k_{27} values to the rate constant ratios of Table III are shown in Table V. Examination of the calculated rate constants in Table V reveals that the reference 8 ratios provide the predicted consistency while the reference 7 ratios do not. Thus, the reference 8 rate ratios yield very nearly the same values of k_{28} and k_{29} by application to the ratios of either k_{27} or k_3. Furthermore, values for k_3 and k_{27} calculated from the reference 8 ratios lie within the standard deviation limits of the independently determined values of these constants, whereas reference 7 values do not meet this criteria. Thus we have succeeded in establishing the validity of the reference 8 rate constant ratios. Moreover, by a method which is independent of primary quantum yield data, we have succeeded in adding additional credibility to the Ford-Endow mechanism. Further confidence in the independently determined k_3 and k_{27} values is also implied by the developed interrelationship between k_3, k_{27}, and k_{28}. The results in Table V further indicate the necessity for additional independent determinations of k_{28} since the values developed from the reference 8 rate ratios are more than a factor of 2 higher than the average of the two independent studies (43,44). The interrelationships and numerical consistency of the constants involved in our discussion warrant the conclusion that these two independent k_{28} values are in error. However, this logic is not entirely foolproof and therefore additional independent studies of k_{28} are necessary.

TABLE V

Calculated Rate Constants Derived from Rate Constant Ratios in Table III

Rate constant ratios investigated	Av literature rate constant applied to rate ratios	k_{28}, $pphm^{-1}hr^{-1}$	k_3, $pphm^{-2}hr^{-1}$	k_{27}, $pphm^{-2}hr^{-1}$	k_{29}, $pphm^{-2}hr^{-1}$
Schuck et al. (8)	k_{27}	8.9×10^3	10.2×10^{-8}	—	2.9×10^{-5}
Schuck et al. (8)	k_3	7.8×10^3	—	1.4×10^{-5}	2.6×10^{-5}
Ford and Endow (7)	k_{27}	4.4×10^3	5.9×10^{-8}	—	8.4×10^{-5}
Ford and Endow (7)	k_3	6.7×10^3	—	2.4×10^{-5}	13.0×10^{-5}
Av literature rate constant values (see Sec. III)		3.5×10^3	$8.9 \pm 1.9 \times 10^{-8}$	$1.6 \pm 0.4 \times 10^{-5}$	—

An alternate method of checking the rate constant ratio data is to compare the k_{27}/k_3 ratio generated by calculating the $k_{27}[M]/k_{28}$ to $k_3[M]/k_{28}$ ratios from Table III with that obtained from the independent values of k_{27} and k_3 from Section III. These independent rate constants yield a k_{27}/k_3 value of 1.8×10^2 while the reference 8 data yield 1.6×10^2 and the reference 7 data a value of 2.7×10^2. Thus, the reference 8 value is 11% lower than the independent value, while the reference 7 value is 50% higher.

The reference 8 findings have obvious implications for other published data involving NO_2 photolysis. A specific case is the generation of rate constants for the reaction of oxygen atoms with hydrocarbons when NO_2 is used as a source of oxygen atoms (see Sec. VII).

VI. NITROGEN DIOXIDE AND ACTINOMETRY

One of the more practical implications of an applicable NO_2 photolytic mechanism is the use of NO_2 as a gas phase actinometer. This is of specific importance to investigators concerned with photochemical air pollution since NO_2 is the light absorber associated with this problem and thus absorbs in the precise wavelength region of interest. The need is further emphasized because there exists no other gas phase actinometer which can be conveniently applied to the 3000–4000 Å region. If we examine the various equations generated from the mechanism, we find three (eqs. 42, 43, and 53) which appear most convenient for application to actinometry. If we accept the approximation that k_a is given by KI_i (see Sec. IV) and insert the $k_{29}[M]/k_{28}$ ratio from reference 8 into eq. 43 the three expressions obtained after solving for I_i are:

$$I_i = \frac{\{0.5\ d(\ln [NO_2])/dt\}_{M=0}}{\phi K} \qquad (54)$$

$$I_i = \frac{\{0.67\ d(\ln [NO_2])/dt\}_{M=1\ \text{atm}\ N_2}}{\phi K} \qquad (55)$$

$$I_i = (k_{1a}[NO]^2[O_2]/\phi K[NO_2])_{M=0} \qquad (56)$$

These equations imply a rather straightforward approach to the measurement of light intensities. Equations 54 and 55 show that I_i is directly proportional to the rate of NO_2 photolysis either in the absence or presence of the third body, M. Equation 56 shows light intensity to be approximately inversely proportional to the equilibrium NO_2 concentration since the NO and O_2 concentrations necessary to produce a finite NO_2 value are quite large (see Sec. V). Calculations show for a light intensity of the order of 10^{15} quanta cm^{-2}sec^{-1} that irradiation of a mixture of 500–1000 ppm each of NO and O_2 will yield an equilibrium NO_2 concentration in eq. 56 of the order of 3–25 ppm. We have used eq. 54 to obtain a value of light intensity at 3500 Å and found its value to be indistinguishable from that obtained using an ortho-nitrobenzaldehyde actinometer. Thus an I_i value of $9.2 \pm 0.4 \times 10^{14}$ quanta cm^{-2} sec^{-1} was found with this latter actinometer as compared with $9.0 \pm 0.2 \times 10^{14}$ quanta cm^{-2} sec^{-1} obtained by use of eq. 54 (8).

The values of ϕ necessary for evaluation of eqs. 54–56 are obtainable from published data (see Fig. 3). The K values are obtained from the published (1) NO_2 absorption coefficients, α. Thus the value of α for NO_2 at 3500 Å is 1.36×10^2 1 mole^{-1}cm^{-1}, and therefore:

$$K = \frac{\alpha\ (2.303 \times 10^3)}{6.023 \times 10^{23}} = 5.21 \times 10^{-19}\ \text{cm}^2\ \text{molecule}^{-1} \tag{57}$$

Using K values in cm^{2}molecule^{-1} and rates in sec^{-1} will yield I_i values in units of quanta cm^{-2}sec^{-1} from eqs. 54 and 55. The units of concentration in eq. 56 can have any convenient definition; however, the time unit of k_{1a} must be in seconds if it is desired to obtain I_i in quanta cm^{-1} sec^{-1} when using the above units of K.

There are, of course, many factors which restrict the use of these I_i equations. The most important of these restrictions apply to the NO_2 concentrations employed in eqs. 54 and 55. We have seen that NO_2 concentration above 10 torr can create problems because of the importance of certain reactions which are unimportant in the fractional torr range (see Sec. V-B). Another reason for restricting the upper limit of NO_2 concentration relates to the M effect of NO_2 and its photolytic products. However, the upper limit of NO_2 concentrations which can be used is determined by the KI_i approximation for k_a. This approximation is only valid for systems which absorb weakly, either because the molecule itself is a weak absorber or because its concentration is low. Since NO_2 is a strong absorber, we therefore make the system weakly absorbing by reducing the absorber concentration. Calculations, for example, show that the percent error introduced in k_a (10) by this approximation reaches the 1% point at 35 ppm in an 8-in. light path. This, therefore, represents the upper limit of NO_2 concentration for this particular system and, of course, presents measuring problems. In our laboratory a 60-liter reaction cell is also the

40-m folded path cell of an infrared spectrophotometer; thus we can measure NO_2 in the 0.1–20 ppm range by its absorption in the 1600–1620 cm^{-1} region. Although this method is convenient, it is not quite as simple as it might sound. One of the major complexities is that the 40-m cells and interior optical bench as received from the manufacturer are completely unsatisfactory for gas analysis, particularly at reduced pressures. Not only do these cells absorb gases at a fantastic rate, but they also are extremely destructive to labile molecules and intermediates. Various methods have been used to circumvent these difficulties. The methods used most often involve inert coating of the optical bench and replacement of the cover can with a glass cylinder. In our laboratory the optical bench and cover can were constructed of low-carbon steel and subsequently coated with porcelain. This is quite satisfactory in terms of making them nonabsorbing and nonreacting, although it is quite a difficult and expensive task. A system which apparently produces satisfactory inert surfaces for the study of fast reactions at a nominal cost and effort involves coating the parts as received with an acrylic paint (53). Entry of irradiation is provided by Pyrex or quartz windows mounted in the surface of the cover can. Having produced a satisfactory inert surface we now calibrate the system. Since NO_2 is a simple molecule, it does not follow Beer's absorption law. To make this matter more complex, the NO_2 absorption is strongly influenced by total pressure as illustrated in Figure 6. Thus calibration curves must be generated as a function of concentration and total pressure. The only

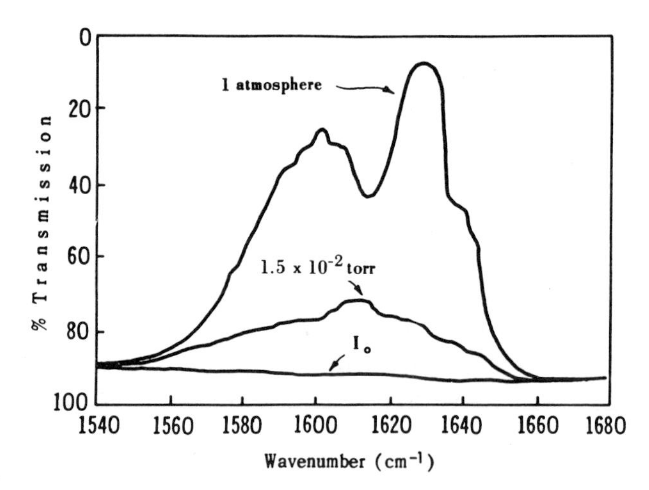

Fig. 6. Infrared absorption of 1.5×10^{-2} torr NO_2 with and without one atmosphere of N_2.

alternative at this time to the use of a long-path infrared spectrophotometer is offered by reacting the NO_2-containing gas with a colorimetric agent (54). The one disadvantage to the use of such an alternative for NO_2 measurement in the ppm range is the gas volume demand and the relatively slow response of the commercially available monitoring instruments which use this colorimetric principle. These, however, are restrictive rather than prohibitive factors; thus Ford and Endow (7) were able to minimize this slow response aspect as well as the volume demand by investigating flow systems which resulted in equilibrium NO_2 concentrations.

Other factors which must be taken into account when using the light intensity equations are the effects of contaminants and, in the case of eq. 55, the effect of different inert gases since the coefficient in this latter equation is a function of third-body efficiency and was determined only for N_2. The extent to which various gases might affect the coefficient in eq. 55 is shown in Table VI. The estimated coefficients were obtained by appropriate recalculation of Holmes and Daniels (11) quantum yield data. Although the values in Table VI are estimates, it is apparent that operation in gases other than N_2 will require redetermination of the $k_{29}[M]/k_{28}$ ratio in order to conform to the gas employed.

TABLE VI

Coefficients for Use in Eq. (55)

Gas	Estimated coefficients (11)
Helium	0.76
Argon	0.63
Hydrogen	0.66
Carbon dioxide	0.56

The amount of permissible contamination depends on the chemical nature of the contaminant. Total inert gas concentrations up to 8 torr (1.1×10^4 ppm) can be present when using eqs. 54 and 56 without producing more than a 1% error in the light intensity measurement. NO concentrations of 1 torr (1.3×10^3 ppm) can be present when using eq. 54 without exceeding a 1% error. For example, irradiation of 20 ppm NO_2 in the presence of 500 ppm NO does not produce a detectable change (8) in the $2\phi k_a$ value obtained by use of eq. 42. Even larger concentrations of O_2 can be present since $k_3[M]/k_{28}$ in eq. 38 is much smaller than $k_{27}[M]/k_{28}$. It should be recognized that these projected limitations refer to specific cases in which only one other gas besides NO_2 is present. For example, if both NO and O_2 are present as contaminants, the concentration limits are determined by the $k_{1a}[NO]^2[O_2]$ term of eq. 38 rather than the rate constant ratio terms. The allowable limits for more complex systems can be determined by evaluating the system using the rate constant ratios from reference 8 (Table III) as applied to eq. 38. It is also apparent from these considerations that I_i expressions can be generated from eq. 38 for any desired degree of contamination or for any NO_2 concentration range. However, the simplicity evident in eqs. 54–56 may be destroyed. This can, in some cases, increase the probable error in the result, thus reducing the usefulness of the derived expression. An interesting application containing implications for light intensity measurements while still retaining the desired simplicity was employed by Bufalini and Stephens (18) when generating values of k_{1a}. They recognized that eq. 52 reduces at equilibrium to:

$$\frac{\phi k_a[NO_2]}{(k_3[M]/k_{28})\ ([O_2]/[NO_2])} = k_{1a}[NO]^2[O_2] \qquad (52a)$$

when operating at atmospheric pressure in air. Although Bufalini and Stephens used this expression to generate values (18) of k_{1a}, it is apparent that it can also be used to obtain values of light intensity or alternately as a further check on the value of the $k_3[M]/k_{28}$ ratio.

The data in Figure 3 indicate the useful wavelength range of NO_2 when used as an actinometer. The information stops at 2700 Å, so we cannot project its use to shorter wavelengths. Above 3700 Å NO_2 may not prove useful because of the rapid drop-off in primary quantum yield, thus leading to possible large errors in the light intensity measurements.

VII. NITROGEN DIOXIDE PHOTOLYSIS IN THE PRESCENCE OF OTHER SPECIES

A. Interaction at Dilute Concentrations in Air

The rate constants discussed in this chapter permit us to outline the important reactions occurring in air which is contaminated with hydrocarbons and nitrogen oxides. This system is most easily understood if the rapid reactions involving nitrogen oxides and the various forms of oxygen are first considered. This scheme can be best understood with reference to the diagram shown in Figure 7. Fast reactions are shown as solid lines while slower reactions are shown by dashed lines. The initiating reaction is the photolysis of nitrogen dioxide to yield nitric

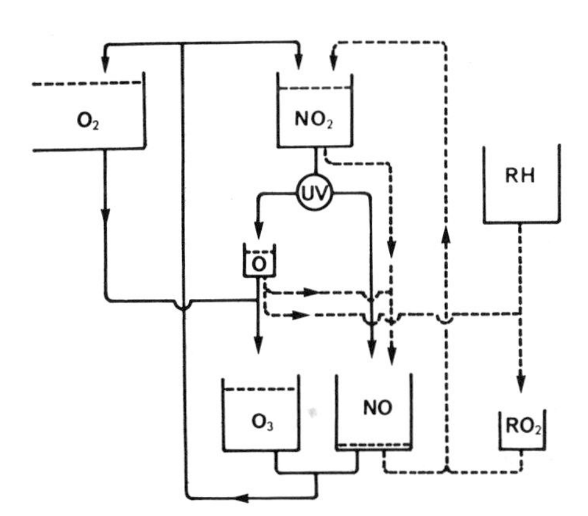

Fig. 7. Representation of interaction occurring in air between oxides of nitrogen and hydrocarbons.

oxide and an oxygen atom (reaction a). It is important to appreciate the speed of this reaction; in full sunlight nitrogen dioxide would have a half-life of about 2 min. This is far faster than any overall change which can be observed when traces of nitrogen oxides and hydrocarbons in air are irradiated. This simply means that nitrogen dioxide is being regenerated almost as fast as it is photolyzed. Subsequent reactions clearly show how this occurs. From the value of k_3 the half-life of oxygen atoms in normal air is found to be 13 μsec. No reaction between oxygen atoms and trace amounts of pollutants can compete with this. The ozone and nitric oxide produced by these two reactions react rapidly with each other to regenerate the initial reactions. If ozone and nitric oxide are each present at 10 pphm concentration, their half-life would be 0.35 min. Since the reactants are regenerated, this constitutes a closed and quite rapid steady state which is not really evident because concentrations do not change. This circulation is shown by the solid lines in Figure 7 and can also be expressed by the equation

$$O_2 + NO_2 \overset{UV}{\rightleftharpoons} NO + O_3$$

The concentrations must obey the steady-state equation

$$k_{31}[NO]\,[O_3] = \phi k_a[NO_2]$$

Since this equation contains three unknown concentrations, there are many possible combinations of concentrations consistent with it. If only nitrogen dioxide is present initially, then irradiation leads to small but equal concentrations of nitric oxide and ozone. If nitric oxide with a small amount of nitrogen dioxide is the starting material, then irradiation produces only trace amounts of ozone but very little over all change in the concentrations; this in spite of the fact that quite rapid formation and reaction is occurring.

Two slower reactions which serve to unbalance this steady state are shown as dashed lines in Figure 7. These two reactions work in opposite directions, one (reaction 28) causing the conversion of nitrogen dioxide to nitric oxide and the other causing the oxidation of nitric oxide to nitrogen dioxide. Although these reactions are slow compared to the steady-state circulation, they do change the balance between nitric oxide and nitrogen dioxide. Under atmospheric conditions the hydrocarbon reactions overcome reaction 28 and there is a net conversion of nitric oxide to nitrogen dioxide. When this is complete, ozone begins to accumulate. Three conditions must be fulfilled for this conversion to be complete: (*1*) there must be sufficient hydrocarbon, (*2*) the hydrocarbon must be of sufficient reactivity, and (*3*) there must be enough time.

B. Interaction of Oxygen Atoms with Hydrocarbons

As noted in Section I the interaction of hydrocarbons with the oxygen atoms produced by sunlight photolysis of NO_2 is considered to be an important factor in photochemical air pollution problems. Assuming the reaction between oxygen atoms and hydrocarbons (RH) is bimolecular we can write:

$$d[\mathrm{RH}]/dt = -k_{\mathrm{RH}}[\mathrm{O}]\,[\mathrm{RH}] \tag{58}$$

Since NO_2 photolysis provides an *in situ* source of oxygen atoms, it is natural that attempts be made to apply this photolysis to generation of k_{RH} values. Using the derived oxygen atom relationship of eq. 47 and the assumption of generating data at low NO_2 conversions in a nitrogen atmosphere, we can rewrite eq. 58 as:

$$d[\mathrm{RH}]/dt = -\frac{k_{\mathrm{RH}}(\phi k_{\mathrm{a}}[\mathrm{RH}])}{k_{28}(1 + k_{29}[\mathrm{M}]/k_{28})} \tag{59}$$

or rearrange to give:

$$(k_{28}/k_{\mathrm{RH}})\ (1 + k_{29}[\mathrm{M}]/k_{28}) = \phi k_{\mathrm{a}}[\mathrm{RH}]/(d[\mathrm{RH}]/dt) \tag{60}$$

In essence, both the Ford-Endow (55) and Cvetanovic (56) studies used variations of eq. 60 to arrive at certain hydrocarbon rate constants. One basis assumption in both investigations cited is that the oxygen atoms react only with NO_2 or with the hydrocarbon. From such an investigation of 1-butene Cvetanovic obtained:

$$(k_{28}/k_{1\text{-Bu}})\ (1 + k_{29}[\mathrm{M}]/k_{28}) = 2.1 \tag{61}$$

By applying the reference 7 value of k_{28} and $k_{29}[\mathrm{M}]/k_{28}$ to eq. 61 a value of 1.6×10^3 pphm^{-1} hr^{-1} was obtained (57) for $k_{1\text{-Bu}}$. However, our findings relative to Tables III and V show this ratio and rate constant to be in error. The effects are mixed, however. Thus the large error in $k_{29}[\mathrm{M}]/k_{28}$ evident in the reference 7 value has very little effect on the $k_{1\text{-Bu}}$ generated from eq. 61. This results from the relatively low (40–50 torr) concentrations of M maintained during the experiments under discussion. Calculations at these low pressures thus provide a value for $1 + k_{29}[\mathrm{M}]/k_{28}$ of 1.11 from the reference 7 ratio and of 1.02 from the reference 8 ratio. Thus operation at these reduced pressures results in only a 10% error in $k_{1\text{-Bu}}$ rather than the 50% error introduced into the Ford-Endow experiments conducted at atmospheric pressure. The largest part of the error is introduced by the k_{28} value. This is not directly related to the reference 7 $k_3[\mathrm{M}]/k_{28}$ ratio used to calculate k_{28} because, as shown in Section IV-C this latter ratio is very nearly correct. The error was introduced by the specific value of k_3 which was introduced into the $k_3[\mathrm{M}]/k_{28}$ ratio. Recalling that k_3 has recently undergone a substantial reevaluation, we recognize that the value of k_{28}, and therefore $k_{1\text{-Bu}}$, is 60% lower than what we currently calculate. Using the more recent k_3 value Cvetanovic would have calculated a $k_{1\text{-Bu}}$ of 3.5×10^3 pphm^{-1} hr^{-1}. Using the reference 8 k_{28} and $k_{29}[\mathrm{M}]/k_{28}$ values in eq. 61 results in a rate constant of 4.1×10^3 pphm^{-1} hr^{-1} for the reaction of oxygen atoms with 1-butene. Applying this corrected $k_{1\text{-Bu}}$ constant to Cvetanovic's list of relative rates produces the rate constants in Table VII for the reaction of oxygen atoms with various hydrocarbons. The relative rates were determined by a technique which involves competition of two hydrocarbons for oxygen atoms; the rates therefore are not affected by the reevaluation of rate constant ratios or rate constants. It is strictly fortuitous that the Ford-Endow rate constant (55) for oxygen atom reaction with *cis*-2 pentene ($1.6 \times 10 \times 10^4$ pphm^{-1} hr^{-1}) is in close agreement with the corrected Cvetanovic constant in Table VII, we say fortuitous because the effect of using a k_{28} value which is low by 60% and a value of $1 + k_{29}[\mathrm{M}]/k_{28}$ which is high by a factor of 50% results in cancellation of errors when introduced into eq. 60. Had Ford and Endow used the corrected value of k_3, their rate constant for this reaction would be higher than reported by a factor of 2.

TABLE VII

Rate of Reaction of Oxygen Atoms with Hydrocarbons

Hydrocarbon	Corrected[a] rate constants $pphm^{-1} hr^{-1}$ Cvetanovic (55,60)
n-Butane	2.7×10^{1}
3-Methyl heptane	3.2×10^{2}
Ethylene	6.5×10^{2}
Propylene	3.9×10^{3}
1-Butene	4.1×10^{3}
Isobutene	1.7×10^{4}
cis-2-Butene	1.4×10^{4}
trans-2-Butene	1.9×10^{4}
cis-2-Pentene	1.5×10^{4}
Cyclopentene	2.1×10^{4}
Tetramethylethylene	7.2×10^{4}

[a] Corrected using reference 8 values for k_{28} and $k_{29}[M]/k_{28}$.

Cvetanovic's corrected $k_{1\text{-Bu}}$ is now in agreement with two direct measures of this constant. Using electrical discharge in oxygen as a source of oxygen atoms, Elias and Schiff (58) obtained 5.3×10^{3} $pphm^{-1} hr^{-1}$, and a similar value of 4.6×10^{3} $pphm^{-1} hr^{-1}$ was recently obtained by Elias (59). These three values lead to an average $k_{1\text{-Bu}}$ of $4.7 \pm 0.6 \times 10^{3}$ $pphm^{-1} hr^{-1}$. Cvetanovic's value of 4.1×10^{3} $pphm^{-1} hr^{-1}$ lies within this 13% standard deviation, thus adding further credence to the k_{28} value generated in reference 8.

C. Projected Studies of Interaction with Hydrocarbons

Having generated a basis for accepting the Ford-Endow mechanism, we can proceed to theoretically analyze the results when, for example, the oxygen atoms from NO_2 photolysis are reacted with olefins. Because it presumably is the simplest case, we will examine this interaction in the absence of a third body. We will assume that the only fate of the oxygen atom is to react with the olefin (RH) or with NO_2 since reaction 28 is the only other oxygen atom consuming path in the absence of oxygen or large concentrations of a third body. Under these conditions our mechanism is described by:

$$NO_2 \xrightarrow{h\nu} NO + O \qquad R_a = \phi k_a[NO_2]\,O \tag{a}$$

$$O + NO_2 \rightarrow NO + O_2 \qquad R_{28} = k_{28}[O]\,[NO_2] \tag{28}$$

$$O + RH \rightarrow \text{Products} \qquad R_{RH} = k_{RH}[O]\,[RH] \tag{58}$$

Inclusion of reaction 58 leads to competition between NO_2 and RH for the oxygen atoms; therefore, we expect the observed rate of NO_2 disappearance to decrease. Note that no provision is made in this mechanism for interaction of the "products" with other molecules in the system. In a recent review Cvetanovic (60) concludes that these interactions with NO_2 are important depending on the particular olefin investigated and the total pressure in the system. The knowledge that such interactions occur should not deter us from examination of the mechanism suggested by reactions a, 28, and 58 since observed deviations between mechanism and experimental data may help delineate these interactions.

The rate of NO_2 disappearance from reactions a, 28, and 58 is given by:

$$-(d[NO_2]/dt) = \phi k_a[NO_2] + k_{28}[O]\,[NO_2] \qquad (62)$$

Using the steady-state assumption, the oxygen atom concentration is given by:

$$[O] = \frac{\phi k_a[NO_2]}{(k_{28}[NO_2] + k_{RH}[RH])} \qquad (63)$$

Insertion of eq. 63 in eq. 62 and rearranging yields:

$$\frac{(d[NO_2]/dt)}{(k_a[NO_2])} = \Phi_{NO_2} = \phi\left[1 + \frac{1}{1 + (k_{RH}/k_{28})\,(RH/NO_2)}\right] \qquad (64)$$

Inspection of eq. 64 shows that in the absence of RH the rate expression reduces to 2ϕ as required by eq. 48. In the presence of RH the Φ_{NO_2} must be less than 2ϕ by an amount which is a function of the k_{RH}/k_{28} and $[RH]/[NO_2]$ ratios. Deviations from eq. 64 will be a measure of further reactions of the "products." If there are no interactions, eq. 64 permits determination of k_{RH}. If product interaction only involves NO_2, then deviations from eq. 64 will be a measure of these reactions providing we know k_{RH} by independent measurement. In the case of 1-butene, for example, we know that k_{RH} has a similar value whether the oxygen atom source is NO_2 photolysis or whether oxygen atoms are generated by electrical discharge in oxygen. Application of eq. 64 to this 1-butene reaction should provide a measure of the possible interaction of NO_2 with the oxygen atom–olefin products. If such interaction is demonstrated, the mechanism can be restated and new rate expressions generated.

Equations a, 28, and 58 also permit derivation of an expression to describe RH disappearance. Thus:

$$d[RH]/dt = -k_{RH}[O]\,[RH] \qquad (65)$$

Substituting the oxygen atom concentration given by eq. 63 yields:

$$\frac{-d[\mathrm{RH}]}{dt} = \frac{k_{\mathrm{RH}}[\mathrm{RH}]\phi k_{\mathrm{a}}[\mathrm{NO_2}]}{(k_{28}[\mathrm{NO_2}] + k_{\mathrm{RH}}[\mathrm{RH}])}$$

which can be rearranged to:

$$\frac{\phi k_{\mathrm{a}}}{\{d(\ln\,[\mathrm{RH}])/dt\}} = \frac{k_{28}}{k_{\mathrm{RH}}} + \frac{[\mathrm{RH}]}{[\mathrm{NO_2}]} \tag{66}$$

Still another relationship which can be established is the $d[\mathrm{NO_2}]/dt$ to $d[\mathrm{RH}]/dt$ ratio. This expression can be generated through the ratio of eq. 62 to eq. 65 followed by insertion of eq. 63. It has the form:

$$\frac{d[\mathrm{NO_2}]/dt}{d[\mathrm{RH}]/dt} = 1 + 2\,\frac{k_{28}}{k_{\mathrm{RH}}}\,\frac{[\mathrm{NO_2}]}{[\mathrm{RH}]} \tag{67}$$

Equation 67 may prove the most convenient since it can be applied without a knowledge of light intensity.

One interesting case occurs if the oxygen atom–olefin reaction results in a single product which then further reacts by bimolecular reaction with NO_2. In this case the form of eq. 66 relating to the olefin rate does not change, whereas the $\Phi_{\mathrm{NO_2}}$ in eq. 64 reverts back to 2ϕ. Projection beyond this point of establishing the method of procedure is not warranted in the absence of actual application. The data from preliminary experiments in this laboratory using *trans*-2-butene as the olefin do not fit eq. 64. In fact, the rate of NO_2 disappearance is 2–3 times the predicted value.

While this may indicate a complex reaction of products with NO_2, it can, for example, also indicate involvement of the olefin as a highly efficient third body in reaction 29.

In conclusion it is noted that some caution is required in interacting the Ford and Endow NO_2 mechanism with other systems. Reflection on the reactions listed in the mechanism (see Sec. III) leads to the conclusion that most of these reactions describe overall effects and that the total number of reactions involved is much greater. As an example, consider reactions 10a and 10b, which evidence suggests are occurring, but which are not required to explain NO_2 photolysis. Interaction of some species with N_2O_5 may therefore demand inclusion of these two reactions in the mechanism.

VIII. SUMMARY AND REMARKS

The evidence at this point is overwhelmingly in favor of the Ford-Endow mechanism and the applicability of the rate ratios of reference 8 to that mechanism. The mechanism permits accurate prediction of overall quantum yields for NO_2 studies, particularly when the NO_2 concentration is less than 10 torr. Insertion of two independent rate constants, k_3 and k_{27}, into the rate constant ratios of the mechanism results in an interconsistency which necessarily means that the k_{28} and k_{29} rate constants thus generated are known within $\pm 22\%$. This is further substantiated by applying the k_{28} rate constant to Cvetanovic's data relating to oxygen atom reaction with 1-butene. This calculation results in a $k_{1-\text{Bu}}$ value which is within a 13% standard deviation of independent measures. Furthermore, the mechanism allows use of NO_2 as an accurate gas phase actinometer in the 3000–3700 Å region. Projected use of the mechanism shows it may be a useful tool for further investigations of oxygen atom–hydrocarbon reactions. The reference 8 findings also show the necessity of a third body in reaction 29.

There are some implications in the experimental details of reference 8 which should be discussed. Thus, it has become traditional in photochemical studies to cast a jaundiced eye at data which was obtained by use of nonmonochromatic light sources and nonuniform light intensities, yet both of these latter conditions were violated in the reference 8 study without noticeably affecting the relationship of the results to systems in which these conditions were closely adhered to. For example, the light source in reference 8 consisted of "black light" fluorescent tubes which radiate a continuum in the 3000–4000 Å region. The intensity peaks at 3500 Å and falls off rapidly above and below this wavelength. In fact, 85% of the irradiating energy is confined to the 3300–3750 Å region; however, this 450 Å spread hardly corresponds to monochromatic light. In addition, the light was not rendered parallel and the intensity within the 8 in. light path of the cell therefore varied by a factor of 10–10^2. To some extent this lack of effect can be attributed to the fact that ratios of photolysis rates were used to calculate the rate constant ratios. Since these ratios of photolytic rates were determined independently of light intensity, we would expect light intensity parameters to cancel out. However, this is not true when using eqs. 54–56 to measure light intensity; yet, here again we see no effect of these "light" parameters. One might note that the use of a weak absorbing system requires only a knowledge of incident light intensity rather than quantity of absorbed light. In such a system, effects usually associated with the fraction of light absorbed are negligible over large changes in absorber concentration. In reality, as pointed out by Calvert and Pitts (51), some of the theoretical effects of light absorption do not take diffusion in gas phase systems into account. A good example of this possible lack of applicability of the light absorption theory is contained in Table A, reference 46. Theory in this case demanded a drastic difference between measured and real rates of reaction because 69–100% of the light was absorbed in the system. However, as pointed out in Section V-B, no such effect could be found in the data.

Another reason why use of a wide bandwidth has no effect in this case is probably associated with the regular manner in which the NO_2 primary quantum yield varies within the wavelength region studied (see Fig. 3). The fact that the particular system under study is not affected by these deviations from accepted practice does not mean we can ignore them; however, it does mean that deviations from these practices are not automatically a basis for criticism of data. Obviously, the use of monochromatic light can provide data which would be difficult if not impossible to obtain without it. A good example of the necessity of monochromatic light is in the generation of quantum yield data shown in Figure 3.

It is a rare occasion when a single investigation such as reference 8 succeeds in providing data with such diverse and encompassing implications, yet to stop at this thought leaves the true meaning of the results hidden. Thus the relationships uncovered would still be hidden in the absence of the detailed experimental data (see Tables I, II, and IV) which led to the mechanism proposed by Ford and Endow in 1957. The magnitude of the facts which were used by Ford and Endow in order to arrive at this mechanism is now very apparent. It is indeed unfortunate that the numerical data which they generated cast suspicion upon their proposed mechanism—a mechanism which now appears to form the basis for correlation of most experimental data relating to reactions involved in NO_2 photolysis.

IX. RECENT DEVELOPMENTS

Since completion of the first draft of this chapter, several developments have occurred which require brief discussion in order to bring progress in this field up to date. These occurrences do not seriously alter the basic content of the chapter. Their main importance is that they tend to modify and in some cases clarify some of the areas discussed.

A. Atmospheric Conversion of NO to NO_2

As discussed in Section I, one important unknown in atmospheric photochemistry is the manner in which NO is converted to NO_2 during sunlight irradiation of polluted atmospheres. In lieu of knowledge, various theories have been proposed to account for this phenomenon. Unfortunately, these theories have not been readily amenable to experimental testing and thus the conversion of NO to NO_2 has remained a puzzle. Quite recently, however, a reaction between NO and one of the end products of hydrocarbon photooxidation has been investigated which may account for this rapid conversion. The compounds involved in this conversion are the peroxyacyl nitrates (PAN's), which are a family of phytotoxicants and eye irritants found in photochemical air pollution mixtures. The PAN's have the general formula

$$\mathrm{R}\overset{\overset{\displaystyle O}{\|}}{\mathrm{C}}\mathrm{-OONO_2}.$$

From a concentration viewpoint peroxyacetyl nitrate (PAN) appears to be the most prominent member of this family found in polluted atmospheres (61).

One of the most convenient methods of synthesizing certain members of the peroxyacyl nitrate family is by photolyzing (with 3000–4000 Å light) the parent alkyl nitrite in air or in an atmosphere of oxygen. It was during one such photochemical study involving irradiation of ethyl nitrite in oxygen that an unexpected rapid destruction of the product PAN was observed (62). Removal of irradiation did not immediately affect the rate of destruction and it was observed that the reaction was accompanied by a decrease in NO and an increase in NO_2. Subsequent studies (63) of nonirradiated mixtures of NO and PAN in absence and/or presence of O_2 and N_2 confirm the existence of a reaction between NO and PAN which appears to be approximately independent of the presence of a third body and whose rate constant is 10^3 greater than the rate constant in air, for the reaction between NO and O_2 to form NO_2 (Section III-B). A first approximation of the initial mechanism can be envisioned as:

$$\mathrm{NO+CH_3}\overset{\overset{\displaystyle O}{\|}}{\mathrm{C}}\mathrm{-OON}\begin{matrix}\diagup \mathrm{O}\\ \diagdown \mathrm{O}\end{matrix} \rightarrow \mathrm{2NO_2+CH_3}\overset{\overset{\displaystyle O}{\|}}{\mathrm{C}}\mathrm{-O\cdot} \qquad (68)$$

A tentative value of 9.6 $ppm^{-1}\,hr^{-1}$ has been found for the rate constant in terms of NO_2 formation. It is tempting at this point to interpret reaction 68 as being similar to the Trautz mechanism (Sec. III-B) in which NO reacts with the NO_3 portion of the PAN molecule. Much, however, remains to be investigated before a reaction mechanism can be justified. The important point at this time is the discovery of a reaction which converts NO to NO_2 at a rate which makes it important to atmospheric photochemistry. Furthermore, this reaction indicates that PAN and probably its homologs are more than end products of atmospheric photochemical reactions. These compounds are, in fact, fascinating intermediates with roles similar in kind if not magnitude to the role played by ozone.

B. Leighton's Review

As previously pointed out, the most comprehensive review concerned with atmospheric photochemistry was published by Dr. P. A. Leighton in 1961 (10). A review of the literature since that time shows that this publication is still recognized as the most current review of the state of knowledge in this area. In addition to emphasizing the thoroughness of the treatment, this latter recognition tends to show how few new facts have been added to our knowledge since 1961. However, it is becoming increasingly apparent that we have to some extent not understood the importance of Leighton's careful analysis. In one specific case involving the photochemistry of NO_2 this has led to a misinterpretation of his conclusions. It is important that this misinterpretation be recognized, since it involves reactions which are basic to the entire atmospheric photochemical phenomenon. Such recognition may serve to stimulate new research, and help us recognize the difference between fact and theory. It is indeed unfortunate that our lack of knowledge concerning the complex atmospheric reactions has caused us to theorize often and extensively—to the point where we now frequently confuse fact with theory. Lack of such knowledge frequently has led us to overemphasize reactions which the facts in Leighton's review clearly show cannot be important contributors to atmospheric photochemistry.

The reactions and predicted results thereof which have been misinterpreted are reactions a, 3, and 31, of the NO_2 kinetic scheme. The rates of these reactions are well known and it was Leighton who clearly pointed out that these three reactions must dominate atmospheric photochemistry. These three reactions which result in a rapid cycling of NO_2 must play a dominant role since NO_2 is the only known effective light absorber present and because the rates of these reactions are at least 10^2 greater than any other known reactions. These facts have important implications which Leighton recognized. Thus, as discussed in Section VII-A, atmospheric ozone levels during sunlight irradiation of polluted atmospheres are determined by

$$[O_3] = \frac{\phi k_a[NO_2]}{k_{31}[NO]} \tag{69}$$

It has frequently but incorrectly been stated that the relationship in eq. 69 only applies in the absence of hydrocarbon. In reality, Leighton's analysis shows that eq. 69 holds *whether or not* hydrocarbons are reacting with the O atoms and O_3 formed in reactions a and 3. The rates of these hydrocarbon reactions are such that they cannot disturb the relationship denoted by eq. 69. Within the limits of instrument sensitivity, the validity of eq. 69 can be shown to hold for observed concentration of O_3, NO, and NO_2 in Los Angeles Basin polluted atmospheres. One of the more interesting possibilities stemming from the general applicability of eq. 69 relates to light intensity measurements. Thus if the concentrations of O_3, NO, and NO_2 at any given location are known, it should be possible to calculate incident light intensity at that location. The extreme speed of the equilibrium in eq. 69 relative to atmospheric mixing factors assures that incident light intensity would be measured. There are many advantages to such a measurement of light intensity. Thus NO_2 is the only identified important absorber in our polluted atmospheres and therefore the use of eq. 69 would give a direct measure of the light which is important to subsequent reactions. Furthermore it would do this without the errors associated with light passing through cell windows, etc. Unfortunately the use of eq. 69 for this purpose must await the development of much more reliable and sensitive instruments for measuring the concentrations of O_3, NO, and NO_2. Instruments are needed which can measure these three gases in the 10^{-3} ppm range rather than those presently available which frequently have difficulty measuring accurately in the 10^{-1} ppm range.

C. Modifications in NO_2 Kinetic Scheme

In Section VII-C, certain possible interactions involving photolyzing NO_2 systems and hydrocarbons were discussed. These projected interactions were based on the correctness of the Ford-Endow mechanism as detailed in Section II. However, it should be clearly understood that this latter mechanism is not the only set of reactions which can account for the behavior of NO_2 during photolysis. Thus in spite of the logical and quantitative agreements afforded by use of the Ford-Endow mechanism, we must be prepared to question and modify the theory on the basis of additional experimental findings. Certain unresolved aspects of the mechanism are worth noting. Thus Leighton concluded several years ago (64) that the existing evidence concerning the effects of inert gases on NO_2 photolysis strongly suggests a primary quantum yield which is third-body dependent. This is, of course, in direct contrast to the Ford-Endow mechanism which is based on a primary quantum yield which is independent of third-body effects. (See Section II, reaction a.) Yet, modification of the Ford-Endow mechanism to account for this third-body dependence does not produce large changes in the form of the overall rate expression (eq. 38) and therefore does not produce significant changes in the rate constant ratios shown in Table III. Thus it is not possible by this type of mechanism change to generate evidence for or against third-body dependence of primary quantum yield. However, it is apparent that such mechanism modifications will seriously alter the pathways of interaction with hydrocarbons as discussed in Section VII-C. Extensive studies of such interactions may in fact provide additional information concerning validity of the Ford-Endow mechanism.

Another related aspect of the mechanism which requires further investigation concerns the formation and destruction of NO_3 (reactions 29 and 9). Examination of the details of overall reaction 1a of the mechanism (Sec. III-B) shows that reaction 40 appears to be identical with reaction 9. Since a reaction may only be used once in a mechanism, this apparent doubling up of reactions would appear to be an error. However, as Leighton (10) has pointed out, there is some evidence which suggests that reaction of oxygen atoms with NO_2 (reaction 29) yields a nitrate structure

```
O
 \
  N—O
 /
O
```

for NO_3, while reaction of O_2 with NO (reaction 39) yields a peroxy structure (OONO). Thus on this basis, reactions 9 and 40 are not repetitious reactions since each refers to reaction of NO with different isomers of NO_3. Subsequent experimentation could conceivably indicate that the same NO_3 isomer is produced by reactions, 29 and 39. In this event a serious discrepancy in theory will become evident since the Trautz mechanism (Sec. III-B), when applied to the ppm concentration range, demands that most of the NO_3 formed must decompose to O_2 and NO, while the Ford-Endow mechanism demands that a sizable fraction of the NO_3 formed must react with NO to form NO_2. As a matter of interest, neither reaction 29 nor 9 need be included in the mechanism, providing we can show that the primary quantum yield is third-body dependent. Thus at the moment there exists some question concerning the importance of reaction 29 during photolysis of NO_2 in the ppm concentration range. Reflection shows that reactions 29 and 9 were selected for inclusion in the mechanism in order to explain third-body effects, assuming a primary yield which is independent of collisional deactivation. Thus there is nothing inherently absolute about the inclusion of these two reactions. As indicated previously, the Ford-Endow mechanism should not be regarded as an end unto itself. Rather it should be recognized as a semiempirical device which can be useful in efforts to extend our knowledge in this fascinating field.

D. Role of Singlet Oxygen

Another mysterious aspect of atmospheric photochemistry relates to hydrocarbon attack. Research in the laboratory has indicated that as much as 50% of the hydrocarbon disappearance cannot be accounted for by attack of oxygen atoms and O_3 generated in reactions a and 3. Recently it has been suggested (65) that singlet oxygen makes a contribution to hydrocarbon destruction. It was further suggested that this species arises by an energy transfer process involving absorption of sunlight by sensitizing compounds and subsequent collision with O_2 to form singlet oxygen. By itself this is an interesting theory well worth investigation, particularly since recent evidence indicates that singlet oxygen is not readily deactivated by collision. However, reflection on the atmospheric NO_2 cycle as outlined in Section IX-B reveals the possibility of another equally interesting source of singlet oxygen. One of the reactions in this cyclic mechanism (reaction 31) is between NO and O_3 to form NO_2 and O_2. The question at this point is, what is the state of the O_2 product? If it occurs as the excited singlet, we would indeed have an active species which might be as important as oxygen atoms in terms of hydrocarbon attack. The evidence concerning the state of the O_2 product in reaction 31 is inconclusive. Thus reaction of O_3 with NO in the torr concentration range yields excited NO_2 and ground-state oxygen. However, in this concentration range this result would be expected even if singlet oxygen were initially formed, since this latter species is known to readily transfer its energy to NO_2. Thus these experiments require repetition in the ppm concentration range. The advantage of singlet oxygen from reaction 31, instead of the previously discussed sensitized hydrocarbon energy-transfer system, is that half-life and concentration considerations become much less important. This is a result of the cyclic nature of the NO_2 photolytic process which would assure an ever-present supply of singlet oxygen. For example, the half-life of the oxygen atom from reaction a is only 10 μsec, yet oxygen atom attack on hydrocarbon is considered to be very important.

Although this singlet oxygen discussion is interesting, it is well to keep in mind that at the moment no experimental fact supports these theories. Experimentation within the coming year should permit an assessment of the role of singlet oxygen. One of the more satisfying features of these singlet oxygen speculations is that here at least we have a theory which appears amenable to experimental testing.

E. Reactivity of NO

During the discussion of the relationship of oxides of nitrogen to air pollution in Section I, it was stated that NO is a "relatively nontoxic gas." The term "relatively" implies comparision with other gases and in this particular context the comparison is generally to NO_2 toxicity. On the basis of mice exposures to several hundred ppm of NO it appears that NO_2 is five times as toxic as NO. Interpretation of NO exposure in these concentration ranges is difficult, however, because of the ready conversion of NO to NO_2 by way of reaction 1a (see Secs. I and III-B). Thus at the 320 ppm NO concentration employed in the noted toxicity study, approximately 30 ppm/min of NO was converted to NO_2. Thus the contract time of this gas mixture with the mice would have to be shorter than 10 sec in order to keep the conversion of NO to NO_2 below 1%.

It has recently been determined that exposure of vegetation to NO is similar in some respects to exposure to NO_2. An atmosphere containing 10 ppm NO or an atmosphere containing approximately 6 ppm NO_2 will result in a 65% reduction in the ability of plants to assimulate CO_2. However, this growth suppression symptom cannot in reality be directly compared because the effect of these two gases on vegetation is quite different. Exposure to NO_2 results in a gradual destruction of certain plant cells and a resultant reduction in CO_2 assimilation which can be reversed only by growth of new cells. On the other hand, exposure to NO results in no detectable damage to plant cells and an immediate reduction in CO_2 assimilation which is completely reversible when the NO exposure is discontinued (66).

From a chemical viewpoint one might expect that NO, because of its high reactivity with many compounds, would be more toxic to biological systems than NO_2. However, in spite of the difficulties associated with

NO exposures, the preceding discussion clearly does not support such an expectation. It is also apparent that we have much to learn concerning the mode of interaction of NO and NO_2 with biological systems.

References

1. T. C. Hall, Jr., and F. E. Blacet, *J. Chem. Phys.*, **20**, 1745 (1952).
2. D. Neuberger and A. B. F. Duncan, *J. Chem. Phys.*, **22**, 1693 (1954).
3. R. G. W. Norrish, *J. Chem. Soc.*, **1929**, 1604, 1611.
4. E. A. Schuck, E. R. Stephens, and J. T. Middleton, *Arch. Environ. Health*, **13**, 570 (1966).
5. E. F. Darley, C. W. Nichols, and J. T. Middleton, *U.S. Dept. Agr., Farmers' Bull., State of California*, **55**, 1, p. 11 (1966).
6. E. R. Stephens, *Infrared Phys.*, **1**, 187 (1961).
7. H. W. Ford and N. Endow, *J. Chem. Phys.*, **27**, 1156 (1957).
8. E. A. Schuck, E. R. Stephens, and R. R. Schrock, *J. Air Pollution Control Assoc.*, **16**, 695 (1966).
9. H. W. Ford and S. Jaffe, *J. Chem. Phys.*, **38**, 2935 (1963).
10. P. A. Leighton, *Photochemistry of Air Pollution*, Academic Press, New York, 1961.
11. H. H. Holmes and F. Daniels, *J. Am. Chem. Soc.*, **56**, 630 (1934).
12. (a) M. Bodenstein, *Z. Physik. Chem.*, **85**, 329 (1913); (b) O. K. Rice, *J. Phys. Chem.*, **64**, 1851 (1960).
13. M. Bodenstein, *Z. Angew. Chem.*, **31**, 145 (1918).
14. M. Bodenstein, *Z. Electrochem.*, **24**, 183 (1918).
15. J. C. Treacy and F. Daniels, *J. Am. Chem. Soc.*, **77**, 2033 (1955).
16. A. P. Altshuller, I. R. Cohen, S. F. Sleva, and S. L. Kopczynski, *Science*, **138**, 442 (1962).
17. J. D. Greig and P. G. Hall, *Trans. Faraday Soc.*, **62**, 652 (1966).
18. J. J. Bufalini and E. R. Stephens, *Intern. J. Air Water Pollution*, **9**, 123 (1965).
19. R. L. Hasche and W. A. Patrick, *J. Am. Chem. Soc.*, **47**, 1207 (1925).
20. F. B. Brown and R. H. Crist, *J. Chem. Phys.*, **9**, 840 (1941).
21. W. A. Glasson and C. S. Tuesday, *J. Am. Chem. Soc.*, **85**, 2901 (1963).
22. M. Šolc, *Nature*, **209**, 706 (1966).
23. H. S. Johnston and L. W. Slentz, *J. Am. Chem. Soc.*, **73**, 2948 (1951).
24. G. Lunge and E. Berl, *Z. Angew. Chem.*, **19**, 857 (1906).
25. G. Lunge and E. Berl, *Z. Angew. Chem.*, **19**, 881 (1906).
26. E. Briner, W. Pfeiffer, and G. Malet, *J. Chim. Phys.*, **21**, 25 (1924).
27. G. Kornfeld and E. Klingler, *Z. Physik. Chem.*, **B4**, 37 (1929).
28. M. E. Morrison, R. G. Rinker, and W. H. Corcoran, *Ind. Eng. Chem. Fundamentals*, **5**, 175 (1966).
29. J. H. Smith, *J. Am. Chem. Soc.*, **65**, 74 (1943).
30. M. Trautz, *Z. Elektrochem.*, **22**, 104 (1916).
31. G. B. Kistiakowsky and G. G. Volpi, *J. Chem. Phys.*, **27**, 1141 (1957).
32. F. S. Klein and J. T. Herron, *J. Chem. Phys.*, **41**, 1285 (1964).
33. A. A. Westenberg and N. de Hass, *J. Chem. Phys.*, **40**, 3087 (1964).
34. M. A. A. Clyne and B. A. Thrush, *Proc. Roy. Soc. (London), Ser. A*, **269**, 404 (1962).

35. P. Harteck and S. Dondes, *J. Chem. Phys.*, **27,** 546 (1957).
36. F. Kaufman, *Proc. Roy. Soc. (London), Ser. A*, **247,** 123 (1958).
37. E. A. Ogryzlo and H. I. Schiff, *Can. J. Chem.*, **37,** 1690 (1959).
38. S. W. Benson and A. E. Axworthy, Jr., *J. Chem. Phys.*, **26,** 1718 (1957).
39. F. Kaufman and J. R. Kelso, *J. Chem. Phys.*, **40,** 1162 (1964).
40. J. A. Zaslowsky, H. B. Urbach, F. Leighton, R. J. Wnuk, and J. A. Wojtowicz, *J. Am. Chem. Soc.*, **82,** 2682 (1960).
41. W. M. Jones and N. Davidson, *J. Am. Chem. Soc.*, **84,** 2868 (1962).
42. S. W. Benson and A. E. Axworthy, Jr., *J. Chem. Phys.*, **42,** 2614 (1965).
43. L. F. Phillips and H. I. Schiff, *J. Chem. Phys.*, **36,** 1509 (1962).
44. F. S. Klein and J. T. Herron, *J. Chem. Phys.*, **41,** 1285 (1964).
45. J. N. Pitts, Jr., J. H. Sharp, and S. I. Chan, *J. Chem. Phys.*, **40,** 3655 (1964).
46. R. G. W. Norrish, *J. Chem. Soc.*, **1929,** 1158.
47. R. G. Dickinson and W. P. Baxter, *J. Am. Chem. Soc.*, **50,** 774 (1928).
48. F. E. Blacet, T. C. Hall, and P. A. Leighton, *J. Am. Chem. Soc.*, **84,** 4011 (1962).
49. J. King, California Institute of Technology Jet Propulsion Laboratory, unpublished data.
50. W. A. Noyes, Jr., and P. A. Leighton, *The Photochemistry of Gases*, Reinhold, New York, 1941, p. 200–202.
51. J. G. Calvert and J. N. Pitts, Jr., *Photochemistry*, Wiley, New York, 1966.
52. F. H. Verhoek and F. Daniels, *J. Am. Chem. Soc.*, **53,** 1250 (1931).
53. J. N. Pitts, Jr., University of California at Riverside, unpublished data.
54. B. E. Saltzman, *Anal. Chem.*, **26,** 1949 (1964).
55. H. W. Ford and N. Endow, *J. Chem. Phys.*, **27,** 1277 (1957).
56. R. J. Cvetanovic, *J. Chem. Phys.*, **30,** 19 (1959).
57. S. Sato and R. J. Cvetanovic, *Can. J. Chem.*, **36,** 279, 970, 1668 (1958).
58. L. Elias and H. I. Schiff, *Can. J. Chem.*, **38,** 1657 (1960).
59. L. Elias, *J. Chem. Phys.*, **38,** 989 (1963).
60. R. J. Cvetanovic in *Advances in Photochemistry*, Vol. 1, W. A. Noyes, Jr., G. S. Hammond, and J. N. Pitts, Jr., Eds., Interscience, New York, 1963, p. 117.
61. E. F. Darley, K. A. Kettner, and E. R. Stephens, *Anal. Chem.*, **35,** 589 (1963).
62. E. A. Schuck, manuscript in preparation.
63. E. A. Schuck, E. R. Stephens, and M. A. Price, manuscript in preparation.
64. P. A. Leighton, unpublished data.
65. A. U. Khan, J. N. Pitts, Jr., and E. B. Smith, *JEST*, **1,** 656 (1967).
66. O. C. Taylor, Report to 1st International Congress of Plant Pathology, London, England, July 22, 1968.

The Formation, Reactions, and Properties of Peroxyacyl Nitrates (PANs) in Photochemical Air Pollution

EDGAR R. STEPHENS
Statewide Air Pollution Research Center
University of California, Riverside

I. Occurrence and Importance of the Peroxyacyl Nitrates (PANs) 119
II. Formation of PANs 121
A. Mechanism 121
B. Kinetics 122
III. Synthesis and Purification of PANs 123
A. Methods of Preparation 123
B. Chromatographic Purification 127
IV. Spectra of the PANs 130
A. Infrared Spectra 130
B. Ultraviolet Spectrum 134
C. Mass Spectra 134
D. NMR Spectra 136
V. Analysis of the PANs 137
A. Infrared 137
B. Chromatography 137
C. Hydrolysis 139
D. Elemental Analysis 140
VI. Reactions of the PANs 141
A. Stability 141
B. Hydrolysis 142
References 145

I. OCCURRENCE AND IMPORTANCE OF THE PEROXYACYL NITRATES (PANs)

The peroxyacyl nitrates are a homologous series of organic nitrogen compounds that are formed when sunlight acts on air which is polluted with trace concentrations of organic compounds and nitrogen oxides. They are of particular interest because of their biological activity. Exposures to concentrations measured in parts per hundred million (pphm by volume) can cause visible damage to agricultural corps (1,2). Concentrations in the parts per million to parts per billion range can cause eye irritation (3,4,24). Concentrations of 100–200 ppm are lethal for mice in a 2-hr exposure (5). These evidences of activity, besides their intrinsic importance, also suggest that lower concentrations may have significant but not readily visible effects, particularly in long-term exposures of plants, animals, and humans. At the present time the value of loss of agriculatural crops ascribed to the peroxyacyl nitrates is estimated in the millions of dollars, and its characteristic damage symptoms have been found throughout the United States and in other parts of the world (6). This is not surprising when it is remembered that auto exhaust contains both ingredients necessary for peroxyacyl nitrate formation. The ubiquitous automobile emits these ingredients into the air, and subsequent sunlight photolysis produces peroxyacyl nitrates.

Peroxyacyl nitrates, abbreviated PANs (plural), is the family name given to these compounds. Although there was considerable uncertainty as to the proper structure for several years after discovery, the accumulated evidence, to be described in subsequent pages, leaves little doubt that this name, corresponding to the structure

$$R\overset{\overset{\displaystyle O}{\|}}{C}OONO_2$$

properly presents this family of compounds. The radical R may be methyl ($CH_3\cdot$), forming the first known member of the series which is regarded as a mixed anhydride of peroxy acetic and nitric acids and is called peroxyacetyl nitrate (PAN). This and the next two members of the series have been prepared, purified, and studied. The second member, in which R is ethyl ($CH_3CH_2\cdot$) is called peroxypropionyl nitrate, (PPN). The third member, peroxybutyryl nitrate (PnBN) with the *n*-propyl radical as R, was studied to a very limited extent. Peroxyisobutyryl nitrate (PiBN) with isopropyl as R was prepared but never purified or studied in any detail. Recently an extensive study of the compound with R as phenyl to form peroxybenzoyl nitrates (PBzN) has been reported (24). This member of the series produced eye irritation at concentrations about 200 times lower than formaldehyde. Since PAN is about twice as strong an irritant as formaldehyde, this makes PBzN about 100 times more powerful than PAN. This nomenclature will be used throughout this chapter. In some of the older literature these same compounds are referred to as "Compound X," peroxyacetyl nitrate, or peroxyacyl nitrites (7–11). Family members in which R is some other radical (e.g., hydrogen, cyclopentyl, etc.) may exist. The one carbon compound:

$$H\overset{\overset{\displaystyle O}{\|}}{C}OONO_2$$

which would be called peroxyformyl nitrate has never been isolated or identified even though some attempts have been made in the author's laboratory. Its existence is doubtful at best. It may be too unstable to survive, or it may not be formed.

Although widespread occurrence of PANs can be expected from the widespread distribution of its precursor, auto exhaust, measurements of its concentration have been made in only a few locations. A regular program of measurement is underway at the Statewide Air Pollution Research Center in Riverside, California, making use of a gas-chromatographic method (see Sec. V.B.) The concentration has usually been below 50 ppb (parts per billion, by volume) at this location, which is about 60 miles east (inland) from the center of Los Angeles. Moderate air pollution of the photochemical type is a fairly common occurrence here, and, judging from the concentrations of carbon monoxide and acetylene, such air pollution episodes represent auto exhaust diluted about 10,000-fold with air. The presence of PAN at many other locations can be inferred from the wide-spread occurrence of its characteristic plant damage symptoms (12).

II. FORMATION OF PEROXYACYL NITRATES

A. Mechanism

In some of the early work on irradiation of dilute mixtures of nitrogen oxides with organic compounds the peroxyacyl nitrates (or Compound X, as it was called then) nearly always seemed to be a reaction product (7). Gasoline vapors, auto exhaust, pure hydrocarbons, and aldehydes all produced this compound. More detailed investigations have shown that olefins (with the exception of ethylene), aromatics (with the exception of benzene), and aldehydes (with the exception of formaldehyde) all produce these compounds (3,13). Saturated hydrocarbons have never been shown unequivocally to produce PAN. Since saturated hydrocarbons react slowly and the PANs are none too stable, it is possible that some PAN is produced.

The complete mechanism of hydrocarbon reaction which leads to formation of the PANs has not been fully clarified. The final step is almost certainly the following:

$$\mathrm{R\overset{\overset{O}{\|}}{C}OO\cdot} + NO_2 \rightarrow \mathrm{R\overset{\overset{O}{\|}}{C}OONO_2} \tag{1}$$

The mechanism of formation of the peroxyacyl radical ($R\overset{O}{\overset{\|}{C}}OO$) from the organic starting material is speculative, but it seems probable that the radical R and the carbonyl carbon must be present and joined in the starting material. *cis*-2-Butene was found to yield PAN when photolyzed with nitrogen oxides in air. This olefin has two carbon atoms on each side of the double bond so that formation of the two-carbon PAN is to be expected. Likewise, irradiation of *cis*- or *trans*-3-hexene with nitrogen oxides yields primarily the three-carbon compound PPN (14). Tetramethylethylene $(CH_3)_2C{=}C(CH_3)_2$ produced only PAN even though rupture of the double bond leads to a three-carbon fragment. Formation of PPN from this fragmeut would require extensive rearrangement of the fragment, which apparently does not occur. These observations may be generalized and extrapolated by saying that formation of new carbon–carbon or carbon–hydrogen bonds does not occur, probably because of the overwhelming excess of molecular oxygen.

The fact that PAN is formed from various methyl benzenes has been cited (13) as evidence that ring rupture does occur when such hydrocarbons are oxidized. The phenyl derivative (PBzN) is formed either from aromatic olefins (styrene and its derivatives) by double bond rupture or from benzilic hydrocarbons by side chain oxidation.

B. Kinetics

In addition to their importance as plant toxicants and eye irritants the peroxyacyl nitrates are of unusual interest to chemists. They were unknown in chemistry until they were detected in synthetic reaction mixtures designed to simulate polluted air (7). The close combination into one molecule of two structures of low stability (nitrate group adjacent to peroxide link) produces a very fragile and reactive molecule.

The elementary reaction which leads to formation of the PANs is almost certainly reaction 1, which can be described as a free radical trapping reaction. In fact, the identification of PANs provides good evidence for the presence of the peroxyacyl radical.

When dilute mixtures of nitric oxide and olefin in air are irradiated, an induction period precedes the formation of PAN (see Fig. 1) (14). During this time interval the nitric oxide is being converted to nitrogen dioxide. Preferential reaction of peroxyacyl radicals with nitric oxide (instead of nitrogen dioxide) will account both for conversion of the nitric oxide and for the delay in PAN formation:

$$CH_3\overset{O}{\overset{\|}{C}}OO\cdot + NO \rightarrow CH_3\overset{O}{\overset{\|}{C}}O\cdot + NO_2 \qquad (2)$$

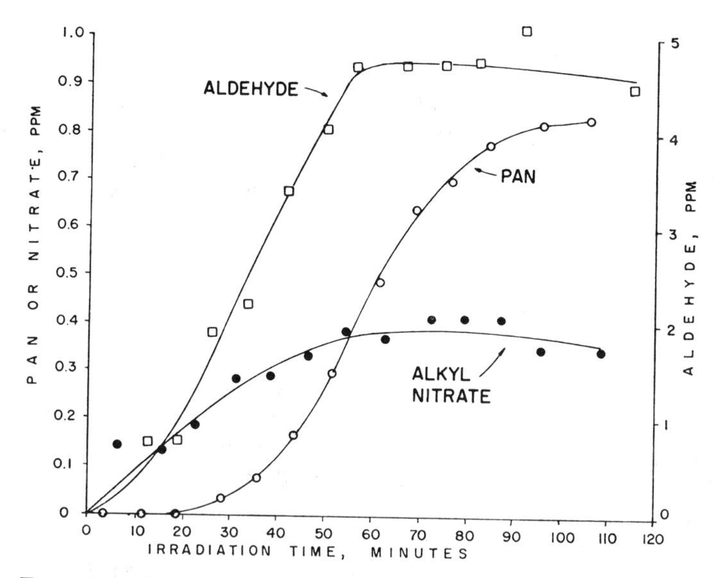

Fig. 1. Formation of peroxyacetyl nitrate and other products by photolysis of 5 ppm each of *cis*-2-butene and nitric oxide in air.

When it was recognized that atmospheric oxidation of hydrocarbons played a key role in photochemical air pollution, the reasonable suggestion was made that oxidation inhibitors might be added to prevent the process. The paradoxical fact, however, is that excellent free radical traps (the nitrogen oxides) are already present and, in fact, are reacting significantly with free radicals. But one of these reactions (eq. 2) simply transforms one radical into another and in addition converts an innocuous substance (nitric oxide) into a substance which is more toxic, highly colored, and a photoinitiator (nitrogen dioxide). The radical trapping reaction 1 converts free radicals into the highly toxic peroxyacyl nitrates.

III. SYNTHESIS AND PURIFICATION OF THE PANs

A. Methods of Preparation

A surprising variety of reactions have been found to yield measurable quantities of PANs. Four different methods have so far been used to prepare quantities of PANs for research:

1. Photolysis of mixtures of nitrogen oxides (either nitric oxide or nitrogen dioxide) with organic compounds in air or oxygen (3,8). This simulates the conditions under which the compounds are formed in polluted air. Biacetyl (2,3-butanedione) and *cis*-2-butene were both used as starting material with nitrogen dioxide or nitric oxide at concentrations of a few hundred ppm each in air or oxygen.

2. Photolysis of alkyl nitrite vapor in oxygen. This is the best method for synthesis of PAN or PPN starting with ethyl nitrite or *n*-propyl nitrite. It has also been used to prepare PBzN from benzyl nitrite.

$$CH_3CH_2ONO + 2O_2 \rightarrow CH_3\overset{\overset{O}{\|}}{C}OONO_2 + H_2O \tag{3}$$

$$CH_3CH_2CH_2ONO + 2O_2 \rightarrow CH_3CH_2\overset{\overset{O}{\|}}{C}OONO_2 + H_2O \tag{4}$$

This did not prove an effective method for synthesis of PnBN from *n*-butyl nitrite. This procedure will be described in some detail in following pages.

3. The dark reaction of aldehyde vapor with nitrogen pentoxide. This reaction was reported by Tuesday (15), who prepared the first three members of the PAN series at low concentration in a long-path infrared cell. He reported that traces of oxygen were necessary for the formation of PAN. Nitric acid and nitrogen dioxide were other products of the reaction. This method appeared to be the best method for preparation of PnBN and has also been used for the preparation of PBzN.

There appeared to be no reason why this reaction would not work at high concentrations, so it seemed to offer an advantage over the photolytic processes which must be carried out with concentrations in the hundreds of ppms. This method was evaluated in the author's laboratory using a simple flow system to react ozone with nitrogen dioxide to form nitrogen pentoxide.

$$2NO_2 + O_3 \rightarrow N_2O_5 + O_2 \tag{5}$$

The pentoxide was then reacted directly with aldehyde in the same flow system to form the PAN.

$$CH_3CHO + N_2O_5 + O_2 \rightarrow CH_3\overset{\overset{O}{\|}}{C}OONO_2 + HNO_3 \tag{6}$$

The effluent was flushed through a 10-cm infrared cell to monitor PAN formation. This was fairly successful using reactant concentrations of a few tenths of a per cent (after mixing). Figure 2 shows the PAN yield in one series of experiments. The striking feature of this plot is the rapid drop in PAN yield for nitrogen dioxide concentrations in excess of

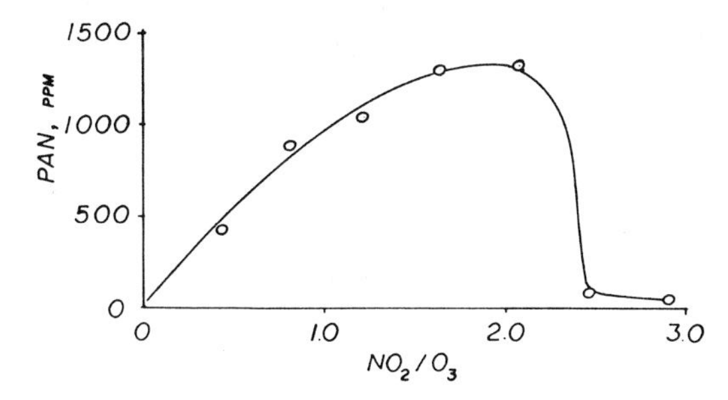

Fig. 2. Formation of PAN by reaction of acetaldehyde, ozone, and nitrogen dioxide CH_3CHO, 0.5–0.76%; O_3, 0.48–0.73%.

that required for reaction with ozone. Apparently excess nitrogen dioxide in some way inhibits formation of PAN, perhaps by inhibiting the dissociation of nitrogen pentoxide.

$$N_2O_5 \rightleftharpoons NO_2 + NO_3 \tag{7}$$

The maximum PAN yield in the data of Figure 2 was about 20–25% Yields above 80% were obtained with low aldehyde concentration and large excesses of the other two reactants.

4. Peracetic acid can be nitrated in various ways to yield PAN. A standard nitration mixture of nitric and sulfuric acids produced small, but clearly identifiable, amounts of PAN when treated with peracetic acid. Treatment of peracetic acid with silver nitrite also produced PAN (9). Although these reactions are of the theoretical interest, they do not appear to be useful methods of synthesis.

All these procedures produced more or less impure products which could be purified by preparative gas chromatography. The procedure used will be described in Section IIIB.

The routine procedure currently in use at the University of California at Riverside for the preparation of peroxyacetyl nitrate makes use of a photochemical reactor to oxidize ethyl nitrite as a dilute mixture in gaseous oxygen (reaction 3) (16). This equipment is shown in Figures 3 and 4. The reaction vessel proper consists of two 5-ft sections of 6-in. diameter borosilicate glass "drain line" connected with a U-bend of the same material and terminated with reducing joints. With the gas mixing manifold, air pump, and freeze trap the system has a volume of about 56 liters and is about 7 ft in length. Each straight section of the reactor is irradiated by a bank of six 40-W blacklite fluorescent bulbs. Fans are used to cool the reactor surfaces.

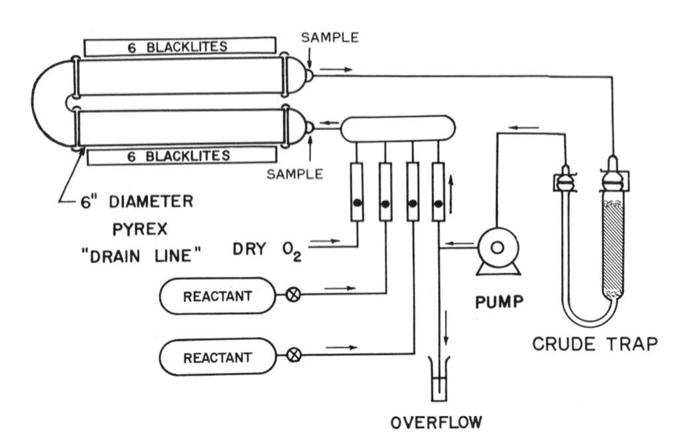

Fig. 3. Photochemical reactor for preparation of the peroxyacyl nitrates (from *J. Air Pollution Control Assoc.*, ref. 16).

Fig. 4. Preparative photoreactor. Ultraviolet lamps folded away from reaction tube.

The circulating pump, which must be of the sealed type, circulates the reaction mixture at about 7 liters/min so as to yield a residence time of about 8 min. The pump is placed downstream from the freeze-out trap so that unstable PAN need not pass through it. Compression heating and contact with the pump diaphragm (neoprene) would certainly cause a loss of PAN. The freeze-out trap is 25 mm in diameter and 125 mm long and is packed with 30/42 mesh C-22 firebrick which is coated with Carbowax E-600. It is chilled with a Dry Ice–acetone mixture.

The reactor is operated dynamically by venting a small flow at the overflow to compensate for the addition of fresh reactants. Ethyl nitrite is dispensed into the reactor from a vapor mixture in nitrogen contained in a war-surplus breathing oxygen cylinder (35-liter volume). Commercial ethyl nitrite is a National Formulary product obtained from Mallinckrodt Chemical Company; it contains about 15% ethyl alcohol. Ethyl nitrite is so volatile that about 18 ml of a 25-ml portion can be vaporized by connecting it directly to the evacuated tank. After pressurization to 60 psig with nitrogen the ethyl nitrite concentration is about 30,000 ppm (v/v). This mixture is metered into the reactor at about 90 ml/min along with 200 ml/min of oxygen. After dilution into the circulation flow, which is approximately 7 liters/min, this incoming stream contains about 350 ppm of ethyl nitrite. Sampling ports on the reactor inlet and outlet covered with rubber septa are provided for syringe sampling. The effluent contains about 250–300 ppm of PAN while the influent contains almost none.

B. Purification

The gas chromatograph used for purification of PAN is made almost entirely of borosilicate glass (see Figs. 5 and 6). Each leg of the U-shaped separating column is 15 mm in diameter and about 450 mm long. The U bend itself is large bore capillary. The column is packed with 30/42 mesh C-22 firebrick coated with 20% of Carbowax E-600. Helium is used as carrier gas at 800 ml/min of which about 10% is fed to the thermal conductivity detector while the remainder passes to the collection system. This is done so the collected material does not pass over the hot filaments of the detector where decomposition of the highly unstable PAN might occur. The flow is split by having a section of 0.5 mm i.d. glass capillary about 1 in. long in the main flow and a section ten times as long in the line to the detector. An oil manometer is used to measure pressure just upstream from the capillaries. This serves to monitor flow so long as the capillaries remain unobstructed and the downstream side of the stream is kept at ambient pressure. A column bypass flow is also tapped in at this point to provide a check on the purity of the column effluent. Both the column and crude trap are fitted into the flow system with O-ring joints.

The photoreactor is normally operated for a full 7- to 8-hr day, but the freeze-out trap is changed every 3–4 hr. The trap is kept in the

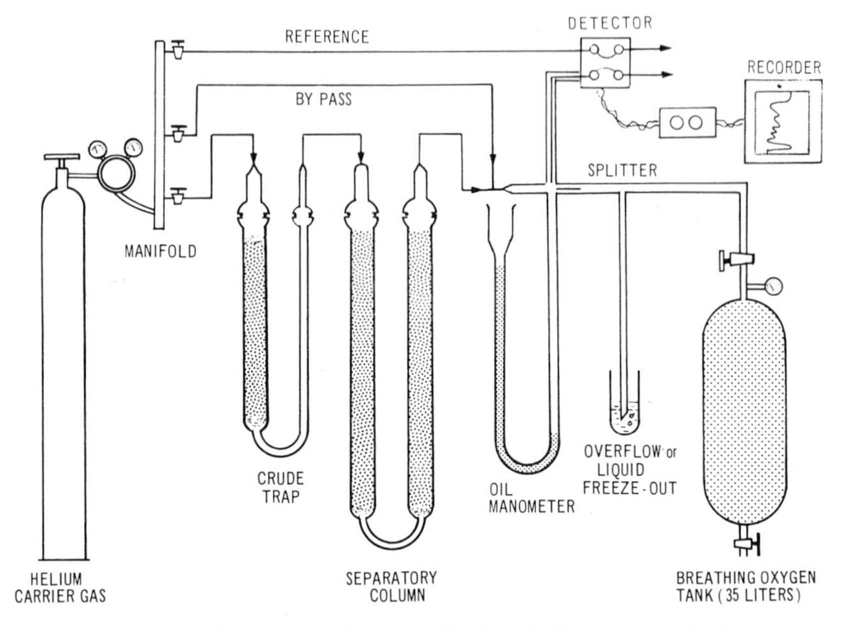

Fig. 5. Preparative chromatograph for purification of the peroxyacyl nitrates.

Fig. 6. Preparative chromatograph. Traps (in small Dewar flasks at right) for freeze-out of liquid PAN.

Dry Ice coolant during transfer from the reactor circuit to the chromatograph. After the desired flow is established, the column effluent is checked for purity by switching momentarily to the bypass. The bypass flow can be made equal to the column flow by adjusting it to give the same pressure on the oil manometer. Normally the detector shows less than 1% of full scale deflection at the sensitivity used for the separation. After switching the flow back to the column, the freeze trap is thawed by immersion in water at room temperature. A typical chromatogram is shown in Figure 7. Impurity peaks, principally of acetaldehyde, methyl lnitrate, and ethyl nitrate, emerge before the PAN and are discarded. The product PAN begins to emerge after about 22 min. The various peaks in the chromatogram are identified by flushing a 10-cm infrared cell with the column effluent and recording the spectra. The product PAN can be frozen out of the carrier gas stream with a Dry Ice–alcohol mixture to yield a liquid product. The pure products are water white mobile liquids. The pure liquid PANs are extremely explosive (see Sect. VI. A) and should be handled with great care. A safer and simpler procedure, if a liquid product is not required, is to bleed the entire column effluent into an evacuated container such as a low-pressure breathing oxygen tank. Using a "T" on the outlet line, an overflow bubbler, and a needle valve on the tank, this can be easily controlled. At a flow of 0.8 liter/min the column effluent can be collected for 43 min in a single 35-liter tank. This, of course, leaves the PAN mixed with helium carrier gas, but for most purposes this is

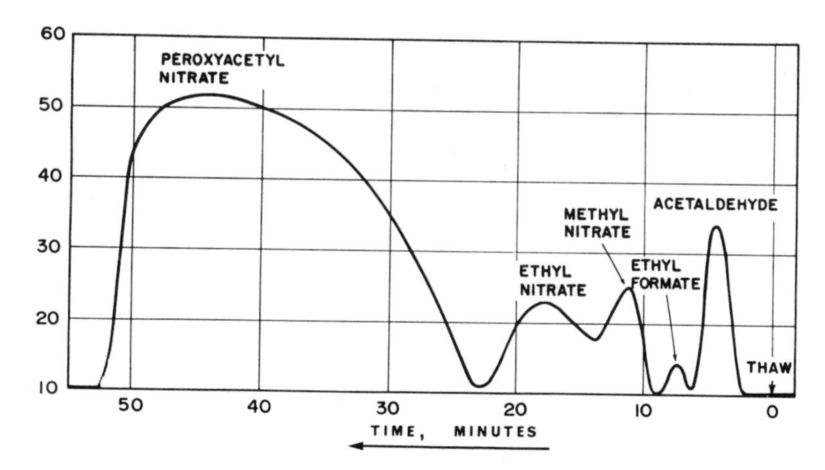

Fig. 7. Chromatogram of peroxyacetyl nitrate purification (from *J. Air Pollution Control Assoc.*, ref. 16).

not objectionable. To dispense the PAN in an experiment the tank is pressurized to about 100 psig with nitrogen. This produces a concentration of about 1000 ppm which can be conveniently measured by flushing a 10-cm infrared gas cell and recording the infrared spectrum (see Sect. IV). If PAN is vaporized directly from the liquid state into an evacuated tank, care should be taken not to pressurize too rapidly with diluent gas. Compression heating can cause decomposition or even explosion. During 8 hr. of photolysis the reactor consumes about 0.053 moles of ethyl nitrite and produces enough PAN to fill a 35-liter tank at 100 psig and 1500 ppm. This requires 0.0165 moles of PAN and indicates an overall yield of 30%.

IV. SPECTRA OF THE PANs

A. Infrared Spectra

The infrared spectrum of peroxyacetyl nitrate shows five strong bands and a number of weaker ones in the 2.5–15 μ wavelength range. A spectrum of the vapor diluted in helium is shown in Figure 8. Nicksic, Harkins, and Mueller (17) have reported that the spectrum in carbon tetrachloride solution is very similar to this vapor spectrum. The vapor spectrum shows some similarities to those of the alkyl nitrates and alkyl nitrites, but the usual strong nitrate band near 6.1 μ is missing, and a strong carbonyl band appears at 5.75 μ along with an unusual band at 5.45 μ. The resemblance to the spectrum of acetyl nitrate, $CH_3C(=O)-ONO_2$, is much closer (see Fig. 9). Although this spectrum is

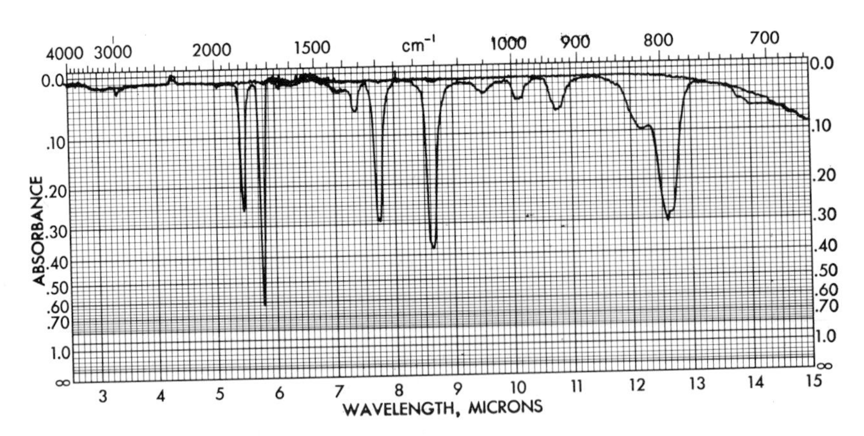

Fig. 8. Infrared spectrum of peroxyacetyl nitrate, 10-cm path. (NaCl windows) Perkin-Elmer Model 137B.

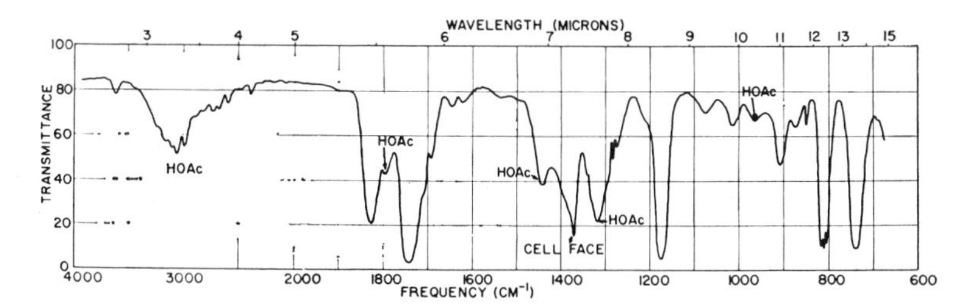

Fig. 9. Infrared spectrum of acetyl nitrate (from *Proc. Am. Petrol. Inst.*, ref. 3).

badly contaminated with acetic acid bands and some cell window absorptions (arising from reaction between the compound, water vapor, and the sodium chloride windows of the cell), the main features of the acetyl nitrate spectrum are clearly discernible. The fact that this spectrum resembles that of PAN so closely and yet is definitely different shows that acetyl nitrate and PAN are closely related in structure but not identical. The acetyl nitrate spectrum shows counterparts to three of the five major bands of the PAN spectrum (5.75, 8.6, and 12.6 μ). The 12.6 μ band of acetyl nitrate shows a branched structure which is completely missing from the corresponding band in the PAN spectrum. The 5.5 μ band in the acetyl nitrate spectrum is at a somewhat longer wavelength than its counterpart in the PAN spectrum. The strong 7.7 μ band in the PAN spectrum is not present in the acetyl nitrate spectrum, but the latter has a strong band at 13.9 μ which is very weak in the PAN spectrum. The weak bands between 9 and 12 μ are different in the two spectra.

The infrared spectrum of PPN is shown in Figure 10. Four of the five major bands of the PAN spectrum are present also in this spectrum. The strong PAN band at 8.6 μ is not present in the PPN spectrum, which suggests that this band arises in the methyl group adjacent to a carbonyl.

The determination of the absorptivities of the PANs presented special difficulties arising from their lack of stability. Ultimately two independent methods were developed which gave results in reasonable agreement (18). In each case very dilute concentrations of the compound were measured into a long-path multiple reflection cell containing air at 1 atm pressure. The spectrometer used was a Perkin-Elmer single-beam double-pass monochromator with a sodium chloride prism. This system is described in reference 19. Wide slits were used to obtain sufficient energy (350 μ at 7.68 and 8.60 μ, and 800 μ at 12.61 μ. In one method, a pressure–volume (PV) technique (see Table I), a

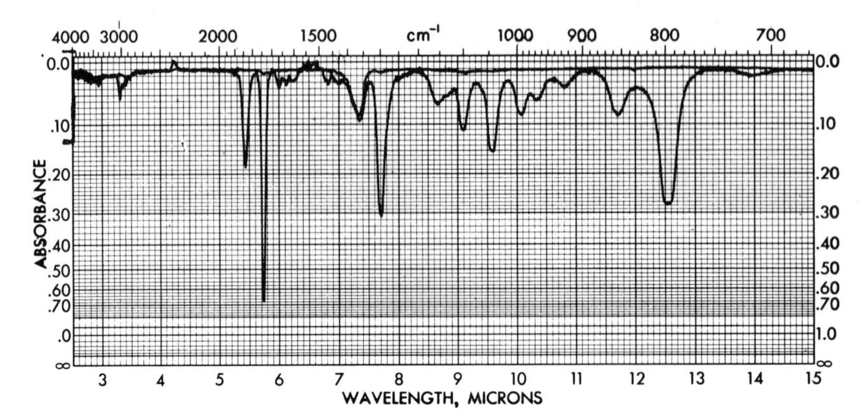

Fig. 10. Infrared spectrum of peroxypropionyl nitrate, 10-cm path. (KBr windows) Perkin-Elmer Model 137B.

glass bulb of known volume was filled to a measured pressure with the vapor. This was then pressurized with nitrogen and swept into the long-path cell. With the pressures and volumes used, this yielded concentrations of a few parts per million which were sufficient for measurement at 120-m path length. The vapor pressure of PAN at room temperature is comparable to that of water (15–20 mm Hg). This technique yielded reproducible results from PAN and PPN, but the vapor pressure of PBN was too low for application of this technique with the equipment available at that time.

In the second method liquid samples (5 μl) were allowed to vaporize from a lambda pipette directly into the long-path cell. To calculate the concentration it was necessary to estimate the liquid densities as 1.2 g/ml and to take the molecular weights according to the formulas given in the table. This technique was successful for PPN and PBN but failed for PAN because the sample exploded while being vaporized into the long-path cell. The data for PPN show satisfactory agreement between the two techniques.

Air at atmospheric pressure has absorption bands of water vapor at the 120 m path which obscure some of the major absorption bands of the PANs. To cover these regions spectra were run in 10-cm gas cells. Since the concentrations were obtained from long-path cell data, these do not constitute independent determinations of absorptivity. A mixture of PAN in nitrogen was analyzed by diluting 100- and 200-ml samples into the long-path cell. This mixture was then flushed through a 10-cm cell. The spectrum was run on a Perkin-Elmer double-grating

TABLE I

Infrared Absorptivities (a) of the Peroxyacyl Nitrates[a]
($a \times 10^4$ ppm^{-1}m^{-1} as vapor in 1 atm of air)

	PAN	$CH_3\overset{O}{\overset{\|}{C}}OONO_2$				
Wavelength, μ	12.61	10.74	8.60	7.68	5.76	5.44
120 m; 5 detn. by PV						
Mean	10.3		13.9	11.4		
Mean dev.	0.4		0.4	0.2		
10 cm using conc. from 120 m	10.1	1.8	14.3	11.2	23.6	10.0
	PPN	$CH_3CH_2\overset{O}{\overset{\|}{C}}OONO_2$				
Wavelength, μ	12.58	9.58	(8.60)[c]	7.68	5.76	5.44
120 m; 5 detn. by PV						
Mean	11.0	4.9				
Mean dev.	0.7					
120 m, liquid; 2 detn.	10.5	4.8	(1.1)[c]	10.1		
10-cm cell	10.5[b]	5.3		11.7	23.0	6.8
	PnBN	$CH_3CH_2CH_2\overset{O}{\overset{\|}{C}}OONO_2$				
Wavelength, μ	12.58	9.63		7.7	5.75	5.44
120 m, liquid 3 detn.						
Mean	11.7	5.8				
Mean dev.	0.3	0.5				
10-cm cell	11.7[b]	7.1		14.8	34.2	8.9
	PBzN	$\phi\overset{O}{\overset{\|}{C}}OONO_2$				
Wavelength μ.	12.67	10.13	8.17	7.65	5.75	5.54
120 m, ref (24)	9.5	17.8	15.6	8.8	22.8	6.9

[a] Reprinted from *Analytical Chemistry,* **36,** 928 (1964).
[b] Taken equal to the absorptivity found at 120 m.
[c] No peak.

spectrophotometer at a slit program of 10.00. The absorptivities obtained in this way are shown in Table I. They are comparable to those obtained in the long-path cell in spite of the large changes in concentration, path length, and resolution. The spectra of PPN and PnBN were also run in 10-cm cells; in this case adsorptivities were calculated assuming the value at 12.58 μ to be the same as that found at the 120-

m path. The infrared spectrum of PBzN reported by Huess and Glasson (24) is quite similar to these alkyl homologues although it does show differences (see Table 1). They also reported that the spectrum of benzoyl nitrate resembled that of PBzN but was distinctly different.

For routine measurement of concentration in nitrogen of helium dilutions, a Model 137B Perkin-Elmer sodium chloride prism spectrometer and 10-cm cell are normally used.

B. Ultraviolet Spectrum

The ultraviolet spectrum of a sample of PAN vapor in air is shown in Figure 11, which is an average derived from two spectra. These spectra were recorded in a 10-cm gas cell using a Cary Model 14 recording spectrophotometer. The most important feature of this spectrum is its weakness in the sunlight region about 3000 Å. This, of course, limits the rate at which PAN could be photolyzed in sunlight. Accurate estimates of the maximum possible rate of absorption of sunlight cannot now be made since the absorbance in this region is so weak. Concentrations used to obtain the data for Figure 11 are at least half the saturation value at room temperature, so a longer path would be required for a better estimate of this absorbance. Huess and Glasson (24) reported finding several absorption bands below 3000 Å for a solution of PBzN in hexane.

C. Mass Spectra

When the chromatographic purification of the PANs was first developed, a time-of-flight mass spectrometer was connected to sample

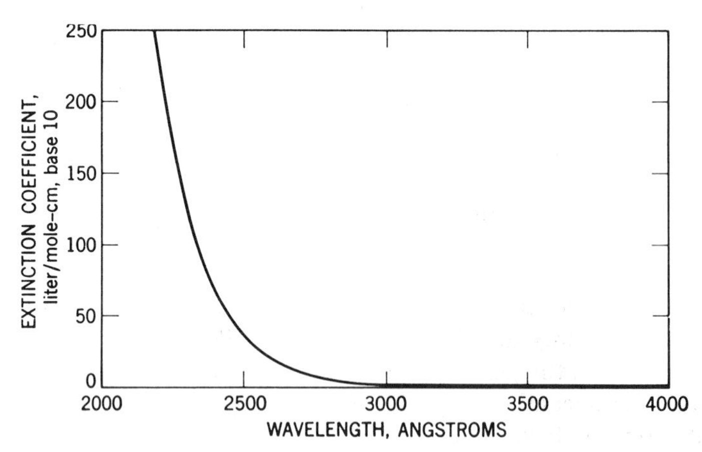

Fig. 11. Ultraviolet spectrum of peroxyacetyl nitrate vapor diluted with 1 atm air.

the column effluent. In this way spectra of PAN, PPN, and PnBN were recorded as summarized in Table II. Two earlier attempts to record mass spectra failed to produce comprehensible spectra probably because the sample decomposed before reaching the ionization region, perhaps in the heated inlet system of the spectrometer. Although the time-of-flight spectra failed to show parent mass peaks, they are believed to represent true spectra of the PANs since they do not show the patterns expected from the known decomposition products. The lack of parent peaks, although not really surprising, was a great disappointment, since these would have provided nearly unambiguous molecular weights which would have permitted a choice between possible structures. The major ions (Table II) are the acyl, the alkyl, the NO_2^+ and the NO^+. Some other ions are evident. Although such fragmentation patterns are consistent with the peroxyacyl nitrate structure, they do not prove that structure, the proof of which must rest on other grounds.

The mass spectrum of PBzN (24) shows a corresponding set of ions. By reducing the ionizing voltage Huess and Glasson were able to detect the parent peak at 183 m/e in the PBzN mass spectrum.

TABLE II
Mass Spectra of PAN Homologs

	PAN		PPN		PnBN	
M/e	Height	Ion	Height	Ion	Height	Ion
71			—		100[a]	$CH_3CH_2CH_2CO$
57	—		100[a]	CN_3CH_2CO	—	
46	17	NO_2	34	NO_2	35	NO_2
45	1		4		12	
44	5	CO_2	7.4	CO_2	24	CO_2
43	100[a]	CH_3CO	0.4		141	C_3H_7
42	3	CH_2CO	—		29	C_3H_6
41	1	CHCO	—		88	C_3H_5
39			—		41	C_3H_3
38			—		12	C_3H_2
30	13	NO	26	NO	29	NO
29	3.6		97	C_2H_5	29	C_2H_5
28	—		4	$C_2H_4(N_2)$	24	$C_2H_4(N_2)$
27	—		79	C_2H_3	100	C_2H_3
26	—		20	C_2H_2	18	C_2H_2
15	39	CH_3	5.3	CH_3	18	CH_3
14	6	CH_2	4.1	CH_2	—	
13	2	CH	1.2	CH	—	
12	0.6	C	1.0	C	—	

[a] Reference peak.

D. Nuclear Magnetic Resonance

Nuclear magnetic resonance spectra of solutions of peroxyacetyl nitrate in carbon tetrachloride and in benzene were recorded by Nicksic, Harkins, and Mueller (17). The proton spectrum (the only one studied) shows a single resonance ascribable to a methyl group. The resonance position of PAN is compared with that of some related compounds in Table III. Nicksic et al. point out that the PAN resonance is 3.5 cps downfield from that of acetyl nitrate while the peracetic acid resonance is 4.0 cps downfield from acetic acid. They did not feel that this was conclusive proof of the correctness of the peroxyacetyl nitrate structure, but they did feel it was consistent with that structure.

TABLE III

Some Nuclear Magnetic Resonance Proton Lines

(Resonance position in cps downfield from reference line (tetramethyl silane). Dilute solutions in carbon tetrachloride except as noted. 60 mc spectrometer)[a]

Compound	cps	Compound	cps
$\underline{CH_3}\overset{\overset{O}{\parallel}}{C}OCH_3$	121	$\underline{CH_3}\overset{\overset{O}{\parallel}}{C}OONO_2$(PAN)	137.5
$\underline{CH_3}\overset{\overset{O}{\parallel}}{C}OH$	124	$\underline{CH_3}\overset{\overset{O}{\parallel}}{C}OONO_2$(in C_6H_6)	78
$\underline{CH_3}\overset{\overset{O}{\parallel}}{C}\underline{CH_3}$	125	$\underline{CH_3}\overset{\overset{O}{\parallel}}{C}Cl$	160
$\underline{CH_3}\overset{\overset{O}{\parallel}}{C}OOH$	128	$CH_3\overset{\overset{O}{\parallel}}{C}O\underline{CH_3}$	205
$(\underline{CH_3}\overset{\overset{O}{\parallel}}{C})_2O$	132	$CH_3\underline{CH_2}ONO_2$	274
$\underline{CH_3}\overset{\overset{O}{\parallel}}{C}ONO_2$	134	$CH_3\underline{CH_2}ONO$	288

[a] After Nicksic, Harkins, and Mueller, ref. 17.

The NMR spectrum provides a new way to analyze PAN solutions. Its unique advantage is that any methyl compound can be used for calibration, since all protons respond equally.

V. ANALYSIS OF THE PANs

A. Infrared

The infrared spectrum (see Fig. 8) has been the primary standard of measurement for PAN since its discovery. Since PAN is normally handled in the gas phase, a 10-cm gas cell is commonly used for analysis with the adsorptivities of Table I. A concentration of about 35 ppm produces an absorbance of 0.005 at 8.6 μ in a 10-cm path. This is the detectability limit. With a multiple reflection cell having a 120-m path the same absorbance corresponds to 0.03 ppm (see Fig. 12). This limit leaves much to be desired because accurate measurements cannot be made at this concentration level; the equipment is large, expensive, and cumbersome, and a very large sample is required. But 0.03 ppm of PAN is capable of causing plant damage, so a method for measurement at this level was most desirable.

B. Gas Chromatography

Gas chromatography and electron capture detection appeared on the scene at just the right time. This combination proved to be nearly ideal for the measurement of PAN at ppb (10^{-9}) levels in ambient air samples. The only shortcoming of the method is that the instrument must be standardized with known PAN samples, and these must be prepared by quantitative dilution from concentration levels which are measurable by some other method such as infrared. Small quantities of PAN for calibration of gas chromatographs can be prepared by photolyzing ethyl nitrite vapor in oxygen in a 10 cm infrared gas cell. After flushing the cell with oxygen small amounts (about 50 μl) of the vapor from above ethyl nitrite are added at about 15 minute intervals while photolyzing with blacklight fluorescent lamps. The PAN which forms in the cell can reach concentrations of over 400 ppm. These can be readily measured using the infrared and portions diluted 5- or 10-thousand fold to obtain suitable calibration mixtures. The alkyl nitrates also formed need not interfere. Sunlight might serve as well as blacklight lamps for the photolysis.

The original report on the electron capture method for PAN analysis (20) described the use of a glass sample valve which was later found to be unnecessary, and stainless steel valves are now used exclusively. With one metal valve loss of PAN was traced to a silver-soldered joint which apparently caused decomposition of the PAN. In the method in current use a modified "Pestilyzer" electron capture chromatograph (Varian Aerograph Company) is used. Modifications include pro-

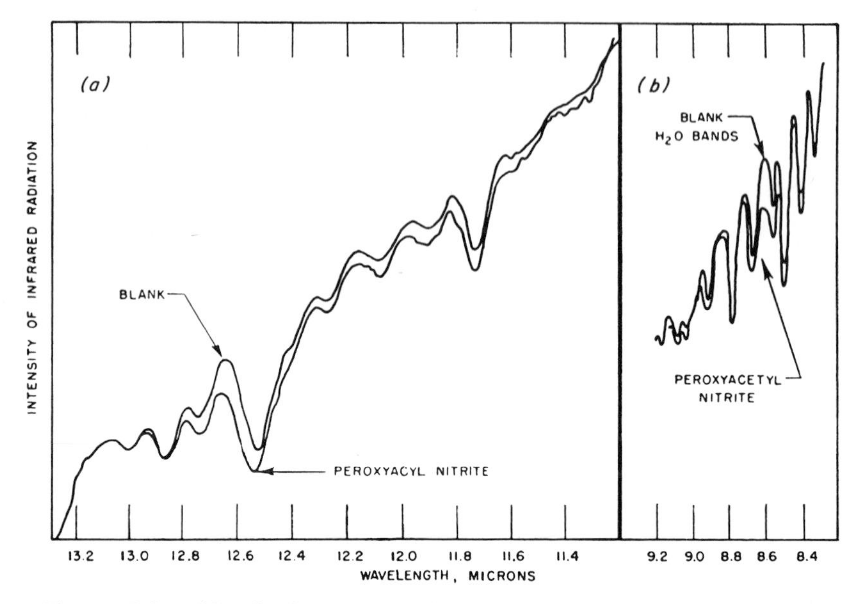

Fig. 12. Infrared bands of peroxyacyl nitrates in ambient air at South Pasadena (from *Proc. Am. Petrol. Inst.*, ref. 9). (*a*) 11:30 A.M., September 26, 1956 (slit 1200 μ, path 408 m); (*b*) 12:00 noon, September 18, 1956 (slit 300 μ, path 444 m).

vision for thermostating at room temperature (25°C) by balancing heating with a thermoelectric cooler. A standard 6-port slide-type stainless steel valve is used and a potentiometer for varying the constant dc detector cell voltage. As with any chromatographic analysis, the column itself is the crucial element. For this analysis 3-mm diameter Teflon tubes 23–46 cm long were packed with 5% (w/w) of Carbowax 400 on Chromosorb. Acid-washed Chromosorb (treatment with dimethyl dichlorosilane) is the best substrate so far tested. Nitrogen gas flowing at 30–36 ml/min serves as carrier; no purge gas is used. A typical chromatogram is shown in Figure 13. This was produced from direct injection of a 3-ml sample of ambient air from Riverside, California. This procedure has now been automated (25) by fitting a solenoid to operate the gas sample valve. The solenoid, an air pump which is used to flush the sample loop, and the recorder chart drive motor are all operated by a timer on a fifteen minute schedule. The first few months of operation in Riverside revealed PAN concentrations up to 58 ppb. Persistence of PAN overnight was a fairly common occurrence.

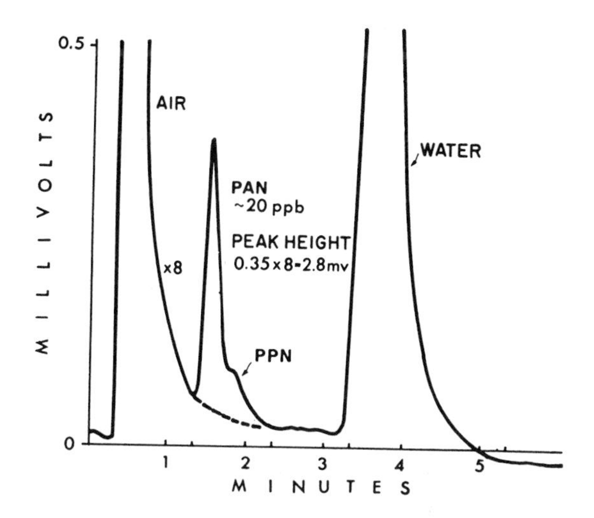

Fig. 13. Gas chromatogram, electron capture detector, showing detection of PAN and PPN in ambient air. Outside air, 11:00 A.M., 2.4 ml sample loop, 30 ml N_2/min, 9×1/8 in. Teflon column, 5% Carbowax 400 on Chromosorb W.

Similar levels of PAN have been reported by Hill and Tingey (27) in a study conducted in the Salt Lake Valley. Mayrsohn and Brooks (26) reported concentrations ranging up to 214 ppb in a study conducted in Los Angeles County.

To measure PBzN in laboratory systems Huess and Glasson (24) used a 25 cm × 3 mm glass column packed with 3.8% S.E. 30 on 80/100 mesh Diatoport S. The retention time was eight minutes at 50°C.

C. Hydrolysis

The discovery that alkaline hydrolysis of PAN yields nitrite ion on a mole for mole basis opened the possibility of a sensitive analytical method independent of infrared analysis. Such a method was described in a recent report (21). In brief, the sample to be analyzed is absorbed in aqueous potassium hydroxide which hydrolyzes PAN to form nitrite ion. The nitrite ion is then determined colorimetrically using Saltzman's reagent. A comparison of this method with the infrared method is shown in Table IV. The agreement is quite good. The method suffers interference by nitrogen dioxide and perhaps other nitrogen compounds, but it should be useful for calibration of other instruments such as gas chromatographs.

TABLE IV

Comparison of Infrared Analysis with Hydrolysis and Determination of Nitrite Ion

Runs	Infrared spectrum ppm	Hydrolysis and determination of nitrite ion (average) ppm
2	1580	1570
2	1460	1465
2	860	865
2	570	540
4	510	510
2	100	99
2	50	48
2	20	20

D. Elemental Analysis

Early attempts to obtain an elemental analysis for PAN were motivated by the desire to prove the structure but were thwarted by the compound's instability. From the beginning it was clear that an acetyl group is present as well as oxidized nitrogen. All possible structures between "nitroso acetyl" $CH_3\overset{\overset{O}{\|}}{C}NO$ and peroxyacetyl nitrate were considered at one time. The real uncertainty lay in the number of oxygen atoms. A direct method for measuring oxygen in the molecule was reported by Nicksic, Harkins, and Mueller (17). They prepared solutions of PAN in carbon tetrachloride and in benzene and analyzed them for methyl groups associated with PAN by nuclear magnetic resonance spectroscopy and for total oxygen by neutron activation analysis. This permitted determination of the number of oxygen atoms per methyl group as shown in Table V. There was considerable scatter in the results, and the average is a bit higher than the five expected for the peroxyacetyl nitrate structure. If the proposed structure were grossly in error, the number of oxygen atoms per methyl might be nonintegral, but it should certainly be constant. Since the results for each solution were repeatable, it seems probable that there were unknown oxygen-containing impurities in some of the samples which made the oxygen results high. In one sample, acetone was revealed by the NMR spectrum, and its contribution to the oxygen subtracted from the total. Huess and Glasson recently reported an elemental analysis of PBzN which was in satisfactory agreement with the calculated composition.

TABLE V

Oxygen Content of Peroxyacetyl Nitrate[a]

Solvent	Oxygen (NAA)		Methyl (NMR), moles/liter	O atoms per CH_3 group
	Wt %	Moles/liter[b]		
CCl_4	0.47	0.234	0.082	5.7
CCl_4	(0.88)[c]	(0.437)[c]		
		0.416	0.175[d]	4.8
CCl_4	2.78	1.38	0.493	5.6
C_6H_6	3.14	0.863	0.260	6.6

[a] After Nicksic, Harkins, and Mueller (17).
[b] Wt % × (density/32) = mole/liter
[c] Before correction for 0.083 moles/liter of methyl as acetone.
[d] At PAN resonance.

VI. REACTIONS OF THE PANs

A. Stability

The major experimental problem in working with the PANs is their low stability. During development of the procedures now in use about a dozen explosions occurred. Those involving samples of the liquid were extremely violent even when only a drop or two of liquid was involved. On one occasion a liquid sample measured from a 5-μl pipette exploded and destroyed considerable glassware even though part of the PAN had vaporized before the remainder exploded. Since the initiating cause of several of these explosions was not apparent, the policy was adopted of handling the liquid as if it might explode at any time and protection was provided so that injury would not result. Gradually the steps in the preparation and purification procedure which involved handling the liquid were eliminated. Under these practices (Sec. III) neither crude nor pure liquid is handled, and no explosions have occurred.

If a drop or two of PAN liquid in equilibrium with its own vapor at room temperature (without diluent gas) is suddenly pressurized with a diluent gas such as nitrogen or air, an explosion of great violence will occur. Presumably compression heating of the vapor initiates decomposition which quickly accelerates into a detonation which propagates into the liquid. On one occasion a 1-liter sample of vapor at 8 mm Hg pressure exploded while nitrogen gas was being bled in to bring the pressure to 1 atm (3). This explosion did not break the glass flask, but was evidenced by a sharp sound and the slow appearance of nitrogen dioxide in the flask. In addition to nitrogen dioxide, the infrared spectrum of the products showed the presence of about 15.5 mm Hg of carbon monoxide and about 2.9 mm Hg of carbon dioxide. This total of 18.4 mm Hg is close to the 16 mm Hg expected for decomposition of a two-carbon molecule. The most serious explosion occurred when a low pressure oxygen tank containing a high concentration of PAN vapor in helium carrier gas was stored in a cold room at about 0°C over a weekend. Apparently a significant fraction of the PAN liquified and drained into the gauge and fitting during this time. An explosion which shattered the gauge and fitting occurred while the tank was being attached to a nitrogen line.

Slow decomposition of PAN vapor diluted in nitrogen has proved to be highly erratic. Since no loss over periods of days or even weeks has sometimes been observed, it appears that decomposition is strongly influenced by the nature of the surfaces of the containers. In some cases the loss of PAN in low-pressure oxygen tanks has been matched by the appearance of methyl nitrate and carbon dioxide in approximately equimolar amounts.

$$CH_3\overset{\overset{O}{\|}}{C}OONO_2 \rightarrow \left[\begin{array}{ccc} & O & \\ & \| & \\ & C & \\ / & & \backslash \\ CH_3 & & O \\ | & & | \\ O & & O \\ \backslash & & / \\ & N & \\ & | & \\ & O & \end{array} \right] \rightarrow CH_3ONO_2 + CO_2 \qquad (8)$$

The ring structure in brackets is suggested as a plausible intermediate in this decomposition. In other cases decomposition apparently led to acetic acid as one product.

B. Hydrolysis

The rather extensive studies of the effects of PAN on plants and animals will not be reviewed here. From the structure it is predictable that PAN might act chemically in several different ways. In connection with his studies of the effect of PAN on biochemical materials, Mudd (22) reported the formation of a short-lived oxidant activity when PAN vapor was bubbled through water. Such solutions are capable of oxidizing biological reductants, but the oxidizing power decays in a few minutes standing at room temperature. Mudd also reported that nitrite ion was a product of this treatment. These observations were not fully appreciated until Nicksic, Harkins, and Muller reported that hydrolysis of PAN by aqueous base produced acetate ion and nitrite ion in quantitative yield (17). This result was surprising in that this oxidizing agent produced such an easily oxidized product as nitrite ion. This result was later clarified when it was found that molecular oxygen was also produced in quantitative yield (23) by the reaction, which may be written:

$$CH_3\overset{\overset{\displaystyle O}{\|}}{C}OONO_2 + 2OH^- \rightarrow CH_3\overset{\overset{\displaystyle O}{\|}}{C}O^- + O_2 + NO_2^- + H_2O \qquad (9)$$

The amount of acetate ion was determined by nuclear magnetic resonance spectroscopy of solutions in deuterium oxide. The nitrite ion was determined by polarography and by a colorimetric method making use of Saltzman's reagent. These results are summarized in Table VI.

TABLE VI

Alkaline Hydrolysis of Peroxyacetyl Nitrate to Yield Acetate Ion and Nitrite Ion [5% NaOH in D_2O][a]

	PAN Concentration, ppm		
Sample	Infrared, vapor	From acetate by NMR	From nitrite by colorimetry
I	450	420	423
II	450	426	380
III	450	417	442
IV	450	430	442

[a] After reference (17).

Gas chromatography was used for the oxygen evolution experiment. A 2.5 m by 5 mm i.d. column containing Molecular Sieve 5A was operated at 35°C with a thermal conductivity detector. This system separates oxygen from nitrogen but not from argon. A U-tube packed with 4-mm diameter glass beads was used for the hydrolysis. The beads were covered with 1% aqueous potassium hydroxide solution. For the blank a trap containing activated charcoal was placed in front of the hydrolysis tube. This served to remove the PAN but not traces of argon or of free oxygen in the PAN–N_2 mixture. For calibration, measured amounts of air were added to the blank. The initial mixture was analyzed by infrared and found to contain 1.45% PAN. The effluent from the hydrolysis contained 1.46% oxygen.

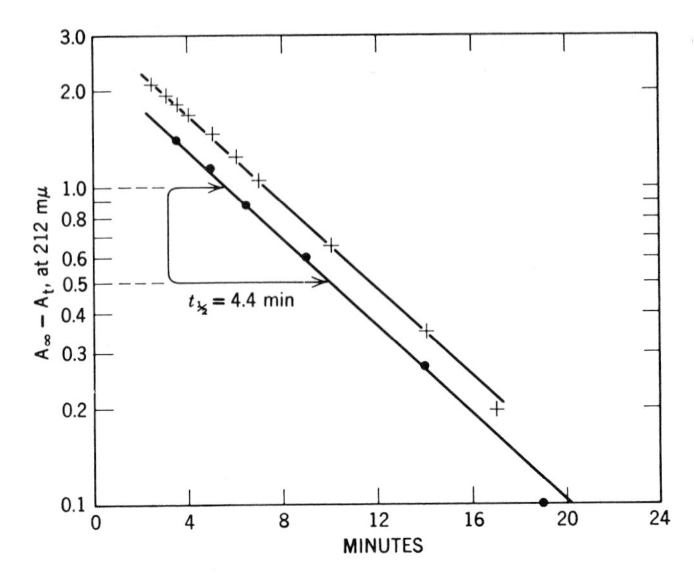

Fig. 14. Formation of nitrite ion by hydrolysis of peroxyacetyl nitrate. $t_1/_2 = 4.4$ min (×–×–×) Phosphate buffer, pH 7.2 (Fig. 2, ref. 22). (–•–•–•–) Tris buffer, pH 7.8 (Fig. 1, ref. 22).

These results renewed interest in Mudd's observation of the oxidizing power of PAN "solutions." The formation of nitrite ion was monitored by its absorption at 212 mμ in the ultraviolet. The plots given in Figure 14 show that the precursor of the nitrite ion decays in a first-order process. Mudd also found that the formation of nitrite ion was more rapid at higher pH, but the nature of the buffer also influenced the rate. Since liquid PAN does not readily mix with water, it seems probable that hydrolysis occurs at the same time as solution. It may be that hydroxyl attack on the carbonyl carbon is followed by expulsion of pernitrate ion.

$$CH_3\overset{\overset{\displaystyle O}{\|}}{C}OONO_2 + OH^- \rightarrow CH_3\overset{\overset{\displaystyle O}{\|}}{C}OH + O_2NO_2^- \quad (10)$$

This ion may then decompose to give nitrite ion and molecular oxygen:

$$O_2NO_2^- \rightarrow O_2 + NO_2^- \quad (11)$$

The pH dependence reported by Mudd may be due to formation of pernitric acid.

$$H^+ + O_2NO_2^- \rightleftharpoons HO_2NO_2 \quad (12)$$

Nicksic et al. (17) titrated solutions of PAN in methanol with base and found only one equivalent was required. This was explained when it was found that methyl acetate was formed:

$$CH_3\overset{\overset{O}{\|}}{C}OONO_2 + CH_3O^- \rightarrow CH_3\overset{\overset{O}{\|}}{C}OCH_3 + O_2 + NO_2^- \qquad (13)$$

The methyl acetate was detected by nuclear magnetic resonance spectroscopy. Huess and Glasson (24) were also able to convert PBzN quantitatively to methyl benzoate by reaction with methyl alcohol in basic solution. This appears analogous to the acetylation of sulfhydryl groups reported by Mudd (22):

$$CH_3\overset{\overset{O}{\|}}{C}OONO_2 + \text{—SH} \rightarrow \text{—S}\overset{\overset{O}{\|}}{C}CH_3 + O_2 + HNO_2 \qquad (14)$$

Mudd also discussed the relationship of this reaction to the observed effects of PAN on biological systems.

Acknowledgment

The author is indebted to many colleagues who have participated in the study of the chemistry of PANs reported here: P. L. Hanst, R. C. Doerr, W. E. Scott, J. K. Locke, E. A. Cardiff, F. R. Burleson, M. A. Price, E. F. Darley, and K. A. Kettner.

References

1. E. F. Darley, C. W. Nichols, and J. T. Middleton, *Calif. Dept. Agr. Bull.* **55,** 11 (1966).
2. E. F. Darley, W. M. Dugger, J. B. Mudd, L. Ordin, O. C. Taylor, and E. R. Stephens, *Arch. Environ. Health,* **6,** 761 (1963).
3. E. R. Stephens, E. F. Darley, O. C. Taylor, and W. E. Scott, *Proc. Am. Petrol. Inst. Sect. III,* **40,** 325 (1960); *Intern. J. Air Water Pollution,* **4,** 79 (1961).
4. E. A. Schuck, E. R. Stephens, and J. T. Middleton, *Arch. Environ. Health,* **13,** 570 (1966).
5. J. T. Middleton, L. O. Emik, and O. C. Taylor, *J. Air Pollution Control Assoc.,* **15,** 476 (1965).
6. J. T. Middleton, *Arch. Environ. Health,* **8,** 19 (1964).
7. E. R. Stephens, P. L. Hanst, R. C. Doerr, and W. E. Scott, *Ind. Eng. Chem.,* **48,** 1498 (1956).
8. E. R. Stephens, W. E. Scott, P. L. Hanst, and R. C. Doerr, *Proc. Am. Petrol Inst. Sect. III* **36,** 288 (1956); *J. Air Pollution Control Assoc.,* **6** (3), 159 (1956).
9. W. E. Scott, E. R. Stephens, P. L. Hanst, and R. C. Doerr, *Proc. Am. Petrol. Inst. Sec. III,* **37,** 171 (1957).
10. E. R. Stephens and E. A. Schuck, *Chem. Eng. Progr.,* **54,** 71 (1958).
11. E. R. Stephens, P. L. Hanst, R. C. Doerr, and W. E. Scott, *J. Air. Pollution Control Assoc.,* **8,** 333 (1959).
12. J. T. Middleton and E. F. Darley, *Bull. World Health Organ.,* **34,** 477 (1966).
13. E. R. Stephens and W. E. Scott, *Proc. Am. Petrol. Inst. Sect. III,* **42,** 665 (1962).

14. E. R. Stephens, "The Photochemical Olefin–Nitrogen Oxides Reaction," in *Chemical Reactions in the lower and Upper Atmosphere,* Interscience New York, 1961, p. 51.
15. C. A. Tuesday, "The Atmospheric Photooxidation of *Trans-butene*-2 and Nitric Oxide," *Chemical Reactions in the Lower and Upper Atmosphere,* Interscience, New York, 1961, p. 15.
16. E. R. Stephens, F. R. Burleson, and E. A. Cardiff, *J. Air Pollution Control Assoc.,* **15,** 87 (1965).
17. S. W. Nicksic, J. Harkins, and P. K. Mueller, *Atmospheric Environ.,* **1,** 11 (1967).
18. E. R. Stephens, *Anal. Chem.,* **36,** 928 (1964).
19. E. R. Stephens, *Infrared Phys.,* **1,** 187 (1961).
20. E. F. Darley, K. A. Kettner, and E. R. Stephens, *Anal. Chem.,* **35,** 589 (1963).
21. E. R. Stephens and M. A. Price, Paper presented at 8th Conference on Methods in Air Pollution and Industrial Hygiene Studies, Oakland, California, February, 1967.
22. J. B. Mudd, *J. Biol. Chem.,* **241,** 4077 (1966).
23. E. R. Stephens, *Atmospheric Environ.,* **1,** 19 (1967).
24. J. M. Huess and W. A. Glasson. General Motors Corp. Research publication GMR 747. Presented at the 155th National Meeting of the American Chemical Society, San Francisco, Calif., April 2, 1968.
25. O. C. Taylor, E. R. Stephens, and E. A. Cardiff. Paper presented at 1968 meeting of the Air Pollution Control Assoc., St. Paul, Minn., June 1968.
26. H. Mayrsohn and C. Brooks. California Air Resources Lab., 434 S. San Pedro St., Los Angeles, Calif. 90013. Paper presented at the Western Regional Meeting of the American Chemical Society, Nov. 1965.
27. D. T. Tingey and A. C. Hill. *Utah Acad. Proc. V.* **44,** part 1, p. 387 (1967).

Biodegradable Detergents and Water Pollution

THEODORE E. BRENNER

Technical and Materials Division, The Soap and Detergent Association, New York, New York

I. The Water Pollution Problem 151
II. Early History of the Detergent Question 154
III. Summary of Early Findings 155
IV. Search for an ABS Replacement 158
V. Mechanisms of Biodegradation 162
VI. Biodegradability Testing 163
VII. Field Studies 167
A. Field Studies at Sewage Treatment Plants 168
B. Studies on Individual Household Disposal Units 173
VIII. Postconversion Experience 180
IX. European Situation 192
References 193

By definition a detergent is anything that cleans. In its broadest sense this would include such diverse materials as old-fashioned lye soap, alkaline dishwashing compounds, solvent cleaners, and, under some circumstances, even sand. However in the interest of clarity, the term will be limited in this chapter to those cleaning compounds which include as part of their chemical make-up petrochemical (or other synthetically derived) surface-active agents. The surface-active portion of the detergent formulation tends to lower surface tension, which in turn permits dirt particles to be lifted or floated off the article to be cleaned by the wash water. Thus, in large measure, surface-active agents (or surfactants as they are sometimes called) give much of the cleaning power to detergents.

The basic formulation differences between typical soap and detergent general purpose heavy duty products are shown in Figure 1.

In the case of a "soap" detergent, the soap portion itself, which usually is the sodium or potassium salt of a fatty acid derived from natural fat or oil, serves as the surfactant. Generally speaking, a smaller quantity of the synthetically derived surfactants is used in detergent formulations.

In both soaps and synthetic detergents mildly alkaline complex phosphates are incorporated into the formulations to provide softening, to aid soil suspension, to adjust the pH of the washing solution, and to enhance overall cleaning efficiency in several other ways. In addition,

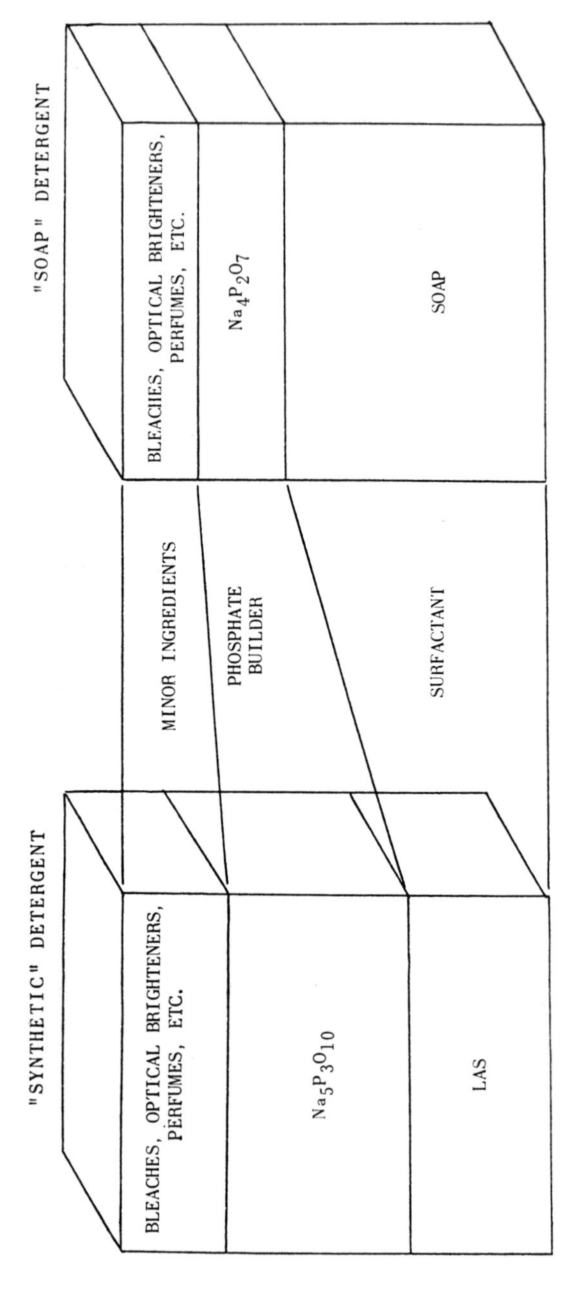

Fig. 1. Representative ingredients in packaged household "synthetic" and "soap" detergents. Reprinted from F. J. Coughlin et al. (1) by permission. Copyright © 1963 by American Water Works Association.

small quantities of other materials, such as soil-suspending agents, optical brighteners, bleaches, and perfumes are added, all of which contribute to the individuality of different products.

The growth in popularity of detergents over the last 20 years has been dramatic. In 1940 somewhat more than 3 billion pounds of nonliquid soap were sold in the United States. Detergent sales volume at that time was a minuscule 30 million pounds. By 1966 the figures were not completely reversed, but 5 billion pounds of detergents were sold as compared to a little more than 1 billion pounds of soap (Fig. 2). Perhaps more importantly, detergents now hold somewhat more than 90% of the total household market. The popularity of detergents stems primarily from their superior cleaning efficiency, particularly when used in hard water areas or under other unfavorable conditions.

One of the key technical factors that led to the successful development of detergents was the application of the surfactant alkyl benzene sulfonate (ABS) to detergent formulations. This material emerged as a prime detergent component because of its ability to be formulated into efficient cleaning products, relatively low cost, ease of formulation, and compatability with other detergent constituents. In 1962, for example, it was reported that ABS probably accounted for some 70% of the surfactant volume likely to be disposed of in waste water (3).

ABS is derived from propylene and benzene, both of which are available in abundant quantities and reasonable price. To manufacture ABS, propylene is first polymerized to a tetramer or pentamer (4). The polymers are then used to alkylate benzene in the presence of an acid catalyst such as hydrogen fluoride, aluminum chloride, or sulfuric acid. After distillation of the alkyl benzene it is then either batch or continuously sulfonated with sulfur trioxide or oleum and finally neutralized with sodium hydroxide to form ABS (Fig. 3).

ABS can be composed of many isomers. These may come from three principal sources: (*1*) the presence of other olefins such as ethylene or butylene in the feed stocks used to make the propylene tetramer and pentamer, (*2*) the various means of polymerization which may be employed, and (*3*) the possibility of fragmentation and isomerization reactions which take place during alkylation of the benzene (5).

One of the key characteristics of alkyl benzene sulfonate is its highly branched alkyl side chain. The significance of this will be discussed in more detail later in this chapter.

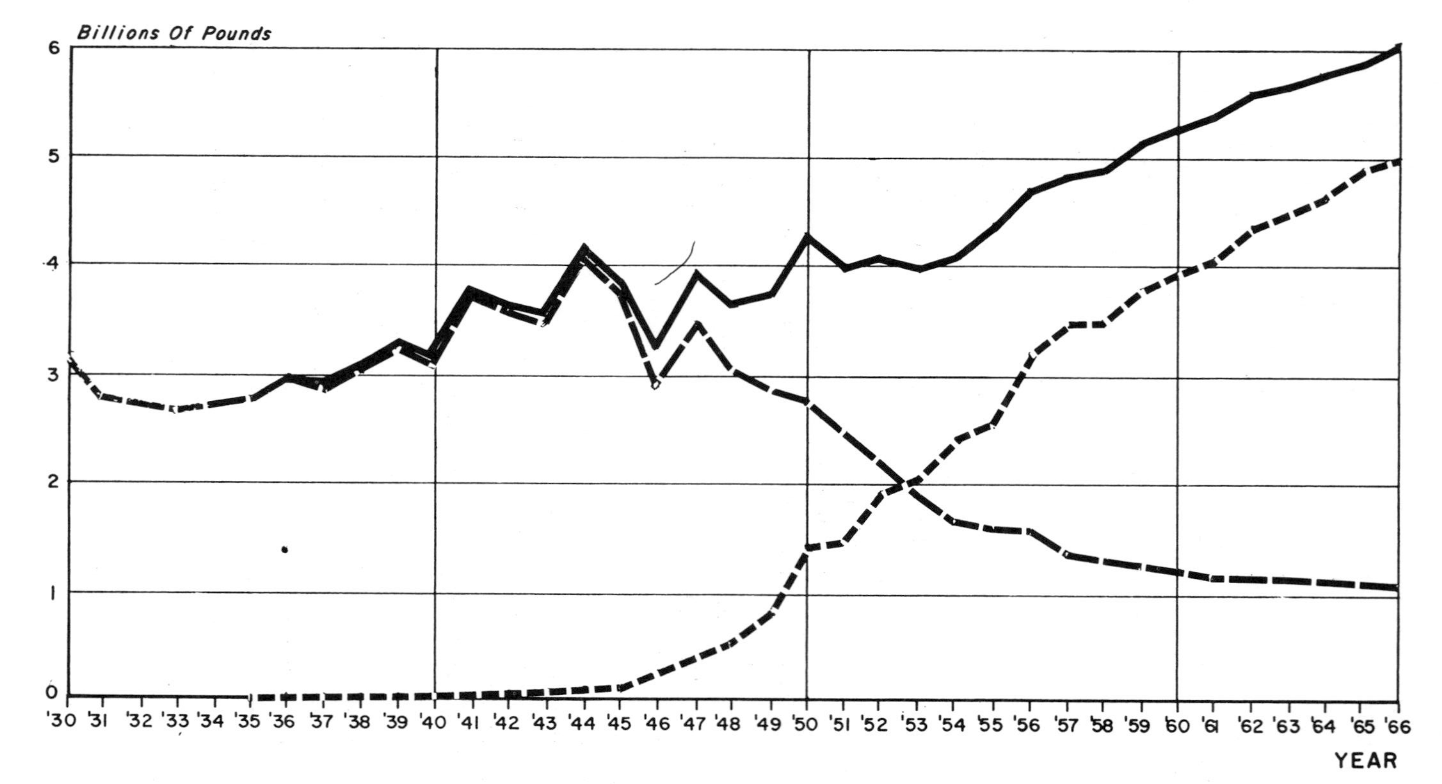

Fig 2. Estimated annual soap and synthetic detergent sales 1930–1966. (——) Total, (- - -) detergent, (— —)soap. From ref. 2 by permission.

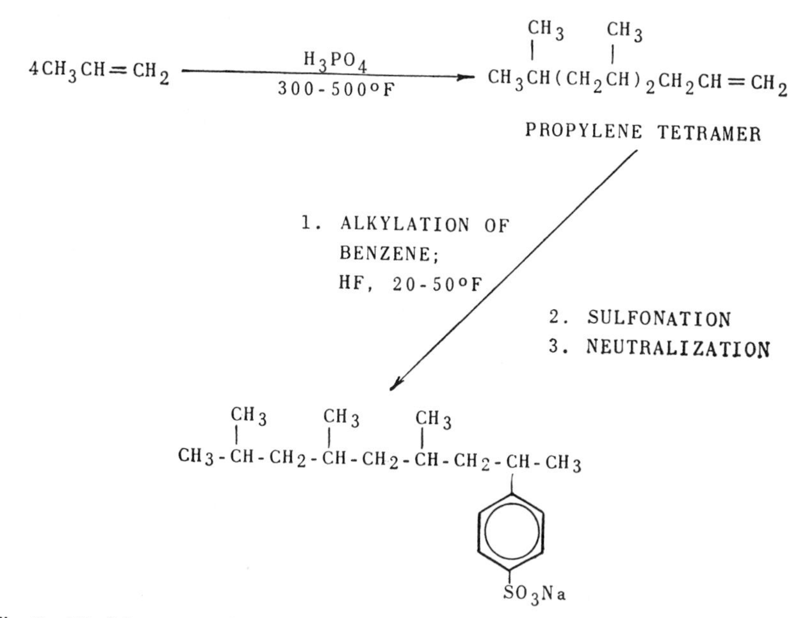

Fig. 3. Alkyl benzene sulfonate (ABS). From D. Justice and V. Lamberti (5) by permission.

I. THE WATER POLLUTION PROBLEM

The passage in 1948 of Public Law 845, the first federal water pollution control act, introduced a new era in water management and water pollution control. Historically water pollution control had been part of local, state, and federal public health programs and was primarily oriented towards the control of waterborne communicable diseases, and, to a lesser extent, to the removal of esthetically objectionable visible waste constituents. Certainly the almost total eradication in the United States of such dread killers as typhoid and cholera must stand as one of the great testimonials to the work done in this field during the 19th and early 20th centuries. Thus, in many local and state governments, the administration of water pollution abatement programs comes under the jurisdiction of the health department. Until very recently (1966), the federal program came under the supervision of the U.S. Public Health Service. Since then, this function has been transferred to a new agency, the Federal Water Pollution Control Administration, under the general jurisdiction of the Department of the Interior.

With the decline of waterborne diseases, greater interest developed in the commercial and esthetic aspects of stream pollution, which in turn led to a better understanding of the chemical, biological, and biochemical forces constantly at work in the dynamic environment of natural waters.

In the 1920's and 1930's it became increasingly evident that streams and other natural waters were being degraded at an alarming rate by the discharge of vast quantities of raw or partially treated municipal and industrial wastes.

While the extent of knowledge about water pollution control measures grew at a rapid pace during the last three decades, this new knowledge was not being put uniformly into practice. Probably the simplest explanation for this was the pressing need for other public works such as schools, highways, hospitals, etc., which took precedence in the expenditure of available tax dollars. Another important and related factor was a generally apathetic public, unaware of the increasing demands being made on our overworked rivers and streams. However, critical water shortages in such areas as populous Southern California, serious droughts in the Northeast, and an overall increased sophistication on the part of the electorate have done much to bring the picture into proper focus. The growing concern over water pollution control led to the enactment of far-reaching legislation at almost all levels of government. Following the passage of the first federal act in 1948, improved and more comprehensive federal laws were enacted in 1956, 1961, 1965, and 1966. At the present time the federal government has authorized the total expenditure of 3.55 billion dollars during the five-year period ending in 1971 for the construction of waste treatment plants, while another 60 million dollars per year is earmarked exclusively for research and development in the area of water pollution control and wastewater reclamation (6).

However, a recent report by the Department of Commerce (7) indicated that in 1964 only 65% of our population (approximately 121 million people) was served by sewers. Of this number, some 70 million people were receiving what was classified as "adequate" waste treatment. The remainder obviously received "inadequate" treatment or no treatment at all. The terms "adequate" and "inadequate" are not further defined, but review of somewhat earlier statistics (1962) compiled by the U.S. Public Health Service (8) presents a rather disquieting picture. The goal of most water pollution control authorities has been the universal application of secondary biological treatment to municipal wastes. Plants of this type would incorporate not only the screening, settling, and chlorination processes usually found in primary plants (and designed to remove only objectionable settleable materials and to control disease), but would also provide for the removal of the bulk of the soluble oxygen-consuming organics that are probably of even greater concern. Within the practical context of today's technology and economics, the most efficient form of secondary treatment involves the use of some form of the activated sludge process. In this process, wastes are maintained in intimate contact with suspended biological life for a defined period of time. Oxygen (supplied either by blowers, compressors, or air entrainment devices) is introduced to the system to assure an active aerobic condition in the biomass and to provide mixing. Soluble organic waste constituents are first adsorbed on the biological cells present and then more slowly assimilated and converted to new cell material.

Generally speaking these processes can routinely remove 90% or more of the biochemical oxygen demand (BOD) in the waste. Yet, in spite of the great need for systems of this type, only a little more than 33 million people were served by activated sludge plants in the U.S. in 1962. Another 23 million used the less efficient trickling filters, while 5 million more made use of a variety of other "secondary" systems, including stabilization ponds, sand filters, etc. Thus in 1962 only a little more than 61 million people had any form of secondary biological waste treatment available to them.

It has been suggested that our present rate of sewage treatment plant construction will barely keep pace with the growth in population. Estimates have been made that if, over the next 25 years, we were to provide secondary treatment nationally (assuming BOD removal in such plants would be about 90%) the BOD load placed on rivers and streams could be reduced by about 40% during this period. However, even with this reduction, after 1990 BOD discharges would continue to increase at about 2–3% annually simply as a result of our growing population (9).

Therefore, our present problem is twofold. First there is an urgent need for the rapid construction of secondary treatment systems throughout the nation in order to stop and reverse the pollution trend. Hopefully funds are now available and interest sufficiently high to accomplish this effectively. Second, there is perhaps an even greater need to develop new methods and systems that can effectively cope with the growing volume and complexity of our wastes in the future. There is every indication that positive action is now being taken to implement the vast research and development effort needed to provide the necessary technology.

It is against this background of inadequate waste treatment and a generally uninformed (about water pollution abatement) public that the question of detergent residues in natural waters first came to light.

II. EARLY HISTORY OF THE DETERGENT QUESTION

The first recorded detergent foam incident occurred in 1947 at Mt. Penn, Pennsylvania, where a heavy blanket of foam was observed on the aeration tanks of the sewage treatment plant (10). This incident had been preceded (a few days earlier) by the widespread distribution of free detergent samples in the community. By 1950 other isolated reports of foaming were being discussed in technical circles and, in some instances, received rather flamboyant press coverage. Among these were incidents at San Francisco's Golden Gate Park Water Reclamation Plant, at the then brand new Hyperion plant serving Los Angeles, California (11), and at the San Antonio, Texas sewage treatment plant. The first detailed report on groundwater contamination appeared in 1958 when an article was published in the *Journal of the American Water Works Association* regarding the situation in the Eastern portion of New York's Long Island (12). Other suspected cases of groundwater contamination, however, went back to the early 1950's.

Although it was well known that many natural organics could create foam in water (decaying leaves are a frequent cause, for example), there was at least enough circumstantial evidence to warrant further investigation. It had been demonstrated that alkyl benzene sulfonate residues could foam at very low levels (1 mg/liter) and that these residues, after use by the housewife, invariably found their way, along with all other domestic wastes, either into groundwater or rivers and streams. The type and extent of treatment which these wastes may have received is open to speculation; nevertheless, a possible cause and effect relationship between detergent residues and some foam incidents was indicated.

In order to study this question in depth and to insure an organized, well directed effort, the Association of American Soap and Glycerine Producers (now The Soap and Detergent Association) established its Technical Advisory Council in 1951. This group was charged with coordinating the industry's own program with respect to this phase of the water pollution problem, cooperating with interested and qualified nonindustry groups and supporting necessary research within the industry and at universities and with other recognized authorities in the field.

Early activities of this group and its subcommittees included development and evaluation of test methods, developing a better understanding of the significance of trace amounts of synthetic detergents on water and waste treatment processes, evaluation of potential toxicity problems, evaluating means and methods of foam control, and many other facets of the problem. Other interested technical bodies initiated their own investigations. Typical of these was the formation in 1954, of the Detergent Subcommittee of the Ohio River Sanitation Commission (ORSANCO). A similar task group under the sponsorship of the American Water Works Association (AWWA) was organized somewhat earlier (10). Many states (and some local governmental bodies) also set up study groups either in compliance with a specific legislative mandate or as part of their overall pollution control responsibility. Maryland, California, New York, and Wisconsin are notable examples.

The combined efforts of all these groups plus the individual programs of the detergent manufacturers and their raw material suppliers contributed significantly to a fuller understanding of the problem by the end of the decade.

III. SUMMARY OF EARLY FINDINGS

During the 1950's many projects were undertaken to provide a fundamental knowledge of the scope and significance of the "detergent problem." Although foaming could be considered as a crude measure of the presence or absence of synthetic detergent residues in water, it did not provide a reliable quantitative or qualitative measure. While methods were available to determine low concentrations of ABS, they needed major refinement. The most commonly used analytical method today is the colorimetric methylene blue technique which was first described by Jones in 1945 (10a).

This method, which depends on the formation of a blue-colored salt when methylene blue reacts with ABS, has gone through several modifications and is now included in the widely accepted *Standard Methods for the Examination of Water and Wastewater* (13). Most of the development work on the procedure took place in the 1950's. This procedure is subject to many positive and some negative interferences however, and data obtained using the method may therefore be misinterpreted. There is a growing body of scientific opinion which holds the belief that results should be reported in terms of methylene-blue-active substance (MBAS) rather than as ABS or similar designation. By so doing, misinterpretation of data would be minimized.

In addition, a referee method was also developed under the sponsorship of The Soap and Detergent Association and is also included in *Standard Methods for the Examination of Water and Wastewater* (14). This procedure eliminates practically all of the positive and negative interferences which may affect the methylene blue technique. While not overly complex in itself, it does require the use of infrared spectrophotometric equipment and is a somewhat lengthy procedure. It involves concentrating the ABS on activated carbon, desorbing, purification of the desorbed material, and ultimate examination of the formed amine complex by infrared spectrophotometric means. The method yields extremely precise results.

The toxicological aspects of water contaminated by detergent residues came in for extensive study during this period. Most of the major findings were utilized in the preparation of the 1962 edition of the *U.S. Public Health Service Drinking Water Standards* (15). In these standards a limit of 0.5 mg/liter of ABS in drinking water was *recommended* (this was not a mandatory limit). In an appendix to these standards, the rationale that led to the adoption of the 0.5 mg/liter limit is developed. The selection was made essentially for nontoxicological reasons (possibility of foam formation at higher levels, indication of serious sewage pollution, etc.). The commentary on safety of water supplies containing these levels of ABS warrants repetition here.

> An ABS concentration of 0.5 mg/l in drinking water, in terms of a daily adult human intake of 2 liters, would give a safety factor of the order of 15,000, calculated on the results of subacute and two-year tests on rats fed diets containing ABS. In these rat studies, it was found that levels of ABS in the diet of 0.5% and below produced no discernable physiological, biochemical, or pathological deviations from normal.
>
> Human experience (6 subjects) with oral doses of purified ABS of 100 mg (equivalent to 2 liters of water containing 50 mg ABS/l) daily for 4 months led to no significant evidence of intolerance.

The matter of ABS toxicity was further evaluated in the definitive report of Woodward, Stokinger, and Birmingham (16) who, in 1964, also concluded that no problem existed at the concentrations at which ABS residues are found. They reasoned that no changes in the 1962 standards were warranted.

In order to determine the actual levels of ABS found in community water supplies throughout the country, the Association of American Soap and Glycerine Producers undertook a sampling program at 32 U.S. cities during 1959 and 1960. This study showed that the highest level found was 0.14 mg/liter (Camden, New Jersey, summer 1959) and that the average concentrations were 0.034 mg/liter, 0.024 mg/liter, 0.015 mg/liter, respectively, for the summer of 1959, the winter of 1959, and the spring of 1960 (17).

Other studies by various groups throughout the country have shown that ABS levels in rivers and streams rarely exceeded 0.5 mg/liter.

Perhaps even more importantly, these studies showed no buildup or increase in these levels over a period of years (18). It became quite clear, therefore, that ABS residues constituted no significant toxicological problem in either raw waters or in drinking water supplies.

In the few instances where high levels of ABS were found, e.g., the oft-cited Chanute, Kansas, situation, where reclaimed waste water was used for an extended period (19), no adverse effects were reported. This gave practical evidence to the validity of the previously developed theoretical and laboratory safety standards.

Extensive studies of fish toxicity also indicated no practical hazard to fish life from the exposure to the levels of ABS residues found in natural waters. Laboratory studies had shown that the median toxicity level (TL_M) of ABS to fish was 3.5–6.0 mg/liter and that the TL_M increased as the ABS underwent partial degradation (20). This fact has been confirmed by the very complete records maintained by the U.S. Public Health Service (now Federal Water Pollution Control Administration).

Based on these data, there is no evidence of fish kills in this country as the result of sewage-borne detergent residues.

Other potential problems that were considered during this period included the effect of detergent residues on water-softening, ion-exchange resins and on waste treatment processes. A particularly persistent question was that of tastes and odors. It had been the opinion of some authorities that ABS residues could contribute to tastes and odors in water supplies. However, after careful study it was found that none of these concerns had any basis in fact—at least at the concentrations actually found in nature (21–24).

The problem, then, was one of foaming. Foam is visible and is esthetically unacceptable in sewage treatment plants and in rivers and streams receiving their outflows. It had been determined that ABS could cause incipient foams at levels of 1 mg/liter or more. At use concentrations, ABS levels ranged between 200 and 600 mg/liter. By the time these wash wastewaters had flowed from the home and had been mixed and diluted by other wastewater in sewers, the concentration reaching the sewage treatment plant would probably be of the order of 2–10 mg/liter and would occasionally be as high as 15 mg/liter. In contrast, in Europe, where water use habits differ from those in the United States, reported ABS levels in wastewater were often in the 20–25 mg/liter range.

A variety of surveys, studies, and reports showed that ABS reduction in an activated sludge plant would be in the range of 50–60%, while trickling filters would effect a somewhat lower removal (25). Hence, sufficient ABS could exist in the effluent of an activated sludge plant to support foaming at the plant outfall, assuming of course, that the other conditions necessary for foam development were present, e.g., turbulence. Obviously, if foaming could occur in the effluent, the chances for foam development in aeration tanks were even greater, due to the higher ABS concentrations and greater turbulence. It could be concluded, therefore, that even with "good" secondary treatment systems, ABS was not as readily removed as were other soluble organic wastewater constituents.

Since much of the discussion about ABS developed over its effect on ground water (particularly as a result of discharges from home septic tanks and cesspools) considerable work was undertaken in this area. The findings are best summarized in the 1963 ORSANCO Detergent Subcommittee report "Components of Household Synthetic Detergents in Water and Sewage" (26), which concluded:

> Where drinking water is drawn from shallow wells near septic tanks or cesspools, ABS and other pollutants may appear in water. ABS occasionally appears under these circumstances in concentrations as high as 1–4 ppm where, in some cases, it may be the only significant pollutant. Often, however, the presence of ABS, apparent because of the foaming it causes, is a signal that other pollutants are present. The hypothesis that ABS facilitates the movement of coliform organisms through saturated sand soil has been disproved by experiments. The development of more readily degradable surfactants may contribute much to pollution abatement where pollution removal is inefficient.
>
> Authorities state that, used in ordinary amounts, detergents will not interfere with the functioning of a septic tank designed, built, and connected properly.

IV. SEARCH FOR AN ABS REPLACEMENT

As early as January 1951, discussions relating to the development of alternate materials to replace ABS were underway. At that time the search for a potential replacement was largely being carried out in individual company laboratories and, at least in the early stages, concerned with establishing a better understanding of the factors affecting ABS biodegradability.

From this early work it was determined that the highly branched alkyl side chain of the ABS molecule (Fig. 3) inhibited its rate of biological degradation.

It was known that the biological processes used in sewage treatment and the microorganisms present in natural waters could rapidly degrade such straight chain materials as soaps and alcohol sulfates. The rate of

ABS breakdown, however, was slower, and this in turn resulted in lower overall removal under the fixed detention times existent in sewage treatment plants. Thus it appeared that if a new surfactant could be developed with an essentially unbranched side chain, many of the previous objections to ABS would be eliminated.

The acceptability of any new replacement for ABS was predicated on four basic hypotheses. They were:

1. The new material must be more biodegradable than ABS.

2. It must have detergency characteristics that were at least as good as those of ABS and it should be compatible with other materials used in detergent formulations.

3. Its price range (and the cost of formulation) must at least be of the same order of magnitude as that of ABS.

4. It must be essentially nontoxic and nonirritating to skin when used in detergent formulations.

New chemical developments (which were at least partially motivated by the search for more biodegradable surfactants) provided the answer to the manufacture of straight-chain materials meeting all four requirements.

The advent of synthetic molecular sieves, Ziegler ethylene polymerization, and paraffin cracking probably contributed more to the availability of readily biodegradable surfactants than any other recent technological developments in the chemical industry. Of the three, synthetic molecular sieves have probably been most important in this situation (27).

Although naturally occurring molecular sieves had been studied for almost four decades, it was not until the early 1950's that they became commercially available. At that time, the development of synthetic, crystalline, aluminosilicate sieves led to the practical application of molecular sieves in the chemical industry (28).

The *n*-paraffins (which today are probably the most-used surfactant feedstock) can be separated from kerosene in either molecular sieves or by complexing with urea (29,30). In both cases, the rejection of branched components is believed to occur essentially through the same adsorption mechanism. For example, in the case of the sieves, an empty but uniformly porous network remains after the water of hydration is removed. When separating *n*-paraffins from kerosene, molecular sieves with pore diameters of about 5 Å are used, and the *n*-paraffins, which have cross-sectional diameters of 4.9 Å are readily adsorbed. The binding force in both molecular sieves and urea complexes develops from the interaction of the outer electrons in the shells of the atoms of the crystal and of the adsorbed molecules. These forces are generally referred to as dispersion or polarization forces. Molecular sieves can be used in several processes, all of which involve adsorption and desorption cycles. These cycles can be induced by pressure differences, temperature differences, gas purging and the use of adsorbable fluids. All of the molecular sieve processes used today employ one or more of these methods in their adsorption–desorption phases (31).

After separation the *n*-paraffins can be converted for use as biodegradable surfactants in detergent formulations by several means. These are shown diagrammatically in Figure 4 and can be classified in the following manner:

1. They can be monochlorinated and reacted directly with benzene. In this process $AlCl_3$ is used as the catalyst. An alternate route involves dehydrohalogenation which yields random olefins that can in turn be used to alkylate benzene. In either case, the alkylate is sulfonated with either oleum or sulfur trioxide and then neutralized (usually with sodium hydroxide). Materials produced by either of these processes have been classified as linear alkylate sulfonate (LAS) and have become the primary replacement for ABS (32) (Fig. 5).

2. *n*-Paraffins can also be converted to random secondary alcohols by an oxidation step and can be used as sulfates, ethoxylates, and sulfated

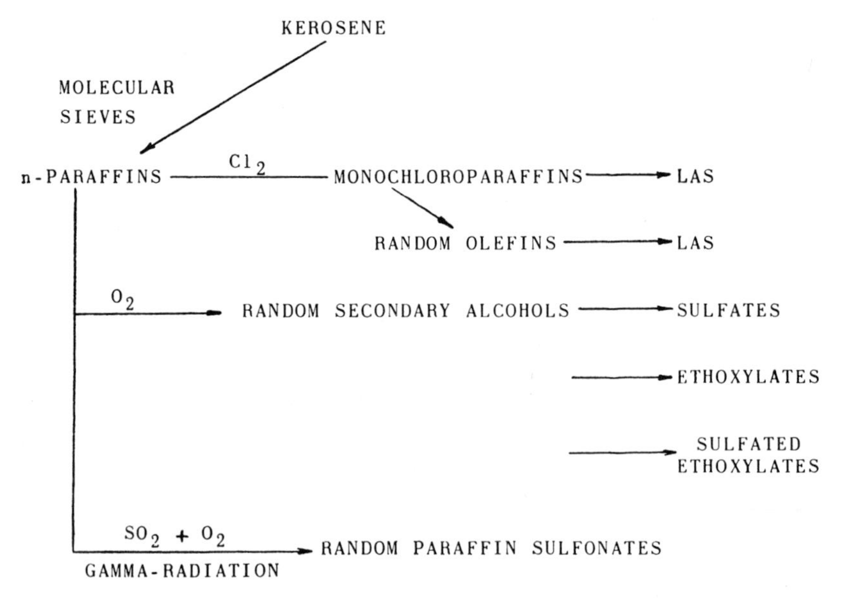

Fig. 4. Biodegradable surfactants from *n*-paraffins. From D. Justice and V. Lamberti (5) by permission.

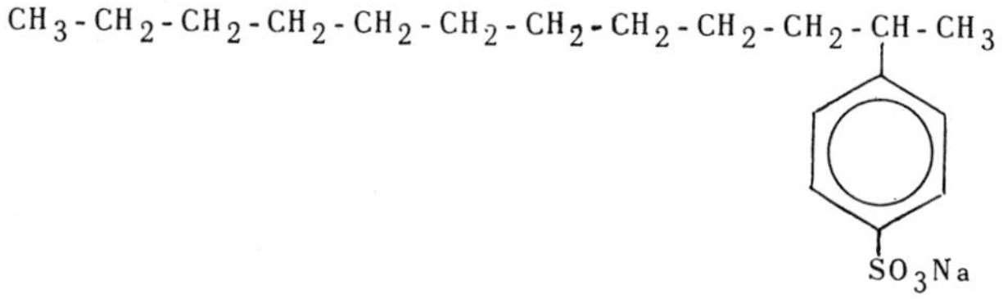
$CH_3-CH_2-CH_2-CH_2-CH_2-CH_2-CH_2-CH_2-CH_2-CH_2-CH-CH_3$
SO_3Na

LAS

CHAIN LENGTHS IN TYPICAL LAS: C_{11}-C_{14}

POSITION OF PHENYL GROUP: RANDOM SECONDARY ALONG EACH CHAIN

Fig. 5. Linear alkylate sulfonate (LAS). From D. Justice and V. Lamberti (5) by permission.

ethoxylates. While highly biodegradable, these materials find application in formulations that generally did not use ABS.

3. Although they do not as yet seem to have been commercially developed, random paraffin sulfonates can also be produced by the sulfonation of *n*-paraffins with sulfur dioxide and oxygen utilizing gamma radiation as the energy source (33).

The Ziegler polymerization of ethylene is probably the second most important development insofar as surfactant technology is concerned and has proven to be particularly suitable for the production of both primary alcohols and alpha olefins. These materials, in turn, are used to produce LAS and other degradable surfactants. In this process, ethylene is built up on Al-triethyl to produce Al-trialkyl. To produce detergent alcohols, the Al-trialkyl is oxidized and then hydrolyzed with an acid. Conversely, to produce alpha-olefins, the Al-trialkyl is subjected to a displacement reaction in the presence of a catalyst (34).

Although at one time cracked wax alpha-olefins appeared to be the most practical approach for the production of LAS, it now seems that this route has primary application in the manufacture of a variety of other biodegradable surfactants and detergent intermediates. These include primary and secondary alcohol ethoxylates and sulfated ethoxylates.

Cracked wax alpha-olefins can be produced by heating selected petroleum waxes to approximately 1000°F at 100 psia in the presence of steam or nitrogen. In this procedure, about 80% of the olefins contain 20 carbons or less and the alpha-olefin content is about 95%.

The derived alpha-olefins can then be converted to useful detergent components by a variety of processes (35).

In the early 1960's it became apparent that the technological and economic practicalities affecting the full scale manufacture of highly biodegradable surfactants had been essentially resolved which made it possible to predict when a total conversion to these materials would be completed. In April 1963 industry spokesmen testifying before a Congressional committee indicated that full conversion to LAS would be possible by December 31, 1965. During the intervening period, new plant construction was accelerated, and by August 1964 the target date for conversion was moved up to June 30, 1965. At that time, all raw material suppliers had completed their changeover (at an estimated cost of $150,000,000) and detergent manufacturers were no longer using ABS in their formulations.

V. MECHANISMS OF BIODEGRADATION

Microorganisms have the inherent ability to utilize organic matter for food, energy, and ultimate conversion to new cell material. Although a wide range of organic compounds may be involved, microorganisms accomplish this degradation process through a limited number of biochemical reactions. The biodegradation of surfactant molecules follows the same general pattern, although some minor variations may exist due to specific characteristics of the predominant surfactant molecule involved. Swisher (36) has described the most important reactions, insofar as surfactant biodegradation is concerned, as being beta-oxidation, methyl oxidation, and aromatic oxidation, and defines these reactions as follows:

Beta-Oxidation. This mechanism is vital to all living matter for it is the process by which fatty acids are degraded. In this reaction the fatty acid chain is oxidized, two carbon atoms at a time, into acetyl groups which can be used for energy. The actual process involves the initial esterification of the carboxyl group with coenzyme A, an organic mercaptan. Following this, two hydrogens are removed to give the alpha-beta unsaturated derivative which ultimately yields the beta-keto derivative. Finally another molecule of coenzyme A adds between the alpha- and beta-carbons, splitting off acetyl coenzyme A and leaving a fatty acid coenzyme A ester, two carbons shorter than the original. The process repeats itself as degradation proceeds along the chain.

Under aerobic conditions, the hydrogen may combine with atmospheric oxygen to form water. Although anaerboic degradation proceeds by the same beta-oxidation mechanism, the hydrogen acceptor would probably be either a carbon compound or a sulfate which would result in the formation of methane or hydrogen sulfide.

Methyl Oxidation. Another, although not as well understood, mechanism is that involving methyl oxidation. In this process the terminal methyl group of a surfactant hydrophobe, for example, is oxidized to a carboxyl group and degradation then proceeds along the beta-oxidation route. Kallio and his co-workers have shown that one important pathway involves an attack by oxygen to produce a primary hydroperoxide. This is subsequently reacted to produce primary alcohol, aldehyde, and carboxylic acid.

Aromatic Oxidation. This process is of importance because it is by this route that the benzene ring is broken down. In the oxidation of alkylbenzene, benzoic acid is a likely intermediate product. In this case, as well as in those involving phenol and other materials, catechol is formed by enzyme-catalyzed oxidation and the benzene ring is split between two hydroxyl groups to yield dicarboxylic acid. After several molecular rearrangements, acetate and succinate groups are formed which are utilized to maintain cell metabolic equilibria.

Swisher (37) has summarized his findings on the mechanisms involved in LAS biodegradation by reporting that initial oxidative attack occurs at one end of the chain remote from the sulfonate group. Once underway, the chain rapidly degrades by beta-oxidation until a terminal carboxyl group is formed either two or three carbons away from the phenyl group. The beta- or gamma-sulphenyl alkanoic acid thus formed is attacked from the other end of the chain, after which oxidation of the molecule proceeds rapidly. He noted, however, that a minor portion of the oxidative process may involve a one-carbon step rather than the more conventional beta-oxidation two-carbon process.

VI. BIODEGRADABILITY TESTING

During the period of development of biodegradable surfactants, literally hundreds of candidate materials were subjected to a variety of rigorous laboratory examinations. In addition to fairly standardized procedures, such as those used to evaluate detergency, stability, and other performance variables, it was necessary to develop meaningful and realistic methods to determine surfactant biodegradability. At first this was a matter of choice on the part of the individual investigator, but it soon became evident that a standardized yardstick against which the performance of many materials could be tested was a necessity.

Over the years considerable work had been carried out in the fields of microbiology and sanitary engineering in evaluating the microbial degradation of a variety of organic compounds. Techniques such as Warburg respirometry and the determination of BOD have been extensively employed to establish "biodegradability." In recent years, batch and continuous bench scale waste treatment units have also been used to establish the fate of various materials undergoing biological degradation. However, of more immediate concern in this case was a test that demonstrated the "treatability" of a surfactant under field conditions rather than its more theoretical "biodegradability."

A number of methods, ranging from the simple river die-away test to the complex continuous activated sludge method, were in use when The Soap and Detergent Association's Subcommittee on Biodegradation Test Methods began its work in 1961. Before developing a method for the evaluation of ABS/LAS biodegradability, more than 1300 laboratory runs were made using the four principal methods under consideration.

The important features of each of these tests are as follows:

River Die-Away. In this method a sufficient quantity of the material under test is added to a liter of fresh river water to achieve a concentration, in most cases, of 20 mg/liter. The solution is placed in a sealed jar and is left quiescent under normal room temperature conditions. Samples are routinely drawn for methylene blue analysis (after mixing) and from these data a biodegradation or die-away curve is developed. This procedure yields information on both the total degradability and the rate of degradability of the material under test. Results obtained, however, can be subject to significant variations depending on the source of the river water and its degree of biological activity (38).

Shake Flask Method. Bacteria which have been preadapted to surfactants are used in this procedure. These microorganisms are inoculated into flasks which contain a chemically defined growth medium, i.e., the base medium, and the material under test. The flasks are aerated by continuous shaking on a mechanical shaker. Shaking is stopped from time to time to permit the removal of aliquot portions of the flask content for analysis. Biodegradation of ABS and LAS is followed by measuring the reduction in methylene-blue-active substance. The test is relatively simple, does not require elaborate or specialized laboratory equipment, and lends itself to reproducibility (39).

Semicontinuous Activated Sludge. Several modifications of this procedure have been used; however, in practically all variations, (and in that adopted by The Soap and Detergent Association) a sewage-derived bacterial inoculum is employed. The inoculum, the material to be tested, and a synthetic sewage used as a food source for the inoculum are placed in a specially designed aeration chamber. The mixture is aerated for an extended period of time (usually 23 hr) and allowed to settle (generally 1 hr). The upper two-thirds of the volume is then removed. Samples from this portion are used for analyses. The liquid in the chamber is brought back to volume by the addition of fresh synthetic sewage and the material under test, and the cycle is repeated. Biodegradation of ABS and LAS is followed by the reduction in methylene blue response under specified conditions. A very high degree of control and reproducibility can be achieved, and the method more closely approaches field conditions than the two previously described systems (40).

Continuous Activated Sludge. Although innumerable variations of this procedure have been used by researchers, at least one version has the authority of law insofar as detergent biodegradability is concerned. The West German government promulgated in 1962 a procedure by which detergents were to be tested. In their method a relatively complex system is employed, including a large storage tank, a 3-liter capacity aeration tank, and a collecting tank. Synthetic sewage, fed at a rate of 1 liter/hr is used as a food source for the bacteria, and the initial surfactant concentration is held at 20 mg/liter. This procedure is unique in two ways: the entire detergent formulation is tested (U.S. practice is usually limited to the surfactant portion) and there is no bacterial inoculation—seeding is dependent upon contact with airborne bacteria. Under German law, a material is deemed biodegradable if its average removal (over 21 days) is 80% or more. A period of adaptation is allowed prior to the 21-day test cycle. Interestingly enough, a credit of 100% biodegradability is assigned to that portion of the formulation which is composed of soap. Thus the synthetic surfactant under test may be less than 80% biodegradable, yet the entire formulation could be acceptable if enough soap were included (41).

After a thorough evaluation of these methods, The Soap and Detergent Association promulgated a two-step procedure incorporating both the shake flask and semicontinuous test methods. The method was released in late 1965.

The shake flask technique is used as the presumptive step in this procedure, and each surfactant must be tested by this method. If a surfactant is 90% or more degraded in the presumptive test, no further testing is needed. If it is not at least 80% degradable, it is considered to be not adequately biodegradable. However, if its biodegradability falls between 80 and 90% in the presumptive test, its biodegradability must be adduced by the confirming test which is the semicontinuous activated sludge test. A material must be degraded at least 90% under this procedure to be considered adequately biodegradable (42).

The procedure and standards are designed to be applicable only to anionic surfactants of the ABS and LAS type and not to total detergent formulations containing varying quantities of these surfactants. Obviously it is much more desirable to control biodegradability of the raw material rather than of the many different finished products using the same raw material. Also, it is possible that certain formulations, such as those of a low surfactant–high inorganic salt content, or those containing bacteriostats or bactericides, could have side effects on the microbial population which would make invalid biodegradability measurements. These effects in the laboratory would have no significance under field conditions because the components disassociate when mixed with other wastes.

Work continues on the development of a universal test method to measure the biodegradability of all nonionic surfactants. While acceptable methods exist for determining the biodegradability of specific nonionics, none of these appear totally applicable for all compounds of this class. The problems so far encountered in nonionic testing have been largely analytical rather than biological in nature. This is partly a result of the many nonionic surfactants in use as well as the complexity in measuring partially degraded nonionic residues. At the present time, there is at least some indication that the use of physical analytical parameters (loss of foam, surface tension, etc.) may offer a practical solution to this problem.

Other biodegradability procedures have also been used for both specific and general purposes.

Renn (43), for example, has used laboratory size rotary tube trickling filters to simulate the performance of full-scale versions of these treatment units. The laboratory unit consists of a transparent plastic or glass tube, mounted horizontally which is rotated slowly about its long axis. As the tube rotates, the inner surface becomes wet, and a film of mixed microorganisms grows on it. The nutrient feed and the surfactant under test flow continuously through the rotating tube. In his studies, Renn operated the unit at a flow of 5 gal/day. Surfactant concentration was 10 mg/liter in the feedstock. This system was developed some 15 years ago and has been used to evaluate the performance of many wastes in trickling filters.

The British Standing Technical Committee on Synthetic Detergents also has published a method for the assessment of biodegradability of anionic surface-active materials (44). After considerable collaborative research, they developed a simple system in which a solution containing 10 mg/liter of the surfactant under test in standard BOD dilution water is inoculated with air-dried activated sludge and gently stirred at 20 ± 1°C in the dark for a period of up to 3 weeks. Samples are removed at appropriate intervals and the residual concentration of surface-active matter is determined. The test is discontinued when a stable residual is achieved. The British concede, however, that there is still a need for a second test which would permit the evaluation of the effect of acclimatization on a material which appears to be of questionable biodegradability by the simple test procedure. Although they feel that the test procedure is applicable to the testing of nonionics, this fact has not yet been fully confirmed and is currently under study. Recent British publications indicate that the procedure, somewhat modified, may have application in determining the biodegradability of some nonionics using thin layer chromatographic techniques and foam measurements (44a,44b).

Bunch and Chambers (45) of the Federal Water Pollution Control Administration's Cincinnati Water Research Laboratory have also developed a "universal" method for gauging the biodegradability of a wide variety of organic compounds. This procedure was designed to be simple in operation and require only readily available and economical laboratory equipment. Its stated range of applicability includes not only surfactants, but essentially all other organic materials as well. In this method 10 ml of settled sewage are used to inoculate 90 ml of BOD dilution water containing 5 mg of yeast extract and 2 mg of the material under test. The test is run in 250-ml Erlenmeyer flasks containing 100 ml of inoculated medium. The flasks are incubated under static conditions at room temperature. Weekly subcultures are made in fresh medium for 3 consecutive weeks. A parallel system is used as a control with a material of known biodegradability. Since this test method is quite new, little definitive information on its scope or applicability is available at this time. Obviously, it is faced with the same analytical limitations as the previously described procedures, and this would be particularly true in the case of nonionic surfactant biodegradability testing.

VII. FIELD STUDIES

Field testing of any new material generally serves two purposes: first, to confirm laboratory findings and, second, to fully evaluate performance under conditions which cannot be readily or realistically duplicated in the laboratory. Because of the tremendous investment being made, it was imperative to assemble comprehensive field data on LAS performance under a wide variety of conditions to supplement previously compiled laboratory information. It was also important that this be accomplished prior to the full conversion to LAS so that both the industry and those officials concerned with water pollution control would be satisfied that LAS would perform as anticipated. These studies derived their support from a variety of sources, including the detergent industry (both raw material suppliers and detergent formulators) and from several governmental bodies. In some cases, these projects were jointly sponsored. (*Note:* Although the research described in this chapter is essentially limited to the water pollution aspects of the detergent industry's conversion to LAS, it should be noted that an even larger effort was needed to test fully LAS-based formulations for performance, stability, safety, etc. This work was carried out almost entirely by individual detergent formulators and their raw material suppliers.)

These field studies can be placed into two general categories, those dealing with LAS performance in sewage treatment plants designed for the treatment of municipal wastes and those evaluating LAS removal in the relatively primitive systems used to handle individual household wastes (i.e., septic tanks, cesspools, and their associated soil systems). Sufficient work was done in each area to warrant separate discussion of the significant findings. Since the more important studies (from an overall pollution abatement standpoint) were those conducted at sewage treatment plants, this research will be discussed first.

A. Field Studies at Sewage Treatment Plants

Studies were conducted at plants employing a variety of processes, including activated sludge units (both conventional and extended aeration types), trickling filters, and oxidation ponds.

In most cases these studies were carried out at small relatively isolated locations. This was required in order to maintain the necessary degree of control both on the tributory sewage flow and the detergent usage in the community served by the treatment system. On the other hand, the communities had to be large enough so that the results would be typical of those encountered in large municipalities.

Generally speaking, a population of 500 or less was found to be best suited for studies of this type. Since many smaller communities are served by extended aeration plants, several of the studies were conducted at sites having these systems because of convenience and availability.

Extended aeration, a variant of the activated sludge process, is characterized by long detention times (often 24 hr or more) and high levels of mixed liquor suspended solids. In theory, all biological sludge produced by synthesis, is consumed by autooxidation within the aeration tank. The contents of the aeration tank are, theoretically, completely mixed.

Almost all the tests followed similar patterns. Since LAS-based detergents were not yet on the market, it was assumed that all methylene-blue-active substances found in plant influents and effluents were ABS. Background data were collected on plant performance in terms of ABS and BOD removal. After sufficient information had been obtained, specially formulated LAS-based detergents were distributed to the contributing population. In the case of some of the activated sludge and extended aeration plant studies, the plants were drained and new sludge was allowed to develop after LAS was in use. This often took several weeks but it minimized the problem of ABS desorption from the activated sludge. Results obtained were uniformly excellent with LAS removal even outpacing BOD removal in some cases. Data obtained in some of these studies are shown in Table I.

The Manassas Air Force Station, Virginia, study was conducted during the summer of 1964. The treatment plant itself was designed to operate on the contact stabilization process; however, by some internal piping changes, it was possible to operate it as a conventional activated sludge plant.

An unexpected situation developed at this site in that a significant drop in influent MBAS values was observed after the changeover to LAS-based formulations was completed. During the background data collection phase of the project, ABS levels had been in the order of 7.5 mg/liter. Immediately after conversion the level dropped steadily until it stabilized at approximately 2.5 mg/liter. Effluent values during this period ranged between 1 and 1.5 mg/liter. Because of the low influent values, however, performance in terms of per cent removal was not too impressive. Interestingly, the effluent values remained at the 1–1.5 mg/liter level, even after the raw sewage had been "spiked" with LAS to bring the influent concentration up to approximately 10 mg/liter (46).

This unexplained drop in influent LAS was not completely understood at the time of the study and could not be accounted for by changes in detergent usage. However, later experience both here and in Europe, would indicate that this resulted from degradation (or some other

TABLE I

Summary of Principal Field Test Results (Note: Not All Data from All Studies are Shown Below)

Location	Process	Material in use	Detention time, hr (where applicable)	Mixed liquor suspended solids, mg/liter (where applicable)	% Removal		
					ABS/LAS	Suspended solids	BOD
Manassas AFS, Va. (46)	Conventional activated sludge	ABS	6–16	3090	54	67	89
Manassas AFS, Va.	Conventional activated sludge	LAS	6–16	3510	85	75	91
Woodbridge, Va. (47)	Extended aeration	ABS	47	—	58–61	—	85–91
Woodbridge, Va.	Extended aeration	LAS	47	—	97.7	—	94.6
Kettle Moraine, Wisc. (48)	Extended aeration	ABS	34.5	8430	90.7	91.1	96.0
Kettle Moraine, Wisc.	Extended aeration	LAS	28.6	3937	96.5	93.6	96.0
Kettle Moraine, Wisc.	Extended aeration	LAS	34.0	3413	51.3	—	46.0
Kettle Moraine, Wisc.	Extended aeration	ABS	39.0	2578	68.3	38.3	84.0
Columbus, Ohio (49)	Extended aeration	ABS	12.8	1820	32.8	—	75.1
Columbus, Ohio	Extended aeration	LAS	13.4	1560	70.2	—	71.6
Columbus, Ohio	Extended aeration	LAS	24.3	2180	93.5	—	73.6
Columbus, Ohio	Extended aeration	ABS	18.5	1560	42.0	—	69.8
New Lisbon, New Jersey (50)	Trickling filter	LAS	—	—	75.5	—	79.5
New Lisbon, New Jersey	Trickling filter	LAS	—	—	80.0	—	83.0
Richmond, Calif. (51)	Oxidation pond (standard rate)	ABS	30 days	—	<40.0	—	—
Richmond, Calif.	Oxidation pond (standard rate)	LAS	30 days	—	93.1	—	—

[a] Reprinted from *The Journal of the American Oil Chemists' Society*, **45**, 433–436 (1968) by permission.

removal mechanism) of the LAS in the sewers leading to the plant. This will be discussed later in this chapter.

The Woodbridge, Virginia, study was conducted at a small mobile-home park which was served by an extended aeration sewage treatment plant. When operating under conditions which would assure high BOD removal, surfactant levels in the effluent averaged 0.5 mg/liter, which is well below the foaming threshold level. Even with extremely high LAS loads (LAS levels were 8–10 times those found in normal, domestic sewage) there was no indication that LAS was not being adequately biodegraded in the system (47).

The Kettle Moraine Boys School is a minimum security correctional institution operated by the State of Wisconsin. The study at this site was jointly sponsored by industry and by the Wisconsin Detergent Study Committee. The total contributory population to the school's extended aeration sewage treatment plant was approximately 420. During preliminary background studies, a high level (90.7%) of ABS removal was observed. It was assumed that this resulted from the high level of mixed liquor suspended solids maintained in the system (8430 mg/liter). At the end of the test cycle, another ABS run was conducted during which the mixed liquor suspended solids were held at a more realistic level of 2578 mg/liter. Under these conditions ABS removal leveled off at slightly less than 70%, a value more comparable to previously obtained data. Under normal operating conditions, LAS removal was of the order of 96%, with MBAS levels in the effluent found to be less than 0.1 mg/liter. During one phase of this study, the oxygen supply to the system was deliberately reduced to observe LAS performance under conditions of adverse plant operation. Even under these adverse conditions, LAS removal was essentially the same as BOD removal (51.3 and 46.0%, respectively) (48).

Another field study involving an extended aeration sewage treatment plant was conducted at an 88-home residential community near Columbus, Ohio. Since both detention time and mixed liquor suspended solids could be varied at this plant, an evaluation of the effect of these variables on LAS removal characteristics was possible. It was concluded that LAS removal efficiency was proportional to aeration tank contact time and that decreases in removal occurred with correspondingly shorter detention times. It was noted that degradability improved with increased mixed liquor suspended solids levels (49).

In order to obtain information on as many waste treatment systems as possible, a study was conducted on LAS performance in a trickling filter. Generally speaking, these systems are less efficient from the standpoint of overall soluble organic removal than are the more sophisticated activated sludge plants. Nevertheless, many of these plants are still being used, particularly in small communities, and probably will be for some time to come. The test site for this study was an institutional home operated by the State of New Jersey at New Lisbon. The total population was approximately 1500, and all wastes were treated in a conventional low rate trickling filter operating at a hydraulic loading of about 60 gal/sq ft per day with a 5% recycle rate and an organic loading of about 10 lb BOD/1000 cu ft per day. LAS was fed to the system in two feed ranges; 0–6 mg/liter and 6–15 mg/liter. As noted in Table I, LAS removals were not as high as those observed in the activated sludge type plants (75.5 and 80%, respectively). However, these findings are not surprising in light of the lower overall BOD removals expected and observed in this lower efficiency system (50).

Although not technically a field study in the sense of the projects previously described, work conducted by McGauhey and Klein at the University of California deserves mention, since it was carried out on a large enough scale to make the results comparable to actual field experience. Their work involved the evaluation of LAS and ABS removal in oxidation ponds. In a 4000-gal capacity, 30-day detention conventional stabilization pond, LAS removal was found to exceed 90%. Under the same conditions, ABS removal was less than 40% (51).

The National Sanitation Foundation recently (September 1966) published a report describing research on the basic performance characteristics of extended aeration package sewage treatment plants of varying design. This work resulted in the development of a standard performance evaluation method and performance criteria which would be applicable to all extended aeration treatment plants. The project was undertaken at the request of the Great Lakes Upper Mississippi Board of State Sanitary Engineers, while prime financial support came from the Federal Water Pollution Control Administration. The twelve full-scale treatment units were supplied by various equipment manufacturers and were installed at a specially prepared test site at the Ann Arbor sewage treatment plant.

A portion of the study included an evaluation of LAS removal under the various test conditions. The plants were operated under a variety of flow and climatic conditions (Table II), and ranged in capacity from 5000 to 16,000 gal/day. Influent MBAS values varied between 5.0 and 6.8 mg/liter, and effluent levels for the different test conditions fell between 0.8 and 1.2 mg/liter. In all cases, these effluent values were close to or below the incipient foaming level of 1.0 mg/liter. Since the

TABLE II

Summary of Results—National Sanitation Foundation Package Plant Study (52)

Number of plants under test	Flow condition	Climatic condition	% Removal	
			MBAS	BOD
10	Steady state [a], full load	Fall–winter	76	90
4	Steady state [a], full load	Spring–summer	84	92
4	Steady state [b], half load	Spring–summer	82	92
4	Subdivision [c], full load	Spring–summer	87	90
4	Subdivision [d], half load	Spring–summer	85	95
4	School [e], full load	Spring–summer	82	82
4	School [f], half load	Spring–summer	85	90

[a] Uniform flow over 24-hr period at full design capacity.

[b] Uniform flow over 24-hr period at one-half design capacity.

[c] Varied flow over 24-hr period to simulate typical subdivision flow pattern at full design capacity.

[d] Varied flow over 24-hr period to simulate typical subdivision flow pattern at one-half design capacity.

[e] Total daily flow applied at uniform rate over 8-hr period at full design capacity.

[f] Total daily flow applied at uniform rate over 8-hr period at one-half design capacity.

plants in the study were fed raw Ann Arbor sewage, there was no opportunity to control the makeup of detergent residues found in the waste. Also, since the methylene blue procedure was used, it is possible that materials other than detergent residues were routinely measured and reported as LAS. Acceptable MBAS removals were observed and gave additional support to the growing volume of data indicating effective LAS removal in waste treatment plants (52).

B. Studies on Individual Household Disposal Units

The situation regarding septic tanks, cesspools, and subsurface soil systems was not as well defined as was the case with sewage treatment plants. This was due in part to difficulties encountered in conducting studies of this type and from an improper understanding of the operation of such systems and their inherent limitations. A septic tank or cesspool is a sedimentation basin in which settled material undergoes anaerobic degradation in the tank. Thus these units operate essentially the same as primary sedimentation basins used in conventional sewage treatment plants. The principal difference lies in the fact that in the conventional primary unit, collected sludge is usually removed for external digestion or other further treatment, while in the septic tank or cesspool it undergoes partial digestion in the settling unit itself (53). The assumption that these primitive systems operate as anaerobic activated sludge plants has no basis in fact. Historically, they were designed to provide a modicum of treatment for isolated rural homesites, but their use, unfortunately, has spread to the point where many of the congested suburban housing areas surrounding some of the nation's major cities are served only by these facilities. The proliferation of their use has been decried by health and regulatory agencies at all levels of government, and steps have been taken in some areas to control further installation of such systems (54). Since they do exist, it was necessary to establish the parameters which affect ABS and LAS removal performance in such units.

Whatever biological action is exerted on the soluble organic fraction of the waste entering these systems takes place primarily in the soil mantle surrounding the cesspool or the percolation field of the septic tank–tile drain field system. This zone of biological activity may be quite narrow or can extend into the soil for some distance, depending on local conditions. In a well designed and operated system, the surrounding soil remains essentially in the aerobic state at all times. Aside from the small amount of soluble organic matter (including surfactant residues) that may be adsorbed on solids and remains in the tank, the bulk of the soluble substrate is removed in the soil. The soil system, therefore, takes on singular importance in considering the performance of these units. In many cases, the rules and regulations controlling the installation of individual household disposal systems are based primarily on tank design considerations, rather than on the more critical soil system.

In an attempt to improve what is obviously an unsatisfactory situation, the Federal Housing Administration, among others, has supported research designed to improve standards and performance of septic tank–percolation fields. Obviously similar work on cesspools would be unwarranted simply because these devices are even more primitive and ineffective than are septic tanks–percolation fields. McGauhey and Winneberger, under federal sponsorship, reported in 1964 (55) on the causes and prevention of failures in septic tank–percolation systems. Their research reemphasized the importance of percolation fields in the overall treatment scheme and proposed several construction and operational procedures that would enhance their performance. Some of the factors considered included the necessity for periodic resting of the bed, the effect of smearing of side walls during construction, the practical aspects of narrowing trench width, and the significance of distribution of filtration media.

While some effort has been made to improve these systems, actual field application of the findings has been limited.

Prior to the conversion to LAS, septic tank–percolation field studies were undertaken at the University of California in order to evaluate the performance of ABS, LAS, and an extremely biodegradable control surfactant in systems of this type (51).

Two test units were constructed, with each septic tank having a capacity of 33.6 gal, giving a nominal detention time of 2 days under test conditions. An above-ground percolation field was also built with a 1 ft wide and 1 ft deep trench.

The trench was filled to a level of 4 in. with crushed rock on which a 4-in. diameter bituminous fiber drain pipe was centered. The trenches were then backfilled with crushed rock and covered with 4 in. of sand.

The tanks were scaled down in width to 1/12 of actual size and represented a longitudinal section of a septic tank. The percolation field was scaled to 1/20 of the trench area usually recommended for the specific soil used in the study.

The septic tanks were loaded on an intermittent basis to simulate actual household flow patterns.

In addition to conventional chemical analytical procedures used to measure the removal of surfactant residues through the system, radioassay techniques were also employed in order to differentiate between overall removal and true biodegradation.

The results obtained in the study are summarized in Table III. It should be noted that the tests were conducted under both normal and failure or "ponded" conditions. In these latter studies, the percolation fields were allowed to be loaded to a point where liquid appeared on the soil surface. This was considered to be tantamount to failure in the system.

In all cases, sufficient surfactant was added to provide a 25 mg/liter initial concentration entering the septic tank. During the first or normal phase of the study, influent BOD, COD, and suspended solids averaged 149, 407, and 143 mg/liter, respectively. BOD analyses were not conducted during this portion of the study since it was assumed that adequate performance data could be accumulated using COD analyses alone.

Several basic conclusions can be drawn from this research. Of paramount importance was the fact that very high LAS removal (approximately 97%) could be achieved in both normally operated and flooded

TABLE III

ABS, LAS, and Control Surfactant Removal in Septic Tank–Percolation Field Systems (51)

		% Removal					
Unit	Condition	ABS	LAS	Alcohol sulfate control	BOD	COD	Suspended solids
Septic tank	Normal	16.3	9.6	60.8	13.1–23.6	20.6–25.2	51.8–58.4
Septic tank and percolation field	Normal	78.2	96.9	99.6	97.8–99.1	89.4–94.2	96.9–99.1
Septic tank	Ponded (failure)	9.8	9.0	63.4	—	21.4–28.1	35.0–45.7
Septic tank and percolation field	Ponded (failure)	54.5	97.1	99.7	—	82.5–95.4	90.8–96.3

septic tank–percolation fields. This overall removal was essentially the same as that observed with the extremely biodegradable control surfactant, alcohol sulfate. ABS removal was significantly lower under both operating conditions and fell as low as 54.5% in the ponded state. Little LAS removal was achieved in the septic tank itself. This finding was not surprising in view of the highly soluble nature of the material. Significantly, both BOD and COD removal were also very low in the septic tank. The higher per cent removal achieved with the control surfactant in the septic tank must be attributed to adsorption on solids in the tank. By radio tracer techniques, 87% of the LAS was actually degraded while only 54% of the alcohol sulfate was degraded under normal conditions. Under ponded conditions these values dropped to 66 and 25%, respectively. In all cases ABS performed considerably more poorly.

Under the general sponsorship of the New York Temporary State Commission on Water Resources Planning (56,57), extended studies were conducted on Long Island during the period 1962–1966. These studies were carried out in cooperation with several county, state, and federal agencies as well as The Soap and Detergent Association.

This work differed from that previously described in that it involved the actual field evaluation of several detergent products in various household waste disposal systems. Six homesites in Suffolk and Nassau counties were selected on the basis of their similarity to the systems commonly used on Long Island.

Both cesspool and septic tank–percolation field installations were included, and four different cleaning formulations were prepared for use by the householders at different times. These included ABS, LAS, and alcohol-sulfate-based detergents and a heavy duty laundry soap. The detergent products were not designed to simulate existing commercial products, but were formulated to provide a relatively uniform surfactant content. Accordingly, these products could not be considered typical from a performance standpoint. In all cases, the test products were given to the individual householders in specially coded identical packages.

The test sites themselves were not necessarily selected on the basis of expected high efficiency, in fact, in at least some cases, poor performance was anticipated. The primary characteristics of the individual sites were as follows:

Site I. Two cesspools (in series), both extending into the groundwater.

Site II. Septic tank followed by cesspool extending into the groundwater.

Site III. Cesspool with 17 ft of unsaturated soil between cesspool bottom and groundwater.

Site IV. Two cesspools with 5 ft of unsaturated soil between the cesspool bottom and groundwater.

Site V. Septic tank with tile drain field with 2 ft of unsaturated soil between the invert of the tile drain field and groundwater.

Site VI. Septic tank with tile drain field with 5 ft of unsaturated soil between the invert of the tile drain and groundwater.

At each site the treatment system served an individual home.

Samples were routinely collected both in the treatment units and in a series of test wells upstream and downstream and at various depths in the groundwater. In addition to methylene blue analyses, a variety of other standard chemical and microbiological analyses were conducted on collected samples which included COD, coliform count, chlorides, phosphates, nitrates, etc. In some cases infrared differential procedures were carried out to determine the actual ABS and LAS content of the samples.

From the beginning the project was beset by serious problems that were not totally resolved throughout the course of the study. These included difficulty in establishing the true course and velocity of groundwater flow, even though practically all classical tracer techniques were employed (including hexavalent chrome, chloride and radioisotopes). Also, since all of these sites had been in use for some time, the presence of fairly high levels of ABS in the soil and in the groundwater tended to mask the effect of changing from one test product to another and made differentiation between various surfactants difficult. In one case, where ABS-based detergents had not been used for 9 months, the predominant surfactant found in the groundwater after this period was still ABS!

No final report has as yet been issued on this study but certain preliminary conclusions have been drawn by the author and others. At site I, little or no removal of either ABS or LAS was observed. This was not surprising in view of the total absence of any opportunity for bacterial action in the soil system since effluent from the cesspool went directly to the groundwater.

Conditions at site II were not much different, although a reduction of about 20% in both ABS and LAS was noted in a test well 40 ft from the cesspool. This may at least be partially attributed to dilution effects in the groundwater.

The most extensive studies in the project were carried out at site III. A large sampling shaft was constructed adjacent to the cesspool to permit the collection of samples immediately below the cesspool and at other points above the groundwater table. In addition to routine chemical methods, radioisotopes were used to follow the course and fate of surfactant residues leaving the cesspool, and tritium was also used to establish groundwater velocity and direction of flow. By previous methylene blue analyses it appeared that somewhat more than 40% of the ABS entering the cesspool was removed by the time it reached a sampling well in the groundwater 5 ft from the center of the cesspool.

Some interesting results were observed during the time ^{35}S tagged LAS was in use. For example, between 84 and 90% of the LAS (measured as total sulfur) was removed by the time it reached downstream sampling wells. Practically all of this reduction took place in the first 18 in. of travel in the unsaturated soil. Actual degradation (as opposed to removal) of the LAS ranged as high as 67% which very closely approached the University of California findings for a septic tank–percolation field. Infrared differential techniques indicated that between 85 and 92% of the surfactant found in the downstream wells was ABS, even though LAS-based formulations had been used for over a year.

Due to sampling problems at site IV, no substantive research was conducted at this location.

At sites V and VI (the septic tank sites) overall LAS removals were approximately 33 and 38%, respectively. However, removal of the extremely biodegradable control surfactant, alcohol sulfate, was not significantly better (40 and 58%, respectively). These relatively low removals can probably be attributed to two factors—the presence of significant residual amounts of ABS and the relative shallowness of the groundwater table which limited the development of a viable aerobic soil microbiological system.

Some questions have been raised in the literature regarding the effectiveness of the conversion to LAS in terms of protecting groundwater resources from contamination by discharges of individual household waste disposal systems (58). Based on the in-depth research described here and elsewhere, several conclusions can be reached.

1. LAS can be effectively removed in adequately designed and operated septic tank–percolation fields.

2. This removal is almost entirely a function of the effectiveness of the active aerobic microbial zone surrounding the point of liquid discharge.

3. Virtually no soluble organic (including LAS residues) removal will occur *in* a septic tank (or cesspool) since these devices serve only for the removal of settleable particulate matter.

4. Where LAS removal is low due to inherent design or operational deficiencies in the treatment system, a serious hazard to the groundwater may exist due to other microbiological and organic contamination.

5. Uniformly high treatment efficiency in systems of this type cannot be routinely expected due to the wide variations in design and construction methods used and in the relatively poor control exercised over their operation.

Thus, while LAS will perform satisfactorily in septic tank–percolation fields, the continued widespread use of these units can neither be encouraged nor condoned as acceptable means of domestic waste treatment.

VIII. POST CONVERSION EXPERIENCE

Although ample laboratory and field test data were available regarding LAS performance in treatment systems, in the soil, and in natural waters, it was not until full conversion had been accomplished on a national basis that it was possible to fully assess the significance of the conversion to LAS.

Three years have now elapsed since ABS was last incorporated into U.S. detergent formulations. During and before the conversion, a variety of survey type programs were undertaken to evaluate LAS performance in use. The most extensive and significant of these studies were undertaken at major sewage treatment plants where laboratory facilities and manpower were readily available.

Much data of this type have been compiled by the industry, although the largest share was collected by sewage treatment plant operators, university researchers, and pollution control agencies. The plants for which the most extensive data are available are those at Columbus, Ohio; Los Angeles, California (Hyperion); Milwaukee, Wisconsin; the Los Angeles County, California plants at Whittier Narrows, Pomona, and Saugus; and the Livermore, California plant. The results obtained at these locations are best summarized graphically and are shown in Figures 6–12. With the exception of the Livermore plant, all those discussed employed the activated sludge process or variation thereof.

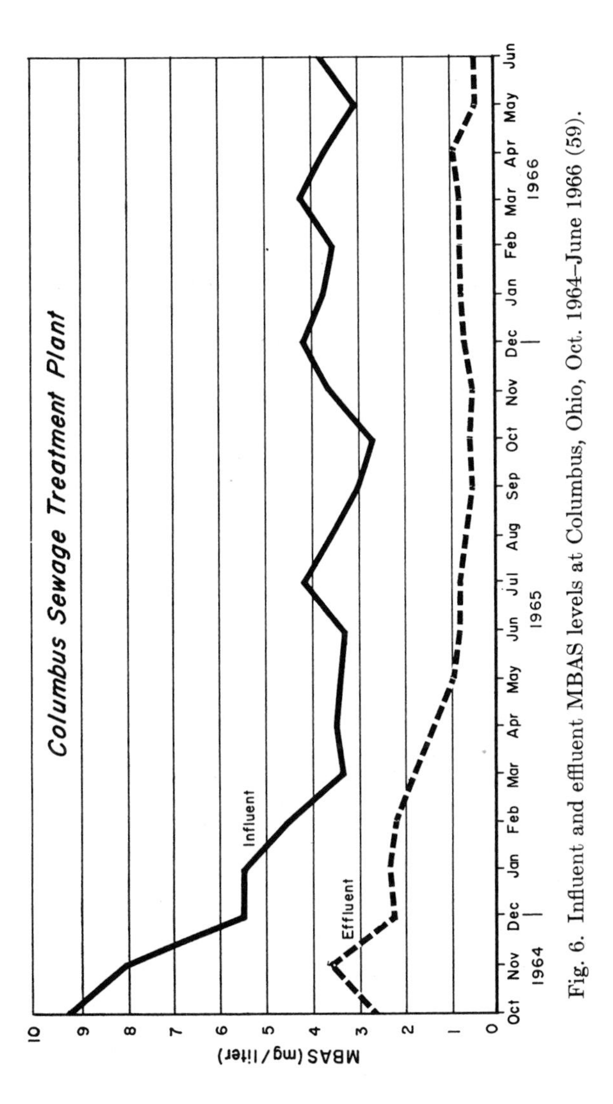

Fig. 6. Influent and effluent MBAS levels at Columbus, Ohio, Oct. 1964–June 1966 (59).

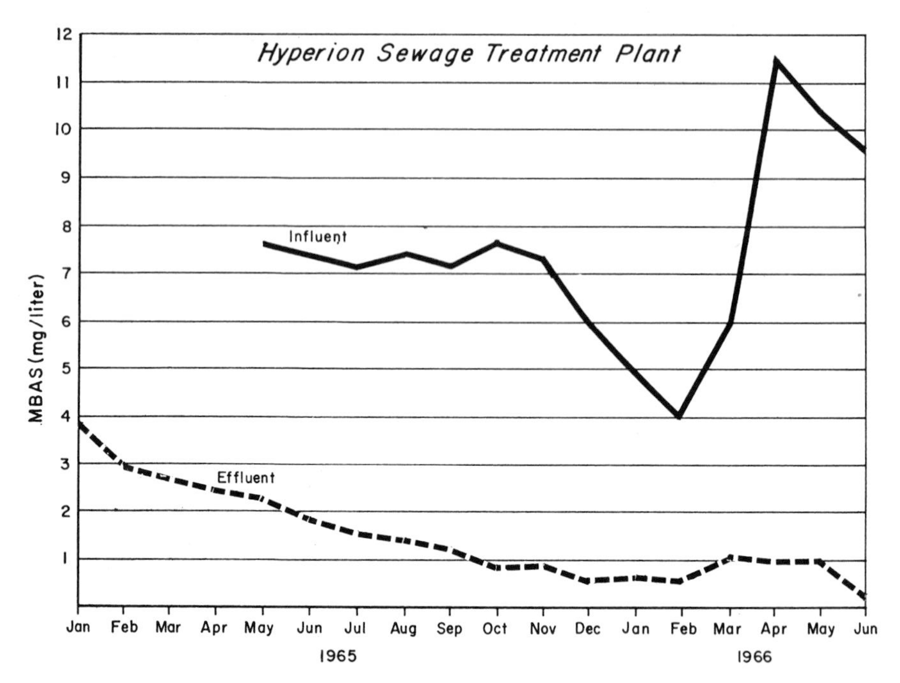

Fig. 7. Influent and effluent MBAS levels at the Hyperion (Los Angeles, Calif.) sewage treatment plant Jan. 1965–July 1966 (63).

Sample collection and analyses began at one Columbus, Ohio plant in October 1964, well before the conversion to LAS was completed (Fig. 6). This plant handles a flow of approximately 45 million gallons per day and has an aeration detention time of approximately 6 hr. Effluent MBAS values at the beginning of the monitoring program were about 3 mg/liter and fell steadily until they reached a level well below 1 mg/liter in August of 1965. They remained at approximately 0.5 mg/liter until the conclusion of the study in early 1966 (59).

In terms of per cent removal, MBAS removal increased from between 60 and 70% at the beginning of the study to about 90% at the end. BOD removals were in the 90% range throughout.

The behavior of influent MBAS levels throughout the course of the study deserves special mention. From an initial high concentration of over 9 mg/liter in October 1964, influent values fell steadily until early 1965 when they stabilized at between 3 and 4 mg/liter. They stayed at this level during the remainder of this study. At first this was attributed to dilution. However, when BOD–MBAS ratios were examined, a steady increase in the ratio was noted, indicating that the effect

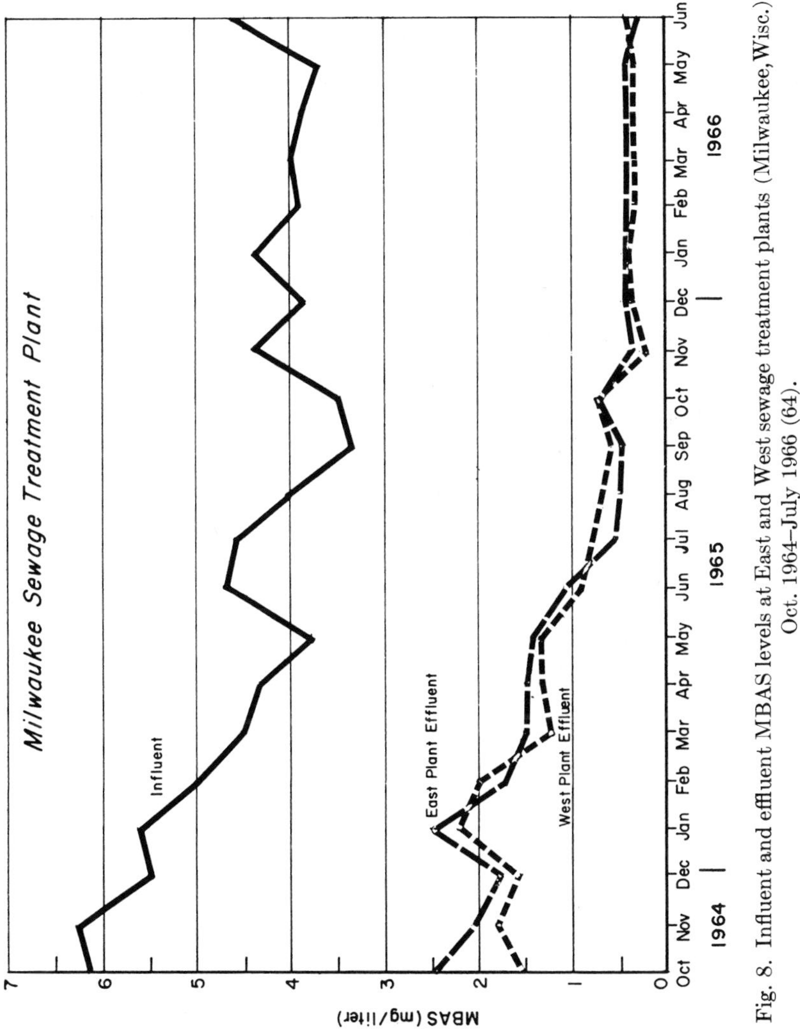

Fig. 8. Influent and effluent MBAS levels at East and West sewage treatment plants (Milwaukee, Wisc.) Oct. 1964–July 1966 (64).

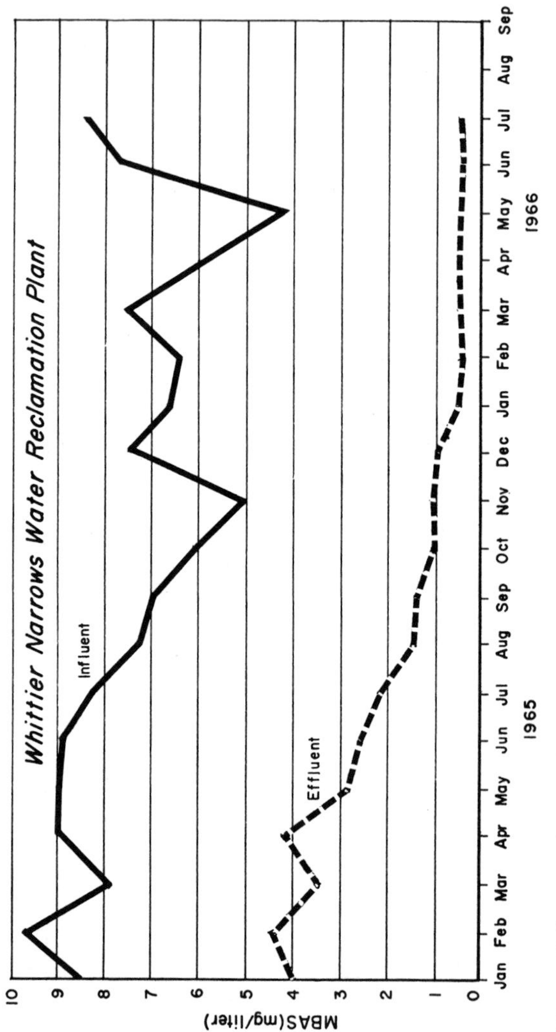

Fig. 9. Influent and effluent MBAS levels at the Whittier Narrows water reclamation plant (Los Angeles County, Calif.) Jan. 1965–July 1966 (65).

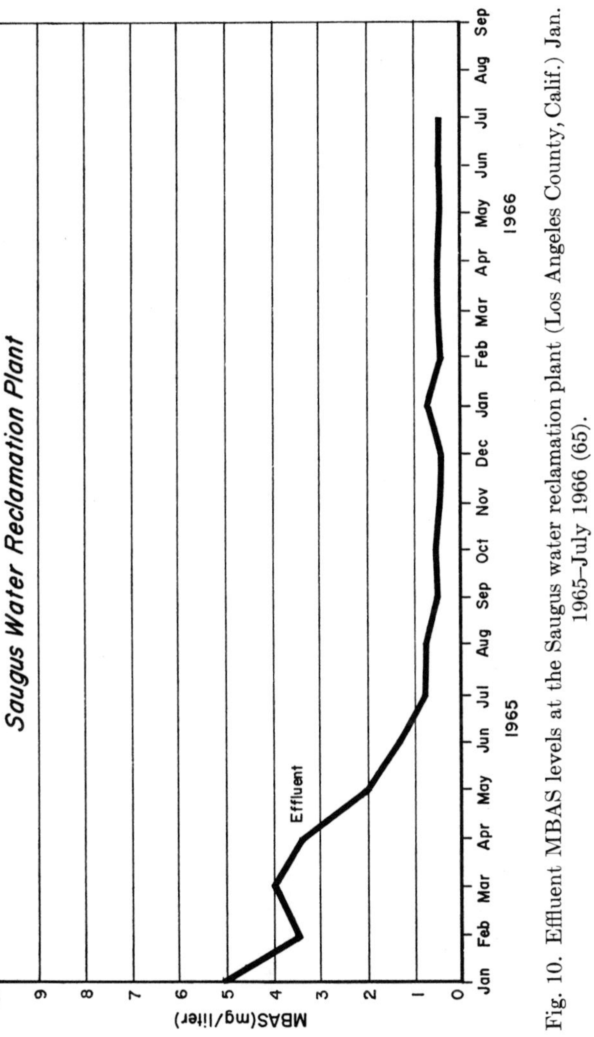

Fig. 10. Effluent MBAS levels at the Saugus water reclamation plant (Los Angeles County, Calif.) Jan. 1965–July 1966 (65).

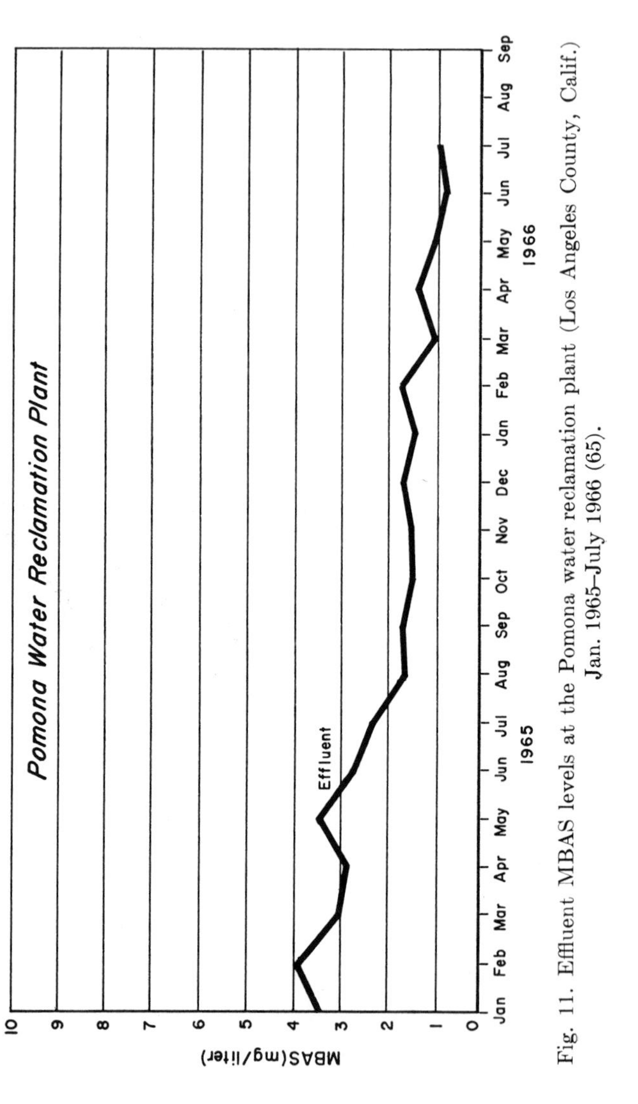

Fig. 11. Effluent MBAS levels at the Pomona water reclamation plant (Los Angeles County, Calif.) Jan. 1965–July 1966 (65).

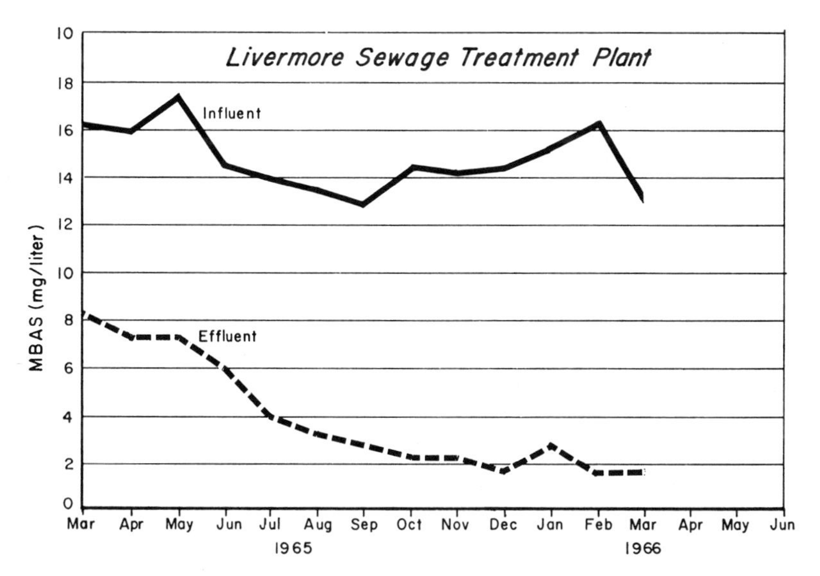

Fig. 12. Influent and effluent MBAS levels at the Livermore sewage treatment plant (Calif.) March 1966 (66).

was not due to dilution alone. As far as could be ascertained, no other external factors were involved. It appeared, at least circumstantially, that some removal or degradation of the LAS was occurring in the sewers prior to reaching the treatment plant (60). The same effect was also seen, to a lesser extent, at Milwaukee (Fig. 8).

This had also been observed in both Germany and Great Britain. Husmann (61) reported that LAS removal in sewers leading to treatment plants in Germany was as high as 24%. A similar reduction had also been noted at the Luton treatment plant in Great Britain (62). Further studies are reportedly underway in Europe to obtain additional information on this subject.

The Hyperion sewage treatment plant, which serves a portion of the city of Los Angeles, handles a waste flow of approximately 100 million gallons per day. The plant generally achieves a BOD removal in the order of 95%. As can be seen in Figure 7, effluent MBAS values dropped from about 4 mg/liter in January 1965 to below 1 mg/liter by October of that year. Percentage removal of MBAS increased from about 70 to slightly over 90% (63).

Data are available for two treatment plants serving Milwaukee, Wisconsin. These plants handle a combined flow of from 150 to 200 million gallons per day. They are unique in that incoming wastewater does not undergo primary sedimentation, but is passed through fine screens instead. Both plants (east and west) demonstrated equally effective LAS removal. Effluents from the east and west plants dropped from about 2.5 and 1.5 mg/liter MBAS respectively, to under 0.5 mg/liter during the period of the study (October 1965–July 1966), ultimately achieving an overall MBAS removal of 90% or better over the last 6 months of the study. BOD removal during this period was consistently in the 95% range. As was previously discussed there was a drop in the influent MBAS values (2–3 mg/liter) observed also at this site (64).

Results from the Whittier Narrows, Pomona, and Saugus Water Reclamation plants operated by the County Sanitation Districts of Los Angeles County are summarized in Figures 9–11. In all three plants, the wastewater, which is primarily domestic in origin, is similar in character and strength. All three plants employ the step-aeration modification of the conventional activated sludge process. The Whittier Narrows and Pomona plants treat incoming wastes in primary sedimentation units while at the Saugus plant raw waste is fed directly to the aeration tanks. Whittier Narrows handled a flow of 15 million gallons per day during the monitoring period while Pomona and Saugus handled a flow of 4–5 million gallons per day and 1 million gallons per day, respectively, during the same period.

At Whittier Narrows there was a steady reduction in effluent MBAS levels from about 4 mg/liter in January 1965 to approximately 0.5 mg/liter in January 1966. Effluent MBAS values remained essentially constant thereafter. During the study period MBAS removal increased from 50–60 to better than 90%. BOD removal approximated 95% throughout the study period (65).

At the Saugus plant, effluent MBAS levels were about 5 mg/liter in January 1965. The effluent MBAS concentration then decreased to about 0.5 mg/liter and stayed at this level for the remainder of the study (65).

At Pomona, effluent MBAS values decreased from about 3.5 to 1.0 mg/liter over the monitoring period (65).

Two waste treatment plants were monitored by staff members of the University of California's Sanitary Engineering Research Laboratory (66). One was the roughing trickling filter–oxidation pond system serving the city of Livermore, California, while the other was a pilot-activated sludge plant operated by the university at Richmond, California.

The Livermore installation was of particular interest since foaming problems had occurred in the past at the treatment plant's receiving stream (Alameda Creek).

This plant handled a hydraulic loading of some 2.5 million gallons per day throughout the monitoring period. The treatment system itself consists of solids grinding, preaeration, and primary sedimentation (1.5 hr). This is followed by treatment in a 110-ft diameter, 4.25-ft deep trickling filter loaded at a rate of 2.84 lb of BOD per cubic yard per day. This in turn is followed by a secondary sedimentation unit with a nominal 2-hr detention time. After chlorination the effluent is discharged to an 18.5-acre oxidation lagoon having a theoretical detention time of 15 days.

When sampling began in March 1965, pond effluent MBAS values approached 8 mg/liter. By October 1965, MBAS levels dropped to 2 mg/liter and have remained at that value or lower ever since (Fig. 12).

Findings at the pilot-activated sludge plant were similar to those at the previously described full scale installations. At this plant (which has an average detention time of 6.5 hr) effluent MBAS values dropped from about 3 to 0.5 mg/liter during the monitoring period. The per cent MBAS removal during this same time increased from 50 to over 90%.

Little definitive data are available on LAS performance in trickling filters or biofiltration plants. This is probably due to the fact that these plants generally serve smaller communities, which often do not have the manpower or facilities needed to develop the extensive data required for meaningful monitoring programs. This, coupled with the fact that great variations exist in filter design and operation (e.g., bed depth, recirculation rate, etc.), makes effective performance comparisons particularly difficult even when data are available. Klein and McGauhey concluded, however, that, based on their observations, MBAS removal efficiency in high rate trickling filters had doubled since the conversion to LAS, but that it has not reached the overall efficiency anticipated in laboratory studies. This, they felt, could have resulted from the peculiar design of the filters investigated or from the slower acclimatization of the filters to LAS. This latter phenomenon had been noted by them in earlier pilot studies (66). Barth and Ettinger (66a) reported on an 18-month study at five sewage treatment plants in the mid-west United States. Although the authors note that these plants cannot be considered as typical waste-water treatment plants, some of their findings are of interest. Four of the plants were of the activated sludge type, while the fifth (Rockford, Ill.) was a high rate trickling filter. In four cases the incoming raw waters were characterized by low MBAS levels, suggesting that either dilution was a factor or that some degradation occurred in the sewers leading to the plants. MBAS removal in the Rockford trickling filter was generally poor; however, overall plant efficiency, as measured by COD removal, was also poor, ranging between 40 and 82% during the study period. In the case of the activated sludge plants, MBAS removal was not as high as expected, but a general correlation was developed by the authors, indicating that the plants that removed COD most efficiently also showed greater efficiencies for MBAS removal.

Information on changes in MBAS levels in rivers and streams is also somewhat limited. This may result from the fact that even when ABS was in use, MBAS levels in streams were extremely low, often less than 0.1 mg/liter, and it is particularly difficult to measure significant changes at these concentrations by the commonly used methylene blue procedure.

However, there is solid evidence that changes are occurring and that MBAS levels have dropped in rivers and streams.

Klein and McGauhey (66) have reported on their studies of Alameda Creek in California. This work began in December 1965 after the conversion to LAS was complete and sufficient time had elapsed to minimize the effect of any residual amounts of ABS.

Alameda Creek was selected for study because foaming problems had occurred on this stream prior to the conversion to LAS. Samples were collected during periods of dry and rainy weather. Combining all data, MBAS concentrations varied between 0.1 and 0.26 mg/liter, levels well below the incipient foaming level. They also reported that there was virtually no foaming tendency in the waters of the creek which was not the case prior to the conversion to LAS.

Another recently published report describes studies conducted on the Illinois River at Peoria by the Illinois State Water Survey (67).

This research is of particular interest since at one time the Illinois River was described as "The largest river in the country that currently contains more than 0.5 ppm ABS" (68).

The river is formed at the juncture of the Kankakee and Des Plaines Rivers at a point some 54 miles downstream from Lake Michigan, and receives the flow of the Chicago Sanitary and Ship Canal. This flow consists largely of treated sewage plant effluent, industrial effluent, and storm runoff from the greater Chicago area. The average time of river flow from Chicago to Peoria is 12.5 days, and the total distance traveled is approximately 154 miles. MBAS monitoring began at Peoria in September 1959. During the period from 1959 through 1965 the average MBAS concentration in the river was 0.56 mg/liter, with a range in monthly average of from 0.42 to 0.87 mg/liter. However, in the 12-month sampling period after the conversion to LAS was completed, (July 1965–June 1966) these values dropped to an average concentration of 0.22 mg/liter with a 0.30 mg/liter maximum value and a 0.13 mg/liter minimum concentration.

In order to minimize the effect of variations in flow, the actual weight of MBAS transported by the river was also calculated. During the period 1959–1965, the average yearly load varied between 13 and 20.6 tons/day, while the comparable value for the 1965–1966 "postconversion" period was 9.0 tons/day. Both in concentration and in total pounds of MBAS, a marked reduction was noted in the Illinois Waterway at Peoria.

Experimentation was also undertaken at this site to determine whether the greater biodegradability of LAS would result in a higher demand on the oyxgen resource of the river. There was no indication of significant differences in dissolved oxygen levels in the river during the "pre" and "post" conversion periods.

G. W. Lawton has also observed reduced MBAS levels in Wisconsin waters since the conversion to LAS (68a). Surface water samples were routinely collected and analyzed, over a three-year period beginning in 1963, at 37 locations in Wisconsin. Generally, the sampling stations were located at the mouths of rivers or at points along the state's border. Significantly reduced MBAS levels were observed in almost all cases following the detergent industry's conversion to LAS in mid-1965. As an example, the Root River, which originates near the southwestern edge of Milwaukee, had an average MBAS content of 1.04 mg/liter during 1963, but by 1966 this concentration had gradually dropped to 0.06 mg/liter or less. A similar pattern was observed in most of the other rivers sampled.

Lawton also noted a marked reduction in the number of well waters containing detergent residues following the conversion to LAS. Only 0.75% of all private water samples tested in 1966 contained more than 0.25 mg/liter MBAS. The author also concluded that treatment of waste waters by conventional means significantly reduced the MBAS content and that this removal approximated BOD removal.

While no comprehensive analysis has been made of the very extensive data collected by the Federal Water Pollution Control Administration on MBAS levels in rivers and streams throughout the country, a general review of this information indicates that levels have dropped. When coupled with the fact that reported foam incidents in natural waters have decreased dramatically (based on reports compiled by The Soap and Detergent Association), this would indicate that MBAS levels have reached a point where they are no longer of practical concern to pollution abatement authorities. The exception, of course, would be those areas where raw or partially treated sewage is introduced to a water course.

A similar conclusion can be drawn relative to the current situation in ground waters. Few reports of foaming well waters find their way into the technical literature or the public press at this writing. This is in marked contrast to the situation that existed as recently as three years ago.

By every visual, scientific, and socioeconomic yardstick, the conversion to LAS has resulted in a marked reduction of the presence of methylene-blue-active substances in sewage treatment plant effluents and in natural ground and surface waters.

IX. EUROPEAN SITUATION

Certain parallels in the American experience can be drawn with developments in some European countries, principally Great Britain and West Germany.

In West Germany, conversion to biodegradable detergents (containing surfactants that meet an 80% degradability standard by the approved test method) was made mandatory by a federal regulation that went into effect October 1964. In Great Britain, as in the United States, the conversion to biodegradable surfactants was voluntary and took effect during the fall of 1965.

Results so far from both countries have been excellent. Husmann (61) reported that in 1955 the MBAS level in effluents of sewage treatment plants of the Emschergenossenschaft and of the Lippengenossenschaft averaged 1.1 mg/liter. In the period 1962–1964, when "hard" surfactants were in common use, this level had risen to 5.4 mg/liter. However, by 1966 the level had dropped back to 1.2 mg/liter. He also reported that the efficiency of MBAS removal in sewage treatment plants during the comparable periods had risen from 25 to 75%.

Husmann further indicated that the overall removal (in sewers and sewage treatment plants) was even higher, due to the degradation of LAS in sewers leading to the treatment plants.

As far as levels in streams are concerned, the total weight of detergents in the Ruhr River has reportedly been reduced from 3060 to 955 lb/day after the conversion to LAS.

Husmann noted, however, that only by the construction of additional biological treatment plants would the full benefit of the conversion be realized in Germany.

In Great Britain, a similar pattern has developed. Levels of MBAS in natural waters have diminished, in some cases dramatically. As an example, in the River Thames at Laleham, where MBAS levels had ranged from 0.22 to 0.42 mg/liter during the period 1954–1965, they dropped to 0.06 mg/liter in the first 6 months of 1966. A similar pattern was observed on the River Lee at New Gauge where levels had varied between 0.21 and 0.53 mg/liter during the 1954–1965 period and diminished to 0.07 mg/liter in 1966 (69).

At Luton, England, where an extensive preconversion field study had been carried out, recent sampling indicated a marked reduction in MBAS levels in the effluent of the sewage treatment plant. As was previously noted, evidence of LAS breakdown in the sewers leading to the plant was also observed (62).

As in the United States, one could summarize the European experience by stating that where adequate sewage treatment facilities exist, LAS will meet the esthetic objectives of water quality without recourse to special treatment.

References

1. F. J. Coughlin, J. B. Green, H. V. Moss and P. J. Weaver, *J. Am. Water Works Assoc.*, **55,** 372 (1963).
2. *Estimated Soap and Synthetic Detergent Sales,* The Soap and Detergent Association, New York, 1967.
3. *Synthetic Detergents in Perspective,* The Soap and Detergent Association, New York, 1962, p. 9.
4. A. H. Lewis, U.S. Pat. 2, 477, 382 (1949), California Research Corp.
5. D. Justice and V. Lamberti, *Chem. Eng. Progr.*, **60,** 36 (1964).
6. P. W. Reed, "1966 Federal Water Pollution Control Legislation," paper presented at 13th Annual Wastes Engineering Conference, University of Minnesota, December 9, 1966.
7. K. L. Kollar and A. F. Volonte, *Distribution of Water and Wastewater Utilities Capital Expenditures in 1964,* U.S. Dept. of Commerce, Business and Defense Administration, December, 1966.
8. A. C. Glass, *Statistical Summary of 1962 Inventory Municipal Waste Facilities in the United States,* U.S. Dept. of Health, Education, and Welfare, Public Health Service Publication No. 1165, 1964.
9. L. W. Weinberger, D. G. Stephan, and F. M. Middleton, *Solving Our Water Problems—Water Renovation and Reuse,* Federal Water Pollution Control Administration, U.S. Dept. of the Interior, August 1966.
10. Task Group Report, *J. Am. Water Works Assoc.*, **46,** 770 (1954).

10a. J. H. Jones, *J. Ass. Offic. Agr. Chem.*, **28,** 398 (1945).

11. *Detergent Report, A Study of Detergents in California*, prepared for The California State Legislature, 1965, p. 18.
12. J. M. Flynn, A. Andreoli, and A. A. Guerrera, *J. Am. Water Works Assoc.*, **50,** 1551 (1958).
13. *Standard Methods for the Examination of Water and Wastewater*, 12th ed., Am. Public Health Association, New York, 1965, p. 296.
14. *Standard Methods for the Examination of Water and Wastewater*, 12th ed., Am. Public Health Assoc., New York, 1965, p. 299.
15. "Public Health Service Drinking Water Standards, 1962," *U.S. Public Health Serv. Publ.*, **956,** 22 (1962).
16. R. L. Woodward, H. D. Stokinger, and D. J. Birmingham, *Archiv. Environ. Health*, **8,** 584 (1964).
17. AASGP Committee Report, *J. Am. Water Works Assoc.*, **53,** 297 (1961).
18. F. J. Coughlin, J. B. Green, H. V. Moss, and P. J. Weaver, *J. Am. Water Works Assoc.*, **55,** 377 (1963).
19. D. F. Metzler, R. L. Culp, H. A. Stoltenberg, R. L. Woodward, G. Walton, S. L. Chang, N. A. Clarke, C. M. Palmer, and F. M. Middleton, *J. Am. Water Works Assoc.*, **50,** 1021 (1958).
20. F. J. Coughlin, J. B. Green, H. V. Moss, and P. J. Weaver, *J. Am. Water Works Assoc.*, **55,** 398 (1963).
21. J. M. Cohen, *Soap Chem. Specialties*, **35,** 9 (1959).
22. J. M. Cohen, *J. Am. Water Works Assoc.*, **55,** 587 (1963).
23. F. J. Coughlin, J. B. Green, H. V. Moss, and P. J. Weaver, *J. Am. Water Works Assoc.*, **55,** 382 (1963).
24. F. J. Coughlin, J. B. Green, H. V. Moss, and P. J. Weaver, *J. Am. Water Works Assoc.*, **55,** 385 (1963).
25. P. H. McGauhey and S. A. Klein, *Public Works*, **92,** 101 (1961).
26. F. J. Coughlin, J. B. Green, H. V. Moss, and P. J. Weaver, *J. Am. Water Works Assoc.*, **55,** 398 (1963).
27. D. Justice and V. Lamberti, *Chem. Eng. Progr.*, **60,** 36 (1964).
28. C. K. Hersh, *Molecular Sieves*, Reinhold, New York, 1961, p. 2.
29. C. K. Hersh, *Molecular Sieves*, Reinhold, New York, 1961, p. 79.
30. W. J. Zimmerschied et al., *Ind. Eng. Chem.*, **42,** 1300 (1950).
31. C. K. Hersh, *Molecular Sieves*, Reinhold, New York, 1961, p. 107.
32. J. Rubinfeld, E. M. Emery, and H. D. Cross III, *Straight-Chain Alkylbenzenes: Structure and Performance Property Relations*, paper presented at the American Oil Chemists Society meeting, New Orleans, April 1964.
33. *Chem. Eng. News*, **41,** 54 (June 17, 1963).
34. D. Justice and V. Lamberti, *Chem. Eng. Progr.*, **60,** 38 (1964).
35. D. Justice and V. Lamberti, *Chem. Eng. Progr.*, **60,** 39 (1964).
36. R. D. Swisher, *J. Am. Oil Chemists' Soc.*, **40,** 650 (1963).
37. R. D. Swisher, *Soap and Chem. Specialties*, **39,** 58 (1963).
38. P. J. Weaver, *Soap and Chem. Specialties*, **41,** 45 (1965).
39. Subcommittee on Biodegradation Test Methods, *J. Am. Oil Chemists' Soc.*, **42,** 989 (1965).
40. Subcommittee on Biodegradation Test Methods, *J. Am. Oil Chemists' Soc.*, **42,** 990 (1965).
41. *Regulation Concerning the Degradation of Detergents in Washing and Cleansing*

Agents, Publication 253/62, Federal Council of the German Federal Republic, 1962.

42. Subcommittee on Biodegradation Test Methods, *J. Am. Oil Chemists' Soc.*, **42,** 988 (1965).
43. C. E. Renn, *LAS Degradation in Trickling Filters*, paper presented at Annual Convention of The Soap and Detergent Association, New York, January 1965.
44. *Supplement to the Eighth Progress Report of the Standing Technical Committee on Synthetic Detergents*, Her Majesty's Stationery Office, London, 1966.

44a. S. J. Patterson, C. C. Scott, and K. B. E. Tucker, *J. Am. Oil. Chem. Assoc.*, **44,** 407 (1967).

44b. S. J. Patterson, C. C. Scott, and K. B. E. Tucker, *J. Am. Oil Chem. Assoc.*, **45,** 528 (1968).

45. R. L. Bunch and C. W. Chambers, *J. Water Pollution Control Federation*, **39,** 181 (1967).
46. J. W. Knapp and J. M. Morgan, Jr., *Proceedings of the Twentieth Industrial Waste Conference, May 4, 5, and 6, 1965*, Vol. 49, No. 4, Purdue University, Lafayette, Ind., 1965, p. 737.
47. C. E. Renn, W. A. Kline, and G. Orgel, *J. Water Pollution Control Federation*, **36,** 878 (1964).
48. P. V. Knopp, L. J. Uhren, M. S. Nichols, and G. A. Rohlich, *Proceedings of the Twentieth Industrial Waste Conference, May 4, 5, and 6, 1965*, Vol. 49, No. 4, Purdue University, Lafayette, Indiana, 1965, p. 745.
49. G. P. Hanna, Jr., W. D. Sheets, P. J. Weaver, and R. M. Gerhold, *Proceedings of the Twentieth Industrial Waste Conference, May 4, 5, and 6, 1965*, Vol. 49, No. 4, Purdue University, Lafayette, Indiana, 1965, p. 725.
50. G. Kumke and C. E. Renn, *J. Am. Oil Chem. Soc.*, **43,** 92 (1966).
51. S. A. Klein and P. H. McGauhey, *The Fate of Detergents in Septic Tank Systems and Oxidation Ponds*, University of California, Berkeley, California Serl. Report No. 64-1, January 1, 1964.
52. National Sanitation Foundation, *Package Plant Criteria Development Part I: Extended Aeration*, Federal Water Pollution Control Administration Demonstration Grant Project WPD-74, September 1966.
53. G. Fair and J. Geyer, *Water Supply and Waste Water Disposal*, 1st ed., Wiley, New York, 1954, p. 899.
54. L. L. Terry, statement in *Cincinnati Enquirer*, November 3, 1963.
55. P. H. McGauhey and J. H. Winneberger, *Causes and Prevention of Failures of Septic Tank–Percolation Systems*, U.S. Govt. Printing Office, Washington, D.C., 1964.
56. *Six Year Progress Report of the Temporary State Commission on Water Resources Planning*, Legislative Document No. 27, State of New York, Albany, New York, 1965, p. 80.
57. *A Short Term Report of the Temporary State Commission on Water Resources Planning*, Legislative Document No. 9, State of New York, Albany, New York, 1966, p. 113.
58. U.S. Dept. of Interior (Geological Survey) press release, September 3, 1965.
59. *Chem. Eng. News*, **45,** 20 (February 27, 1967).
60. G. P. Hanna, Jr., private communication.
61. W. Husmann, *Third International Conference on Water Pollution Research*,

Munich, Germany, 1966 (*Advances in Water Pollution Control*, Vol. I), Water Pollution Control Federation, Washington, 1967, p. 364.

62. Ministry of Technology, Water Pollution Research Laboratory, *Removal of Detergents at Luton Sewage Treatment Works, December 1965–May 1966*, W.P.R. Report No. 1134, Stevenage, Herts., Great Britain, 1966.
63. Private Communication, City of Los Angeles, California, Bureau of Sanitation, Dept. of Public Works, N. B. Hume, Director.
64. Private Communication, City of Milwaukee, Wisconsin, Sewerage Commission, Ray D. Leary, Chief Engineer and General Manager.
65. Private Communication, County Sanitation Districts of Los Angeles County, California, J. D. Parkhurst, General Manager and Chief Engineer.
66. S. A. Klein and P. H. McGauhey, *Effects of LAS on the Quality of Waste Water Effluents*, University of California, Berkeley, California, SERL Report No. 66-5, September 1966.

66a. E. F. Barth and M. B. Ettinger, *J. Water Pollution Control Fed.*, **39**, 815 (1967).

67. W. T. Sullivan and R. L. Evans, *Environ. Sci. Technol.*, **2**, 194 (1968).
68. R. L. Woodward, H. D. Stokinger, and D. J. Birmingham, *Archiv. Environ. Health*, **8**, 585 (1964).

68a. G. W. Lawton, *J. Am. Water Works Assoc.*, **59**, 1327 (1967).

69. E. Windle Taylor, London Metropolitan Water Board, private communication.

Aeroallergens and Public Health*

WILLIAM R. SOLOMON, M.D.
*University of Michigan Medical School,
Ann Arbor, Michigan*

I. Introduction 198
II. Allergic Mechanisms 198
A. A Unified Concept of Allergy 198
B. Allergens 199
C. Types of Allergic Response 199
1. Antibody-Dependent Mechanisms 199
2. Cell-Dependent Mechanisms 201
III. Allergic Populations at Risk 202
A. The Atopic Group 202
1. Definition 202
2. Atopy and Skin-Sensitizing Antibody 202
3. The Familial Aspect of Atopy 204
4. The Prevalence of Atopic Allergy 205
5. The Sequence of Atopic Manifestations 206
6. Allergenic Exposure and the Prevalence of Symptoms 207
B. Allergy in Nonatopic Populations 209
IV. The Nature and Sources of Aeroallergens 210
A. Aeroallergens as Particulates 210
B. Sources of Aeroallergens 210
C. The Transport of Aeroallergens 212
1. Transport Processes in Closed Spaces 212
2. Transport Processes Out-of-Doors 212
V. Aeroallergens in Atopic Disease 216
A. House Dust 217
B. Pollens 218
C. Fungi 220
D. Animal Emanations 221
E. Miscellaneous Allergens 221
VI. Quantitating Airborne Allergens 223
A. Air Sampling Methods 223
B. Pollinosis and Pollen Counts 226
VII. Aeroallergens in Nonatopic Diseases 227
A. The "Farmer's Lung" Group 227
B. Allergic Contact Dermatitis 229
C. Diseases of Unknown Cause 229
References 230

* This work was supported by Research Grant AP-00001 from the National Center for Air Pollution Control, U. S. Public Health Service.

I. INTRODUCTION

As a group, aeroallergens lack those toxic, invasive, and oncogenic properties that are most familiar to students of atmospheric pollution. Yet, the morbid effects of these agents have been observed for centuries (1) and their impact upon human health now is established with unique clarity.

Any attempt to define aeroallergens requires the recognition, within the total exposed population, of specific groups of susceptible persons. Among the members of such groups, however, a broad spectrum of reactivity is common, and the responses of sensitive individuals may vary with time. Furthermore, several distinct mechanisms of tissue injury underlie phenomena that are properly regarded as "allergic." These considerations hold little promise for broad generalizations concerning the effects of aeroallergens upon personal and community health. However, it is entirely feasible to characterize individual agents and to examine their interactions with susceptible human populations.

II. ALLERGIC MECHANISMS

A. A Unified Concept of Allergy

To facilitate this discussion, the general term "allergy" will be shorn of its popular connotations and used in a sense closely approaching that meant by its originator, Von Pirquet (2). Stated simply, allergy is the tendency of an organism to manifest *abnormal, altered reactivity* to a specific chemical grouping, acquired in the course of prior exposure to that grouping. This change in host responsiveness or "acquired sensitivity" may be extreme and frequently confers biological activity on otherwise innocuous environmental agents. In current terminology, "hypersensitivity" expresses this generic state of altered reactivity (3) and may be equated with "allergy" as defined above. Commonly, also, the overall process leading to the allergic state is spoken of as "sensitization" and the inciting agents as "sensitizers."

Tissue changes produced by hypersensitivity mechanisms also underlie effective resistance to many infectious agents. Where the net effect is deleterious and overt disease is produced or augmented, the response is usually regarded as "allergy" rather than "immunity." Since the aeroallergens rarely have inherent harmful properties, allergic responses to them provide no obvious benefit. Still, it is common to speak of allergic mechanisms as "immunologic" in recognition of the value of these same processes to the host when their net effect is protective.

B. Allergens

Chemically specific substances that sensitize man and lower animals are called *antigens* generically and *allergens* when exposure of the sensitive host provokes overt illness. Although all allergens are recognized as "foreign" by organisms responding to them, the several forms of hypersensitivity involve allergens of different characteristic sizes and levels of complexity. Those molecular species involved in asthma and nasal allergy tend to be nitrogeneous with molecular weights over 5000. Farr (4) has presented a concise discussion of the relationship of allergens to hypersensitivity processes.

C. Types of Allergic Response

1. Antibody-Dependent Mechanisms

a. General Properties

Of the varieties of human allergic response, those that involve circulating antibodies are most familiar. These are complex proteins of the globulin class ("immunoglobulins") that occur in blood plasma and may be fixed at tissue sites. Antibody synthesis follows exposure to an appropriate antigen, and the molecules produced carry reactive sites which are complementary to groupings on the antigen that determine its specificity. The binding of specific antigen is a basic feature of antibody-mediated responses. However, it is the *form* of this interaction and its secondary tissue effects that determine the type of allergic process observed. For example, precipitating antibodies are a familiar response to certain antigens in man and lower animals. Large molecular aggregates form when these immunoglobins react with solutions of specific antigen, and the resulting precipitates excite an inflammatory response, especially in the walls of the blood vessels (5,6).

Precipitates do not occur in most responses to aeroallergens. However, complexes of antigens and specific antibodies or the reaction of antigens with tissue-fixed antibodies serve to release or activate secondary agents that produce the changes observed. Typically, the resulting reactions are of an "immediate" type, becoming apparent shortly after the exposure of a sensitive person. These allergic responses are usually manifested by nasal, conjunctival, laryngeal, bronchial, or skin changes. Fatal human reactions (7) are rarely if ever induced by exposure to common aeroallergens and occur predominantly in response to the injection or ingestion of potent sensitizers.

b. Mediators of Tissue Reactions

Several "mediator substances," including histamine, slow-reacting substance (SRS-A), and plasma kinins may be responsible for the tissue changes of "immediate" allergic reactions in man. Histamine is a potent contractor of visceral smooth muscle and acts to dilate small blood vessels while increasing their permeability to the passage of fluid. When histamine is liberated chemically from specific cellular sites in man (8), the effects begin and then recede rapidly; itching, hives, redness of the skin, mild air hunger, and a modest depression of blood pressure are all observed (9). The white blood cells (10) and isolated lung tissue (11) of allergic persons are known to release histamine on exposure to pollen allergens to which the donor is sensitive. SRS-A has been identified by its prolonged constrictive effect on the hollow viscera of several animal species (12). Material with this activity is also released when isolated lung fragments and bronchial segments from pollen-sensitive persons are challenged with the appropriate allergens (12). However, unlike aerosolized histamine, which is a strong bronchial constrictor (13), SRS-A causes only modest decreases in airway patency in asthmatic subjects (14). Similiar attempts to induce asthma with aerosols of plasma kinin have been unrewarding (15). Visceral smooth muscle spasm, dilatation and changes in permeability in blood vessels, and local pain at injection sites are prominent effects of the kinins (16); however, their role in human allergy remains speculative. An additional, complex set of substances, the *complement system*, is activated in some, though not all, immunological reactions (17). Histamine release from washed peripheral white blood cells by pollen allergens occurs without the *obvious* intervention of complement components (18). However, it is premature to extrapolate this observation to allergic reactions occurring in intact humans.

c. Antibodies in Hay Fever and Asthma

Human antibody responses to specific aeroallergens are heterogeneous (19), and studies in animals have shown that the biological effects of different antibody types can vary markedly (20). However, in the situations where sensitization to inhaled materials leads to hay fever and asthma, a well-defined group of heat-labile, nonprecipitating antibodies has been implicated. The remarkable capacity of these agents to fix to skin, respiratory tract mucous membrane, and other tissues for prolonged periods is well known, and they are often spoken of as skin-sensitizing antibodies (SSA). In persons showing asthma and nasal allergy provoked by airborne allergens, specific SSA are usually demonstrable in serum and associated with skin. Similar activity has been found inconstantly in tears (21), nasal secretions (22), and saliva (23), although its carrier in these fluids has not been completely defined. It is assumed that antigenic groups reaching sensitive tissue sites are bound by SSA and that the resulting complexes effect the release of histamine and other mediators of tissue inflamation.

The symptoms of nasal allergy might be explained by postulating the local action of a histaminelike substance. Affected persons experience variable nasal obstruction, sneezing that is often repetitive, and nasal oversecretion. Itching of the eyes, nose, and throat are also prominent, particularly when symptoms recur annually at specific seasons (a condition called "hay fever"—even by physicians). Bronchial asthma may occur alone but often coexists with nasal symptoms. Discrete, spontaneously occurring episodes of air hunger and noisy respiration are typical of asthma, and these alternate with symptom-free periods. Hay fever and *seasonal* bronchial asthma are the result of exposure to aeroallergens in virtually all instances.

The presence of SSA at tissue sites permits skin test reactions to be used as confirmatory evidence of sensitivity in persons whose symptom patterns suggest certain inhalant (and ingestant) allergens. In practice, nonirritating extracts of potential allergens, such as pollens or fungus spores, are injected into the skin or introduced superficially by pricking through a drop of the material. If significant SSA is present locally, a hive and redness, significantly greater than those at appropriate control sites, will appear within 15 min. Skin sites in nonsensitive persons can be made reactive by injecting them with serum containing SSA. Indeed, by infusing relatively large amounts of SSA-rich blood from single allergic donors, it has been possible to *systemically* sensitize normal persons for periods of several weeks (24,25). This observation has made some blood banks reticent to accept donors with respiratory allergy. However, the risk of symptoms in exposed recipients of such blood is extremely small and may be obviated by excluding allergic donors during and for one month preceding their pollen seasons.

2. Cell-Dependent Mechanisms

In some forms of allergic response, no specific antibodies are demonstrable, and specifically sensitive lymphoid cells constitute the only apparent vectors of host reactivity. In a sensitive animal or man, these cells accumulate at sites where antigen has been introduced. Since the resulting tissue changes require 8–24 hr for maximal development, cell-mediated immune mechanisms are often termed "delayed hypersensitivity" in contrast to "immediate" or antibody-dependent processes.

Allergic contact dermatitis is probably the only proven reaction of delayed hypersensitivity that may be induced by aeroallergens in man (see Sect. VII. B). The disease produces areas of thickened, red skin on which small blisters may be apparent; extreme itching is almost always present. With chronicity, the blisters are unroofed, and dryness, crusting, and scaling of the affected skin become prominent. This reaction type is most familiar as the result of *direct contact* with poison ivy by sensitized persons.

The environmental agents that induce contact hypersensitivity are often simple chemicals of low molecular weight. When applied to the skin, they are thought to combine with host protein in the deeper epidermal layers, forming complexes which are antigenic (26,27). The allergic contact reaction essentially always requires direct application of a sensitizer to the skin surface, and this mode of exposure is used diagnostically in performing *patch tests*. In this technique, a suspected offender is applied to the normal skin and covered with a water-resistant material. After 24–48 hr, the test area is examined for redness, swelling, and blister formation; these indicate a state of contact hypersensitivity if a nonirritating test material is used.

III. ALLERGIC POPULATIONS AT RISK

A. The Atopic Group

1. Definition

The association of bronchial asthma, nasal allergy (allergic rhinitis), and certain immediate reactions to foods was made clearly by Cooke and Vander Veer (28). Along with a chronic skin disorder, atopic dermatitis (infantile eczema), these conditions occur together in families and in the cumulative health experience of individuals (29,30). To distinguish these conditions from other forms of allergy, the term "atopic diseases" was applied to them by Coca and Cooke (31), and the familial predisposition that permits their occurrence was designated "atopy" (Greek: a strange disease).

2. Atopy and Skin-Sensitizing Antibody

Perhaps the most fundamental trait of atopic persons is their tendency to produce skin-sensitizing antibodies (atopic reagins) in response to common allergens encountered by inhalation and ingestion. There is evidence (32,33) that the responses of atopic and nonatopic persons to *injected* antigens are comparable. Therefore, the possibility that atopic persons may process antigens abnormally at mucosal surfaces has been raised. Salvaggio et al. have shown that skin-sensitizing antibody is a prominent feature of the response of atopic persons to the intranasal instillation of bovine ribonuclease (32) and dextran (35). Although normal subjects synthesize antibodies with these specificities, they do not produce skin-sensitizing factors.

On occasion, SSA occurs in a setting, apart from that of typical atopic responses, such as human serum sickness (36). Certain injected antigens, such as tetanus and diphtheria toxoids (37) and extracts of the roundworm, *Ascaris lumbricoides* (38), have also been shown to elicit skin-sensitizing antibodies in a large percentage of atopic and normal subjects. When injected with pollen extracts as stable water-in-oil emulsions, both of these groups may also develop specific SSA, *de novo* (39). As a rule, such induced reagins have been transient, and pollen symptoms (pollinosis) have not appeared in persons sensitized in this manner. Similar changes rarely, if ever, follow the prolonged administration of *aqueous* pollen extracts in either atopic or normal persons (40). Clearly, the simple presence of SSA is not an infallible sign of atopy. However, in each of the foregoing examples, *injection* of the antigen is required for sensitization. To date, there is little convincing evidence for the production of SSA in response to commonly inhaled substances apart from the atopic state.

When definite symptoms occur, atopic persons may present three cardinal disease types singly or in combination—nasal allergy, bronchial asthma related to inhalant or ingestant allergens, and atopic dermatitis. In addition, a relatively greater risk of hives, systemic reactions to insect stings (41), and immediate drug reactions (42) have all been reported in atopic persons. There appears to be no greater incidence of allergic contact dermatitis in the atopic group.

Reagins with specificities for common aeroallergens are often demonstrable in persons without other evidence of allergic illness. In some, typical skin and/or respiratory manifestations have occurred previously but have been lost with increasing age. In many such persons, however, neither past symptoms nor prospective study furnish evidence of atopic disease despite positive skin test reactions. Whether these individuals may be considered atopic is a matter of contention although certain considerations favor this view. (*1*) A proportion of such persons, especially children, do ultimately develop symptoms correlating with their long-standing positive skin tests (43). (*2*) Asymptomatic, skin test-positive persons frequently have close relatives with overt atopic disease, suggesting that a latent familial tendency may be present. Unfortunately, this point is difficult to evaluate until there are additional data estimating the frequency of reagins in unselected populations. In small groups of individuals having no personal evidence, but a family history, of atopic allergy, positive skin test reactions to a group of common inhalant allergens were found in 50% by Curran and Goldman (44) and in 30% by Pearson (45). In those with neither family nor personal history of atopy, 10% and 4%, respectively, had at least one positive reaction in these series.

3. The Familial Aspect of Atopy

Whatever the bases of atopic responsiveness, it is this tendency, rather than tangible antibodies or a given illness, that seems to be the hereditable factor. Human atopic reagins do not pass the placenta from the maternal circulation (46). Furthermore, atopic conditions shown by affected offspring and by their atopic parents are often dissimilar (28), although members of some families may be selectively prone to develop asthma (47). Many theories have sought to explain the familial prevalence of atopy after the initial suggestion of simple, dominant, genetic transmission (28) became untenable. This subject has been critically reviewed by DeGara (48). Among the problems that have beset workers in this field are: (*1*) the effects of variable exposure to allergens, especially aeroallergens, upon the expression of atopic symptoms; (*2*) the lack of acceptable criteria permitting symptom-free individuals to be considered atopic; (*3*) the broad age range in which symptoms may first appear; and (*4*) the questionable reliability of information obtained by interviewing subjects. In general, a genetic mechanism has been postulated with both homozygous and heterozygous atopic persons resulting (47,49,51). Despite these uncertainties, some empirical guidelines are available for genetic counseling. A family history of atopic disorders is present in from 40 to 70% of persons having respiratory allergy. Most writers find this incidence greater than that encountered in nonallergic persons, although Ratner (52) has challenged this view. The probability of allergic offspring rises when both parents have demonstrable atopic symptoms; based upon their experience in the eastern United States, Spain and Cooke (53) estimated the chance at approximately 70% in those children with bilateral inheritance. In addition, an earlier onset of symptoms has been described in children with atopy in both parental lines (28). Although the development of concordant allergic symptoms has been observed in identical twins, this has not been the rule (29,54–56).

4. The Prevalence of Atopic Allergy

Data reflecting the morbidity due specifically to aeroallergens are largely unavailable, although estimates concerning *seasonal* asthma and nasal allergy are pertinent. Figures (57) from the National Health Survey for the period 1959–1961 suggest that among children below age 17, 600,000 school days were lost annually due to hay fever and 7.5 million due to asthma. In the same period, 4.2 million days of restricted activity were referable to hay fever, of which almost 900,000 were bed days. Morbidity due to asthma is naturally greater; however, the proportion referable to aeroallergens is less readily determined. In both adults and children, typical bronchial asthma can occur in association, primarily, with infection and airborne irritants and without *demonstrable* relationship to atopic allergy.

The occurrence of asthma and nasal allergy is often difficult to establish, especially in retrospect. Mild transient symptoms are often ignored or ascribed to viral infections or "sinus" disorders even when they appear annually at a definite season. Furthermore, where public awareness of allergic diseases is keen, upper respiratory infections and nasal changes due to environmental irritants may be glibly ascribed to "allergy." The difficulties in classifying obstructive pulmonary diseases are substantial, and symptoms associated with chronic bronchitis and pulmonary emphysema are often reported as "asthma." While diagnostic criteria for these diseases have been outlined (58), even the results of careful physical examinations may not always permit clear-cut diagnoses. Reports of asthma in early childhood are especially difficult to interpret since variable respiratory obstruction often accompanies episodes of infection in this age group. However, an excessive proportion of young children with asthma during bacterial bronchitis appear later to develop obvious respiratory allergy (59). Similar reports linking prior (viral) bronchiolitis with allergic asthma (59,60) have not been regularly confirmed.

Vaughan and Black (61) have reviewed many older surveys of the prevalence of allergic manifestations within samples drawn from population groups in this country. While the diagnostic standards employed were not strictly comparable in these studies, a cumulative prevalence of asthma and nasal allergy approximating 10% was suggested. More recently, Broder et al. (62) have studied all inhabitants of Tecumseh, Michigan (population ca. 9800) above the age of six by a detailed questionnaire and physical examination. They found past or present asthma to be definite or probable in 4.1% and suggested, though less certain, in an additional 5.1%. Nasal allergy was considered highly likely in 6.3% and less certain in 3.4% more. In 80.4% of the asthmatics and in 89.3% of those with nasal allergy, a seasonal worsening of symptoms was reported.

Several recent studies suggest that respiratory allergy may be somewhat more frequent in university populations. Van Arsdel and Moltulsky (51) found past or present symptoms of asthma in 4.7% and hay fever in 14.7% of students surveyed at the University of Washington. Slightly higher rates of 5.7% and 16.6%, respectively, were reported at the University of Michigan (63), and Tips (50) obtained similar data at the University of Notre Dame in Northern Indiana. Differences between these rates and those of nonstudent groups have not been satisfactorily explained.

Sex-specific prevalence rates for respiratory allergy have been comparable in most groups studied. In the Tecumseh survey, for example, cumulative prevalence rates for asthma were 4.0% for males and 4.1% for females, while those for hay fever were 6.3% for both sexes when the "more certain" groups were considered (62). Many reports indicate that both asthma and nasal allergy demonstrate a trend toward earlier occurrence in males, however (48).

Observations in North America suggest that no presently constituted racial group is exempt from inhalant allergy, although definite ethnic differences may exist. The prevalence of hay fever in Negroes has been estimated at from one-third (64) to one-half (65) that of white persons living in the same vicinity. However, the occurrence of bronchial asthma has been reported as more closely comparable for the two races (65, 66). The contention that hay fever and asthma are rare among American Indians is still based upon impressions reported to Thommen (67) by scattered medical workers at schools and reservations. It seems wise to defer judgment on this point also until systematic data from groups of full-blooded individuals are available.

5. The Sequence of Atopic Manifestations

The risk of later asthma, in persons with nasal allergy alone, has been placed at from 25 to 75% by various allergists based upon observation of patient groups. Since these estimates reflect the morbidity experience of persons with rather severe symptoms, the validity of extending them to unselected allergic persons has been questioned. The Tecumseh study provided data pertinent to this matter: of those with both asthma and nasal allergy, 75% had asthma alone initially or developed both manifestations in the same year (68). Furthermore, among those with nasal allergy who were at risk to develop complicating asthma, this had occurred in only 5–10%. These figures suggest that, for prognostic purposes, at least two major atopic groups must be recognized. A majority of persons have relatively mild and remittent symptoms; of these, only a few experience complicating asthma. Among those with nasal symptoms sufficiently persistent and severe to warrant medical consultation, the inherent risk of asthma appears to be distinctly increased.

6. *Allergenic Exposure and the Prevalence of Symptoms*

For unknown reasons, reactivity often develops to one aeroallergen but not to others of apparently similar antigenic potency during roughly equivalent exposure periods; as a result, patterns of sensitivity vary widely from person to person. Those with isolated responses to extremely "strong" inhalant allergens, such as animal danders, are usually regarded as atopic individuals who require especially intense stimulation for sensitization and the occurrence of symptoms. However, the alternative view, i.e., that occasional normal persons may develop sensitivity with very intense exposure, has also been advanced (69,70). There seems little hope of resolving this question until criteria for the atopic state that do not involve skin sensitivity or overt respiratory symptoms are available.

Differences in allergenic exposure have been postulated to explain the reported earlier incidence of ragweed hay fever in boys (71) and the discordant expressions of allergy between siblings, especially twins. The onset of allergic symptoms in persons following a geographic move to an area of more intense allergenic exposure has often been recorded; ragweed pollen has frequently been the offender. We have been impressed, also, with the appearance of pollinosis in personnel following heavy occupational exposure to these particles (72). The involved individuals have been either manufacturing pharmacists or botanists and meteorologists engaged in studying allergenic aerosols.

Differences in the reported prevalence of atopic allergy among various national groups may be related to regional differences in the distribution of aeroallergens. The relative sparsity of anemophilous (wind-pollinated) plant species in tropical areas and the high endemicity of the ragweeds within the Western Hemisphere (73) may both be important in this regard. All the difficulties previously cited (see Sect. III.A.4) have also been encountered in studying atopic allergic disease in foreign groups. In addition, where populations have relatively little medical knowledge, especially if illiteracy is prominent, underreporting of symptoms is bound to occur. This deficit is naturally magnified where the true prevalence of allergy is low, where stoicism is the accepted response to minor illness, and where symptoms, especially those recurring in a given season, are considered a peculiarity of the local climate rather than of individual response. Even in countries where comprehensive health records are kept, statistics related to atopic allergy are not always collected. Furthermore, many published surveys of "allergic disease" have included conditions such as epilepsy, peptic ulcer disease, and psychoneurosis for which the probability of allergic mechanisms is tenuous or nonexistent.

Data for the occurrence of asthma and nasal allergy (as well as for eczema and hives) derived from national groups in Great Britain and northern Europe have been discussed by Williams (74). For asthma, *current* prevalence rates of from 0.4 to 1.4% (approximating 1% overall) have been found. Since most recent reports from the United States relate to *cumulative* prevalence, a ready comparison with European rates is not possible. However, as Williams observes, the prevalence of asthma in the United States seems not to exceed that in European groups studied. Prevalence rates for hay fever in Britain and Continental Europe have seldom exceeded 0.5%, suggesting an occurrence less than one-tenth that experienced in North America. Information concerning inhalant allergy in Africa, southern Asia, and tropical portions of South America is fragmentary (67) or nonexistent. Regarding the occurrence of atopic allergy in Cuba (and other Caribbean islands), Quintero has commented that it is "about the same as in the U.S.A., except for the smaller number of pollen cases" (75).

The suggestion that potent environmental allergens may affect the regional prevalence of inhalant allergy is made most intriguing by observations of immigrant groups (63,76,78). In Maternowski and Mathews' series of 322 foreign and 639 native-born students at the University of Michigan, cumulative prevalence rates for nasal allergy and asthma were comparable (approximately 21%). However, among the foreign atopic students, significantly fewer (33%) had a history of familial atopy than in the native-born group (62%), a discrepancy also noted in other reports. In the atopic groups at the University of Michigan, symptoms (pollinosis) due to ragweed pollen occurred with equal frequency among foreign and native-born students. Ragweed pollen has been a prominent offender in immigrant patients, and considerably less than half of these (63,76,66) have had evidence of atopic manifestations prior to entering this country. These data attest to the sensitizing power of ragweed pollen allergens and suggest that the atopic tendency is present, though latent, in many national and racial groups. Although ragweed pollinosis is often the initial allergic problem in immigrants, additional sensitivities often become apparent in later years. At times the list of offending agents comes to include substances previously well tolerated by patients in their countries of origin. Whether exposure to a potent aeroallergen, such as ragweed pollen, somehow augments the capacity for further sensitization is unknown; however, available data do admit this possibility.

Understandably, the mean age of onset of ragweed pollinosis in immigrants is later than that for native-born affected persons. In the former group most observations suggest that the second, third, and fourth seasons of ragweed pollen exposure have been the periods in which pollinosis has appeared most commonly. However, the development of typical symptoms after much longer exposure periods is well known. Among 60 immigrant hay fever patients reported by Hughes, for example, 28 had their initial symptoms from 5 to 15 years after arriving in Canada (78). The periods of exposure necessary for sensitization to other aeroallergens are generally unknown. Phillips (79,80) has suggested that at least two seasons were necessary for skin sensitivity to sugar beet (*Beta vulgaris*) pollen and spores of a powdery mildew. However, in neither case could preceding inapparent exposure to the allergenic substances in related airborne particles be excluded.

B. Allergy in Nonatopic Populations

With its limited spectrum of symptoms, strong familial tendency, and characteristic skin-sensitizing antibodies, the atopic population constitutes a reasonably coherent unit. Unquestionably, most recognized morbidity associated with aeroallergens occurs within this group. However, additional forms of allergic response to airborne materials, quite unrelated to atopy, also contribute to human illness. Recent studies of these conditions (see Sect. VII) have helped to clarify the tissue interactions involved, but the relationship of additional host factors to prevalence has received little attention. In many of these instances, the risk of inhalant exposure is sharply limited by geography and occupation to highly selected populations. Understandably, no determinants, apart from the duration and intensity of exposure, have become apparent in these groups.

IV. THE NATURE AND SOURCES OF AEROALLERGENS

A. Aeroallergens as Particulates

With few exceptions, airborne allergens are associated with solid particles having average diameters of from 1 to 80 μ. The physical attributes of suspended materials in this size range have been reviewed extensively by Green and Lane (81). To obviate confusion, the practice of designating as "aeroallergens" *both* specific allergenic molecules and the characteristic particles that carry them, should be clearly recognized. The latter are sometimes dusts in the sense of particles formed by attrition from homogeneous solids. More often, as exemplified by pollens, spores, and vegetable fibers, they have a well defined ultrastructure involving many chemical groupings that may act as allergens.

Since specificity of chemical structure is a feature of allergens, it is clear that climatic variables such as temperature, humidity, and sunlight are not *primary* determinants of allergic reactions. However, they readily accentuate preexisting tissue changes and may simulate allergic responses by activating mediators of inflammation nonspecifically (see Sect. II.C.1).

Most particulate allergens are hygroscopic, and this property has made measurements of their specific gravity difficult. Published estimates of particle density for selected pollens (82) and fungus spores (82,83) fall between 0.4 and 1.2 g/cm^3. When airborne, biological particles take on moisture, and those less than 10 μ can serve as condensation nuclei. Certain fungus spores carry small static charges that may contribute to short-range effects (84), but the electrical properties of pollens are unknown.

Since most inhaled allergenic particles exceed 1–2 μ in diameter, significant deposition in the upper air passages is expected. For particles in the size range of pollens, removal by the nose during nasal breathing should be virtually complete; capture by the nasal hairs alone may exceed 40% (85). Lung deposition is very much greater for small spores even though these may grow through hydration in moist inspired air (86). Especially at high wind speeds, the mucous membranes of the eye present a potentially important surface for the impaction of particles exceeding 5–10 μ in size.

B. Sources of Aeroallergens

Since the adverse effects of aeroallergens are best circumvented by avoidance of exposure, the sources of offending agents assume major importance. Determinants of exposure may be related to geography, occupation, or local domestic practices, and effects may be exerted on temporal and spatial scales of varying size. For example, occupational allergy to castor bean dust is well known in industries processing fertilizers and cattle feed. However, localized epidemics of inhalant allergy have also occurred in persons living near such plants and exposed to their highly allergenic wastes (87,88). Perhaps the most practical classification of aeroallergens is by source, and two major categories may be considered: (*1*) those aeroallergens encountered primarily indoors through occupational, educational, and domestic activities and (*2*) those originating from sources out-of-doors. These categories are not exclusive, and representatives of groups such as fungus spores and insect emanations originate in both situations. However, many types fit naturally into one or the other division, as shown in Table I.

TABLE I

Characteristics of Important Aeroallergens in Atopic Diseases

	Origin	
Characteristic	Out-of-doors	Usually indoors or in work environment
Mode of exposure	Natural (in free air)	Associated with work or domestic activity
Pattern of prevalence	Regional with flora and fauna	With local patterns of culture and industrial practices
Period of prevalence	Usually seasonal	Often perennial or related to seasonal activities
Avoidance of exposure	Usually difficult or unfeasible	Often simple and very effective
Microscopic appearance	Often readily recognizable	Often amorphous or not characteristic
Specific allergens	Pollens, fungus spores, algal particles, insect emanations	House dust and kapok, animal danders, fungus spores, insect emanations, vegetable gums, seed proteins (flax, cotton, soy, castor bean and cereal grains)

C. The Transport of Aeroallergens

1. Transport Processes in Closed Spaces

For allergenic particles, transport indoors occurs on a more modest scale than in the free atmosphere; however, the same processes of convection and ventilation are active. Ventilation may be effected by outside wind or artificially by fans. In either case, air speeds of 15–35 ft/min, which provide subjective comfort at room temperature (89), are adequate to circulate most aeroallergens. Convection from mechanical heaters as well as that generated by the human body also fosters air motion indoors. The efficiency of these often impalpable processes in disseminating minute particles throughout a dwelling has been demonstrated by Christensen (90).

Domestic heating and air conditioning arrangements that utilize extensive duct work and basically closed systems pose particular exposure problems. With these, air circulation is brisk, and allergenic particles originating in one area are readily dispersed throughout the volume of the system. In addition, particles tend to accumulate, especially at bends in the conduits, until they are refloated by turbulent eddys. Persons allergic to house dust and animal emanations (see Sect. V) are especially poor candidates for forced air ventilation. When institutional complexes utilize forced air systems, dust and mold contamination from intakes situated in basements and storage areas must be carefully avoided.

Even minor disturbances of indoor air and surfaces due to the movement of human occupants are effective in the transport and reflotation of allergenic particles (91). Dry dusts are readily dispersed in the course of cleaning activities; however, particle clouds resulting when pillows and mattresses are sharply compressed provide a comparable, though less obvious, hazard.

2. Transport Processes Out-of-Doors

Although atmospheric motion largely controls the movements of aeroallergens in nature, human activities commonly do affect the particle content of free air. The plowing of cultivated fields, for example, has been associated with local concentrations of airborne algae (92) and soil fungus particles. Where crops are infected by smut fungi, the spore clouds liberated during combining may affect sensitive individuals exposed occupationally as well as those living at substantial distances. A similar dispersal of biological particles occurs when grass is cut and dry leaves are raked.

All aeroallergens necessarily originate at the earth's surface within a layer of transitional turbulence. A few types are projected into more actively moving air by specialized release mechanisms. Such devices are rare for pollens although the anther sacs of the mulberries, nettles, and related species typically burst when mature, releasing their 14–22 μ grains explosively (82). Varied takeoff mechanisms are more common in the fungi, and these have been lucidly described by Ingold (93). However, most windborne pollens and many fungus spores are merely blown from their attachments by turbulent air currents. This is facilitated when the particles originate from structures on long erect or dependent stalks, a common adaptation in wind-pollinated (anemophilous) plants. Fungus spores dispersed by wind scavenging are also borne on relatively long upright branches and usually originate on vegetation at some distance from the ground. Spore liberation from these organisms increases directly with wind speed (94) and may be augmented by sudden changes in velocity (95). It is generally assumed that wind speed and the initial takeoff of pollen grains are similarly related.

Much less is known about the efficiency of ambient air currents in resuspending pollens and spores following their deposition upon surfaces in nature. Basic to this uncertainty is a general lack of knowledge concerning the adhesive properties and surface characteristics of these particles. Bagnold (96) emphasized that relatively large suspended particles can significantly increase the scouring effect of air currents. His observations also suggest that *inorganic dusts* in the size range of common aeroallergens have prominent adhesive properties. Although few data are available from biological systems, strong attractive forces have been shown between small particles and smooth manmade surfaces (89), effectively resisting reentrainment. For any specific particle type, it is probable that reflotation will be found to vary directly with wind speed, the roughness of specific surfaces, and the relative accessibility of deposited particles to turbulent eddys.

Particulate aeroallergens, having sufficient energy, may pass upward into the zone of well-established turbulence where significant translocation is possible. In this layer, particles are diffused in the horizontal by eddys of varying size and are subjected periodically to vertical forces averaging perhaps 10 cm/sec (97). Both the height of the turbulent layer, usually 500–1000 m, and the degree of disturbance prevailing at a given time vary with the comparative "roughness" of the underlying surface, the wind velocity, and the stability of the atmosphere (82). Besides their unique surface characteristics, urban areas have been shown to have higher surface temperatures than forests and grasslands

(98); these properties further modify mixing processes and ambient particle concentrations, especially at night.

In addition to the vertical components of turbulent motion, convection is effective in carrying suspended particles aloft. As warm air expands, it ascends, and its place at the surface is taken by downward-moving, cooler air. These processes add significantly to the mixing function of the turbulent layer. The volume of this layer is greatest during warm sunny days, when convection is also maximally active, and becomes minimal during cool, clear nights. At any hour, the presence of a temperature inversion, with warmer air aloft, establishes an upper limit for the volume in which mixing occurs.

The loss of particles that have entered the turbulent layer takes place primarily by gravitational settling. For common aeroallergens, this tendency is indicated by their terminal settling velocities (V_T) in still air. As expressed by Stoke's law, V_T is proportional to the density of a given particle and to the square of its equivalent radius. Most pollen grains have terminal settling velocities of from 1 to 10 cm/sec, while those for common fungus spores are between 0.05 and 2.0 cm/sec (82). It is a clear implication of Stoke's law that, for particles of comparable density, the tendency to fall out rises exponentially with particle size. This relationship is a major source of bias in air sampling methods that involve gravitational collection (see Sect. VI.A); it also explains why wind-disseminated particles such as corn pollen (average diameter 95 μ) are seldom recovered at any significant distance from the source plants. Where particles are not added continuously to an air mass, gravitational deposition tends to reduce their concentration, especially in the lowermost layers. These losses vary with particle size and are partially offset by settling and diffusion from above. In addition, the subsidence of particle-bearing air as it cools, particularly after sunset, may cause unexpectedly high concentrations at ground level. Especially when relatively small particles are considered, such concentrations may represent the earlier productivity of sources many miles upwind of the sampling point. Airborne biological particles are known to gain or lose water rapidly, and their relative hydration appears to influence settling in a complex fashion (99). However, it is doubtful that major diurnal variations in particle concentrations near the ground may be ascribed confidently to changes in relative humidity, as has been suggested (100).

There is abundant evidence that long-distance transport of potentially allergenic particles does occur. This has been shown by the finding of pollens and spores over polar regions (101) and oceans (102,103), far from possible source areas. However, the relative contributions made by short- and long-range transport to the concentrations of aeroallergens at any point near ground level are frequently obscure. Gregory has suggested that perhaps 10% of the spores dispersed "escape" local deposition and become available for extended dissemination (104). Undoubtedly, the composition of this fraction, whatever its magnitude, is strongly weighted toward a size range below 10 μ.

In considering aeroallergens out-of-doors, especially pollens, the effects of relatively short-range (several miles or less) transport seem most worthy of attention. Despite much work, it is not yet possible to explain or predict ambient pollen concentrations precisely using existing theories of atmospheric diffusion. Field studies with naturally occurring aeroallergens must contend with temporal variations in source strength and often with sources that are extremely numerous and distributed over large areas. Furthermore, the reduction in dosage risk, as one recedes from any single source, is not simply stated, although it must vary inversely with *some* power of the distance, depending upon existing conditions of wind velocity, atmospheric stability, and the relative strength of long and short eddys. The release of particles at increasing heights above the ground allows equivalent concentrations at progressively greater surface distances from a source (82). This point affects the capabilities of tree pollens to act as aeroallergens despite their frequently large radii. The tendency of any monodisperse cloud to become depleted with distance from its source varies, in general, with particle size. For ragweed pollen grains, Gregory has calculated that a 50% decrease in airborne concentration occurs within a distance of 25 m from the source, under weather conditions commonly prevailing (82).

The complexities of atmospheric processes and of source characteristics discourage generalizations concerning the effects of single weather variables upon particle concentrations. In considering wind as a factor in human exposure to allergenic particles, for example, it is difficult to dissociate the effects of wind speed and direction from those of atmospheric turbulence. With increasing horizontal velocities, reflotation increases, and particle concentrations may be maintained for longer distances before being depleted by vertical mixing and gravitational fallout. However, on windy days, associated changes in turbulence augment particle losses due to impaction and serve to increase mixing, both vertically and in directions normal to the mean wind path. When high wind speeds are associated with strong solar heating, the increased depth of the frictional layer and the convective activity each serve to increase the volume of air in which particles may be diluted.

An overall effect of rain upon particle concentrations is equally difficult to state. McDonald (105) has described the scouring effect of rain drops on allergenic particles and emphasized that this is usually proportional to the total duration and depth of rainfall rather than to its rate alone. Droplet size is an important determinant of capture efficiency, and, in general, this is greatest for rain drops 1 mm in diameter and diminishes when larger or smaller drops are considered. It is noteworthy that 1-mm droplets are the most common components of steady rain, while, in convective showers, larger drops, up to 4 mm in diameter, predominate. In any precipitation, the scavenging effects are greater upon the larger suspended particles, and McDonald has proposed that the probability of capture varies directly with particle density and with the square of particle diameter. Very small spores are probably removed by cloud droplets (less than 0.2 mm in diameter) and by serving as condensation nuclei with subsequent precipitation in the droplets formed (106).

Hirst has suggested that very high air velocities accompany the outwash from impacting rain drops and that this effect, plus the "tapping" disturbance created, can refloat deposited particles (107). These processes may be especially prominent during thunderstorms when the rapid subsidence of particle-laden air further tends to elevate concentrations near the ground. Steady frontal rainfall is commonly accompanied by thermal inversions which tend to retard the dissipation of particle clouds, and diurnal prevalence patterns for ragweed pollen commonly show anomalies that suggest this effect. Pollen release in the ragweeds occurs primarily in the early hours after sunrise, and pollen concentrations peak by late morning with lower levels typically prevailing thereafter. When afternoon rain follows a clear morning, however, it is common for an abnormally high pollen level to persist throughout the day. High atmospheric humidity depresses or prevents the release of most types of pollen and impairs wind scouring of the dry spores produced by many common saprophytic fungi (95). By contrast, the spores of certain other fungi, dispersed primarily in rain splash or by active mechanisms requiring free water (93), may become airborne only in damp weather.

V. AEROALLERGENS IN ATOPIC DISEASE

In persons with asthma and nasal allergy, the observed signs of illness and subjective complaints per se usually provide little help in identifying the responsible allergens. It is common, also, to find skin sensitivity to allergens that produce no effect during normal exposure. Therefore, the circumstances in which symptoms appear furnish the main evidence permitting specific causal agents to be inferred. Of the great variety of airborne materials, a limited number may be mentioned as important determinants in atopic allergy.

A. House Dust

Exposure to domestic dust, especially that from old mattresses, furniture, and books, is probably the most frequent factor precipitating allergic symptoms. House dusts from diverse geographic areas have been shown repeatedly to elicit positive skin test reactions in dust-sensitive persons (108); however, the nature of the common allergens remains controversial. On microscopic examination, vegetable fibers and human epidermal cells are prominent components of house dust. Variable amounts of animal hair, including wool fibers, feather dust, bacteria, fungus spores, food particles, and insect debris, as well as soil particles and pollens originating out-of-doors, may be present. Despite this heterogeneity, there is evidence that degradation products of cotton cellulose may comprise the major allergenic materials (109). Analyses of crude house dust fractions have suggested that the active substances are mainly acidic polysaccharides with small polypeptide moieties linked, perhaps, by covalent bonds (110).

House dust preparations have also been shown to possess endotoxin-like properties (111), probably of bacterial origin. Gram negative organisms are prominent contaminants of raw cotton and have been implicated in febrile episodes observed in textile workers (112). Whether this activity affects the atopic response to house dust has not been determined. The existence of sensitivity to human dandruff has been claimed (113), and parallels have been drawn between responses to this ubiquitous material and to house dust. Other workers have noticed the abundance of a specific mite (*Dermatophagoides pteronyssinus*) in dust samples obtained from humid Dutch homes, and, to a lesser extent, from interiors elsewhere (114). Skin sensitivity to this organism has been demonstrated; however, its distribution and actual clinical importance remain unclear.

Skin and respiratory sensitivity to kapok have been frequently described, and the tendency of these fibers to fragment with age appears to foster aerial dissemination. Kapok provides an inexpensive stuffing for mattresses, pillows, and upholstered furniture and is favored for sleeping bags and life preservers because it is not readily wetted. The allergenicity of kapok seems to develop as it ages and may reside in certain glycopeptides (115). Biodegradation by saprophytic fungi may be the major process producing the characteristic antigens (116). The successful avoidance of kapok by specifically sensitive persons requires that products containing these fibers be clearly and prominently labeled.

B. Pollens

The common allergenic pollens are morphologically distinctive reproductive particles 15–80 μ in size. Since the flowering of most plant species is accomplished annually at a characteristic season, the precise dates at which pollinosis occurs provide useful clues to the particles responsible. It is helpful to consider three major groupings among the windborne pollens—those of trees, grasses, and weeds.

The tree pollens are heterogeneous antigenically and morphologically and, in this country alone, tree species of over a dozen genera have been associated with pollinosis. Whether major differences in pollen allergens commonly occur within genera is not clear, although similarities between related species are often impressive. Especially in the northern tier of states, tree pollens are shed predominantly in the period from mid-March to early June; notable exceptions include pecan pollen in the Southeast (as early as mid-January) and mountain cedar in West Texas and adjacent states (mid-December to mid-February). Although natural wooded areas still contribute much of the tree pollen load, other sources also merit attention. In cities, the choice of shade trees planted is a factor, and ornamentals immediately about homes can strongly influence the exposure of their occupants. Among the popular street trees, ashes, sycamores, birches, oaks, and walnuts all produce potentially allergenic pollens. Skin and mucosal sensitivity to pollens of the yews and junipers may also be demonstrated (72). As a rule, the willows, poplars, mulberries, hollies, yews, and junipers, as well as the white, red, and green ash all have male (pollen producing) and female trees that may be purchased separately from nurseries. Many popular ornamentals are primarily insect-pollinated and shed little windborne pollen; these include the lindens (basswood), locusts, buckeyes, and horse chestnut, magnolias, tulip tree, dogwoods, and mountain ash. The conifers have not been widely implicated as sources of aeroallergens, in part, perhaps, because they are usually concentrated in relatively remote areas. Inhalant allergy and skin sensitivity to pine pollen are rare, although both may occur (117).

In temperate regions, airborne grass pollen is prevalent from late spring to mid-summer, while in tropical and subtropical areas its occurrence may be perennial. Although the common meadow and pasture species are the major contributors, exposure to lawns or to cereal grains in flower may precipitate symptoms. Pollens of the most common northern grasses (the blue grasses, timothy, orchard grass, red top, and sweet vernal grass) have close antigenic similarities. Significant antigenic differences between these and the pollen of Bermuda grass have been suggested; less closely related genera may show additional unique groupings (118).

Herbaceous weeds, especially the ragweeds, are the most important determinants of pollinosis morbidity in North America; elsewhere their importance appears to be relatively minor. Members of the goosefoot (Chenopodiaceae) and amaranth (Amaranthaceae) families are especially significant in the central states and Great Plains where Russian thistle, burning bush, and the water hemps grow abundantly. In the North Central states, hemp pollen is a locally important sensitizer as is the very similar pollen of the cultivated hop in the Pacific Northwest. Ongoing volumetric air sampling studies should help to clarify the place of the ubiquitous plantains (*Plantago*) and of the sorrels and docks (*Rumex*) as sources of aeroallergens.

Ragweeds (*Ambrosia*) of over 40 species inhabit disturbed areas throughout the Western Hemisphere. They are most abundant in temperate parts of North America while, in other regions, their occurrence is restricted or casual. Throughout the eastern United States, the most important species are dwarf and giant ragweed, and in the Great Plains, burweed marsh elder (prairie ragweed) is a major pollen source. In the Southwest, a typical August to late-September ragweed season is recognized and, in addition, desert ragweeds such as rabbit brush and canyon ragweed flower abundantly in spring. A remarkable capacity to colonize disturbed ground has permitted the ragweeds to reach their present condition of widespread abundance. In soil, the seeds of common ragweeds can remain viable for many decades after ragweed growth has been eliminated in the course of plant succession (119). These weeds are familiar inhabitants of urban lots, road cuts, and unfinished parking areas. However, in the Midwest, at least, ragweed growth in grain fields appears to constitute the major pollen source (120). Following seed germination in late May, growth proceeds slowly until the more rapidly developing grains are cut; thereafter, the ragweeds develop without competition. Unless there is regular overturning of the soil or other disturbance of competing plant species, the dominance of the ragweeds cannot be maintained for more than several years. The feasibility of controlling local ragweed by allowing it to be overgrown is of particular importance since other large-scale measures have often proven impractical. Despite the evident effect on health of ragweed pollinosis (121), enthusiasm for urban eradication programs has waned considerably. The inflexible need for yearly application of effective broad-leaf herbicides like 2,4,-dichlorophenoxyacetic acid (2,4-D) has been a major source of disenchantment. Furthermore, estimates of pollen prevalence in the air of large cities have not confirmed the effectiveness of municipal ragweed eradication programs (122). In these instances, it can be assumed that rural sources were the major contributors of pollen recovered in urban air. Considering the importance of immediate local sources, however, ragweeds found close to patients' homes should be actively eliminated.

Plants of several related anemophilous genera, including the sages (*Artemisia*) and cockleburs (*Xanthium*), produce pollens antigenically similar to the ragweeds. This similarity extends also to insect-pollinated members of the large family Compositae, and pollens of dandelions, asters, and goldenrods as well as those of zinnias, chrysanthemums, dahlias, etc., all can elicit symptoms in ragweed-sensitive persons (123). Exposure to the active pollens of these ornamentals, however, usually requires intimate contact with the plants and is readily avoided by informed persons. Symptoms that follow the use of pyrethrum-containing insecticides, by ragweed-sensitive persons, have a similar basis. This material is obtained by extracting dried flowers of the genera *Pyrethrum* and *Chrysanthemum*, both relatives of ragweed.

Immunochemical studies of fractions of short ragweed pollen extracts suggest that these contain a minimum of two distinct human allergens (124). The active materials appear to be large polypeptides; in addition, several workers have reported a skin-reactive, low molecular weight (dialyzable) material (125).

C. Fungi

Fungus spores and fragments of hyphae are frequently the most abundant biological particles in air. In frost-free areas, their occurrence is perennial, while in more northern areas, the growing season is the principal period of exposure. However, indoor situations, marked by high humidity and, at least, traces of organic matter support the growth of fungi at any season. Certain activities, such as cutting grass and raking fallen leaves, as well as exposure to hay, silage, and composting vegetation, will typically worsen the condition of individuals sensitive to fungi.

Microscopic saprophytes of genera including *Alternaria*, *Cladosporium*, *Helminthosporium*, *Fusarium*, *Penicillium*, *Aspergillus*, *Mucor*, *Rhizopus*, and *Candida* have received the greatest attention as sources of airborne allergens. Spores of crop parasites, including rusts and smut fungi, may also precipitate symptoms, especially in agricultural workers (126). Air samples obtained in diverse areas regularly have shown additional spore types, principally of fleshy fungi (Ascomycetes and Basidiomycetes) that do not grow readily on synthetic media. Several workers have observed positive skin tests and mucosal reactions on testing with extracts of some of these spores (127,128).

D. Animal Emanations

The epidermal scales of animals are among the most potent allergens known. In addition to danders per se, proteins of urine, saliva, and lacrimal secretions contribute to the airborne load of allergenic material. When sensitivity develops in persons such as veterinarians, farmers, and laboratory workers, whose exposures are occupational, major hardship often results. In addition to pets, commercial products containing animal hair provide important sources of inhalant expsure. The introduction of synthetic fibers has not eliminated the use of cattle, horse, and hog hair in rug and typewriter pads and their addition to upholstered furniture. Goat hair, such as cashmere, alpaca, and mohair, has gained considerable popularity in modern fabrics. The common use of rabbit and cat hair to line and trim articles of clothing and in stuffed and furry toys still poses a hazard, especially for sensitive children. The interests of dander-sensitive persons require that fur products be clearly labeled as such and that the species of origin be specifically designated.

E. Miscellaneous Allergens

With the exception of the pollens, the airborne reproductive spores of vascular plants, including the mosses, ferns, and club mosses, do not commonly act as aeroallergens. The demonstration of skin and mucosal sensitivity to extracts of green algae (129,130) is of interest since algal cells, often in large numbers, have been recovered from air (92). While bacteria seem to occur regularly in outdoor air (131), their possible allergenicity, as inhalants, has not been documented.

The extent to which insect emanations induce allergic respiratory symptoms is unclear, although skin test reactivity to insect extracts can be demonstrated frequently (132). In certain localities, endemic caddis fly (133) and mayfly (134) populations are known to affect a high percentage of exposed atopic persons. Atmospheric pollution is usually related to the swarming of these insects, the responsible particles comprising hairs, scales, and fragments of previously shed larval pellicles. Isolated instances of occupational exposure to insect emanations have also been recognized (135). In addition, soluble protein materials (sericins) associated with raw and processed silk fibers have proved to be potent sensitizers (136). The antigenic specificities of the sericins appear to differ from those of insect exoskeletal components (132).

Inhalant allergy has often been associated with the processing of grains, both through the direct occupational exposure of atopic persons and, possibly, by industrial contamination of community air (137). Grain dusts comprise a varied allergenic flora, including the spores of rusts, smuts, and saprophytic fungi, as well as grain mites and insect fragments. In addition, respiratory irritation by inorganic dusts, seed chaff, and plant hairs may contribute to the total adverse effects of exposure. In millers and bakers, sensitivity to intrinsic grain proteins more commonly underlies allergic symptoms; wheat, rye, barley, and buckwheat flours have all been offenders. In a prospective study, Herxheimer (138) has traced the acquisition of grain sensitivity by apprentice bakers, noting that 18% have positive skin test reactions after 5 years of work; of these, however, only a fraction have symptoms. Powerful allergens associated with dried castor bean, flaxseed, and cottonseed are also encountered by granary workers as well as by feed handlers and persons processing organic fertilizers. In addition, flaxseed is an ingredient of certain hair preparations and prepared cereals, while cottonseed flour is used widely in the baking trades.

Many other allergenic materials including orris root, vegetable gums, and fish glues can be important offenders by inhalation, especially in industry. These occur as fine dusts and seem to bear allergenic molecules that are relatively complex. True inhalant allergy to gaseous materials or to simple chemical substances is extremely rare. It is known, however, that cooking odors of fish and shellfish can induce asthma in certain persons who develop immediate systemic reactions on ingestion of these same foods. In addition, instances of typical asthma and nasal allergy due to phthalic anhydride (139), tannic acid (140), and the sulfone chloramides, halazone, and chloramine T (141), have been reported. In each case, passively transferable specific reagins were present, and their importance in the observed reaction was assumed. Recently a similar instance was observed in which the offending agent was phenylmercuric propionate (142) used as a disinfecting agent by an institutional laundry. In addition to their irritant effects, possible allergenic properties have also been suggested for certain diamines (143) and for toluene diisocyanate (TDI). Many instances of severe asthma have been recorded in persons exposed to low concentrations of TDI (less than 0.2 ppm) during work in industries producing polyurethane plastics (144). Neither positive skin tests nor circulating antibodies have been demonstrable; however, the finding (145) of TDI-induced, *in vitro* "blast cell" transformation in blood lymphocytes from these workers raises the question of an immunological response (146).

VI. QUANTITATING AIRBORNE ALLERGENS

A. Air Sampling Methods

Blackley's original correlation of his own symptoms with the seasonal recurrence of airborne grass pollen (1) established the value of prevalence data for aeroallergens. Several important objectives of air sampling are apparent: (*1*) to study exposures qualitatively as an aid in diagnosis (*2*) to estimate the level of the allergenic "load" prevailing as a guide in evaluating the results of treatment (*3*) to compare the exposure potential of different environments and (*4*) to evaluate the effects of eliminating sources of aeroallergens. Current sampling techniques are useful, in general, only for particles that may be identified microscopically or that will grow on laboratory media. Unfortunately, many important allergenic dusts admit neither approach, and even recently developed techniques are still applicable primarily to pollens and fungus spores.

Authenticated "known" specimens remain the surest aid in pollen identification; in addition, certain illustrated references may be consulted (147–149). Identifiable fungus spores have been figured by Gregory (82), and other published guides (150–154) will provide help in identifying intact culture mounts. Several general reviews of air sampling methods and equipment are also recommended (82,155,156).

Where the productivity of a discrete source or the level of an occupational exposure must be known, studies at ground level or indoors are often indicated. More commonly, however, sampling is carried out to trace the average exposure of large urban populations to a variety of aeroallergens. For this purpose, one or more centrally located stations are used, and the flat roofs of tall buildings offer particularly favorable locations. A rooftop site should not be flanked by taller structures, and, if elevator housings, etc., are present, samplers should be separated from these by a distance fully three times the height of the obstruction. Aerodynamic considerations and the safety of personnel both discourage the placement of samplers at the roof edge or on window ledges; a central position and firm attachment to the roof are recommended. Porches and balconies are usually poor choices for the placement of equipment, since air mixing is often deficient at these points. A survey of the surrounding area is helpful in identifying strong local sources of aeroallergens, especially pollens. Even when these are absent, however, it is unwise to draw inferences, concerning particle levels at any appreciable distance, from the results obtained at a single sampling point.

Since it was proposed as a "standard" device, the gravity slide or Durham sampler has been widely used for collecting pollens and spores. Full descriptions of this method have been presented (157,158). Basically, the sampler consists of two polished metal disks, 9 in. in diameter, separated by three 4 in. metal struts. The upper plate acts as a rain shield for a horizontal, adhesive-coated microscope slide (1 × 3 in.) mounted 1 in. above the lower disk. This device is still available commercially from the Wilkens-Anderson Co. (4525 W. Division St., Chicago, Ill. 60651) (retail price as of July 1965—$55.00) or it may be built at modest cost. A working drawing may be obtained from the Chairman of the Pollen and Mold Committee, American Academy of Allergy (756 N. Milwaukee St., Milwaukee, Wisc. 53202.)

Standard procedure calls for slides to be exposed for 24-hr periods and for examination at 100–200× magnification. Rain of any appreciable intensity will wash particles from the slides in spite of the shield. Furthermore, hydrophilic coatings (e.g., glycerin jelly) change their adhesiveness with relative humidity and occasionally imbibe enough moisture to run from the slide. Particles are counted during spaced traverses of the slide's width, and results are expressed as particles per square centimeter of slide area.

Gravity slide data provide semiquantitative information that is useful for tracing the seasonal trends of specific airborne particles. However, since deposition is by settling *and* turbulent impaction, both of which vary exponentially with particle radius (see Sec. IV.C.2), smaller pollen grains and spores are severely undersampled in gravity collections. A common reticence to use higher than 100× magnification has also, undoubtedly, contributed a size-related bias.

In recent years, the quantitative limitations of the gravity slide sampler have also received emphasis (159,160). Past attempts (157) to derive particle concentrations per unit volume of air, by applying empirical conversion factors to gravity slide counts, can no longer be defended; the volume of air contributing particles in any time period simply is never known. Furthermore, field comparisons with mechanical volumetric samplers have shown that wind speed and direction serve to modify the "catch" of the gravity slide. Since day-to-day variations in wind velocity and the structure of turbulence are great, *short-term* trends indicated by gravity slides often may be factitious. Despite these limitations, its simplicity, low cost, and independence of power source have maintained the popularity of this device. Some physicians have also emphasized that, until dose-response relationships for aeroallergens are known, more precise information than that obtained from gravity slides has limited usefulness (161). However current this view may be, the effects of known concentrations of allergenic particles are being studied quantitatively in several laboratories at present. It seems possible, therefore, that a precise volumetric approach to airborne allergens may become a *necessity* in the foreseeable future.

Both suction and rotating-arm samplers have proven useful in the quantitative study of aeroallergens; however, no single device will collect the full range of particle sizes with acceptable efficiency. Suction samplers are especially well adapted to the study of particles with diameters less than 10 μ. In most devices, air is drawn in through a narrow intake, and the entrained particles impinge on collection surfaces just behind the entrance orifice. Even with small particles having relatively low momentum, these devices must be wind-oriented to prevent the occurrence of major sampling errors. Two versions of the automatic volumetric spore trap developed by Hirst (162) have become especially popular for sampling in free air. However, the tendency of molecular membrane filters to become overloaded rapidly has severely limited their usefulness (163).

Rotating-arm impactors (164,165) sample particles in the size range of pollens with high efficiency. With these devices, collection is effected by narrow, adhesive-coated surfaces that are whirled through the air at from 1500 to 2500 rpm. Impaction efficiency rises rapidly with increasing particle size, and collection is analogous to that from moving air by narrow vertical cylinders (166). The "flag" sampler provides a simple example of the same principle (164). In this case, a conventional pin, mounted in a glass bearing, is wound with sticky transparent tape to create both the sampling surface and a miniature wind vane. To obtain volumetric concentrations from the (pin) flag sampler, accurate recordings of wind velocity are essential. Three mechanical rotating-arm samplers, the rotorod, rotobar, and rotoslide devices, have also been developed, but only the last (165) has become available in a form permitting intermittant operation out-of-doors for 24 hr. The sources and relative merits of these samplers as well as details of their operation have been summarized by the author (167).

Fungi of many common genera produce spores that are indistinguishable microscopically and must be enumerated, where possible, by the colonies they produce in culture. As a rule, open Petri dishes of semisolid medium are exposed, and many surveys (168) have employed this method. As in the case of gravity slide sampling, larger particles are preferentially deposited, and wind conditions markedly affect the catch (169). In addition, spore viability is a factor that limits sampling efficiency but probably not the allergenicity of the total particle load. A truly "ideal" device for collecting living spores in moving air has not been devised. However, the Andersen sampler (170) can be made wind-oriented, and modifications of this instrument to reduce large particle losses have been described (171,172). For regular sampling in free air, the use of a single culture plate beneath stage 6 of the Andersen sampler has been feasible if very short exposure periods are also employed (167).

B. Pollinosis and Pollen Counts

The practice of publishing gravity slide counts of ragweed pollen is well established in many areas and is accepted eagerly by the public. While published "pollen counts" probably do little harm, their value is questionable, and news media have done little to prevent misinterpretation of these reports by a majority of their audience. Considering the limitations of the gravity slide method previously cited, it is not surprising that discrepancies between pollen data and patients' concurrent symptoms are common. In addition, since the samples are based on 24-hr exposures, each published "count" necessarily refers to a previous time period in which different airborne concentrations may have prevailed. Spatial variations in pollen prevalence are well recognized also, and persons living close to disturbed urban sites or at the edges of cities may expect a relatively heavy exposure.

Individual activities are known to influence the inhaled pollen dose. Remaining indoors has definite protective value for pollinosis patients, and this is augmented when artificial air cooling permits windows to be closed for prolonged periods. Conversely, heavy outdoor exercise and long automobile rides are conducive to heavy pollen exposures. The worsening of pollinosis symptoms by concurrent infection and by other potent allergens is predictable; however, additional factors are less readily characterized. Differences in the response to specific concentrations of ragweed pollen seem to occur in the course of a single pollen season (173,174). Furthermore, the circadian peaks of pollinosis symptoms, especially that of hay fever in early morning hours, often more closely anticipate, than follow, the daily pollen maxima (174). Quite apart from pollen exposure, experience suggests that irritant gases and inert particulates, changes in environmental temperature and humidity, exercise, body position, and emotional factors may all modify respiratory symptoms. Because of these factors, day-to-day variations in reported pollen data have very limited significance for individuals, and a greater public awareness of this point is indicated. It is important, also, that the comparative pollen prevalence at different "hay fever resorts" is not misrepresented by similar crude statistics.

VII. AEROALLERGENS IN NONATOPIC DISEASES

A. The "Farmer's Lung" Group

A variety of organic dusts produce illness with immunological features that do not suggest atopy. The best studied of these is "farmer's lung," in which prominent fever, cough, and breathing difficulty follow within hours after exposure to moldy hay and silage. Chills, headache, and lassitude are commonly associated, and granular or mottled areas appear on the chest X-ray. The changes in lung function are not those of asthma, but consist of nonuniform gas distribution and impairment of alveolar gas transfer, with little evident bronchial obstruction. These defects are reversible in acute situations but may become fixed, with reduced arterial blood oxygen saturation, if exposure to the offending agents continues. Thickening of the alveolar walls, due to an influx of tissue fluids and collections of mononuclear cells (granuloma formation), is prominent, and extensive scarring may follow (175). Acute symptoms are a cause of significant morbidity, and the chronic lung changes have led to severe respiratory disability and even death in agricultural areas of North America as well as in Great Britain and northern Europe.

Precipitating antibodies (see Sect. II.C.1) reacting with components of moldy hay are present in sera of almost all affected farmers (176,177) and in occasional *exposed* normal persons (177) but not in others. Furthermore, cell-free extracts of moldy hay, administered by aerosol, have reproduced the typical acute disease in farmer's lung patients, while others were essentially unaffected (177–179). Evidence has now accumulated that the involved antigens are derived from certain thermophilic actinomycetes, especially *Thermopolyspora polyspora* and *Micro-*

monospora vulgaris (177,179). These organisms flourish when hay and other forage are heated during fermentation by certain fungi and bacteria (180), whose growth is related, in turn, to the moisture content of the stored silage. This association is reflected in a higher incidence of acute symptoms after excessively wet summers; annual rates of from 10.5 to 193.1 cases per 100,000 persons have been reported from England and Wales (181). Although the antibody changes observed in farmer's lung patients are impressive, definite proof that they may contribute to the disease is lacking. Resolution of this problem may be aided by the study of "fog fever," an analogous respiratory disease of horses and cattle fed stored silage, in which precipitating antibodies also have been found (182). Both of these conditions are distinct from "silo-filler's disease" in which damage to small air passages reflects the direct toxicity of NO_2 (183).

A syndrome apparently identical with farmer's lung has followed exposure to moldy corn, barley, and oats, although precipitins to moldy hay and to *T. polyspora* have been less constant in such persons (177). In addition, febrile respiratory illnesses described in bagasse workers, foresters, and pigeon fanciers, appear to be closely related (178). Bagasse (dried sugar cane fiber) has found increasing use in industry, and affected workers have been those engaged in cutting or pulverizing the crude material. The disease (bagassosis) has been reported especially from Louisiana and Puerto Rico; scattered cases have appeared elsewhere (184). Precipitating antibodies to crude, moldy bagasse have been demonstrated in almost all affected patients and in a high percentage of healthy workers exposed to this material (185). Fresh dried cane and pressed bagasse by-products, such as paper, wallboard, and insulation, have relatively minor antigenicity, suggesting that their use does not pose a threat to community health (185). The bagasse components responsible for this illness have not been determined. A respiratory disease resembling farmer's lung has also appeared in sawmill workers exposed to logs infested with the fungus *Cryptostroma corticale* (syn. *Coniosporium corticale*) (186,187). The bark-stripping operation, which may involve massive inhalation of airborne spores, has been an important predisposing factor in this condition. Among hobbyists exposed to pigeons and budgerigars (parakeets), a similar illness is associated with antibodies precipitating with the serum, egg proteins, feathers, and droppings of these birds (188,189). The finding of material related to avian excreta in free urban air (190) raises the possibility that similar illness might occur apart from direct contact with birds. Possibly related though less well studied diseases have appeared in persons working with mushroom compost (191) and tamarind seed (192) as well as following therapeutic exposure to pituitary snuff from bovine and porcine sources (193).

B. Allergic Contact Dermatitis

Of the few contact sensitizers encountered primarily as aeroallergens, pollen oils are probably the most prevalent. The typical skin changes (see Sect. II.C.2) appear seasonally and are localized to exposed areas of the head and neck, upper anterior chest, and forearms. Ragweed pollen has been the principal offender (194), although pollens of burweed marsh elder and certain trees have in a few cases produced this condition (195). Persons with contact hypersensitivity to ragweed pollen commonly react also to pyrethrum (see Sect. V.B) insecticides used as aerosols. However, the smoke of burning plants, including poison ivy, does not appear to carry active material for any appreciable distance (196). Many contact sensitizers that are usually encountered as solid objects and liquids are also important, occupationally, as dusts and mists; paraphenylenediamine, nickel, cobalt, and chrome salts as well as parathion and malathion are familiar examples.

C. Diseases of Unknown Cause

Aeroallergens have been invoked as possible determinants in many additional illnesses. Their role has been most strongly questioned in endemic, obstructive respiratory disease such as that seen in the Tokyo-Yokohama area and in New Orleans. At present, specific allergens have not become apparent in either locality, although many affected persons have shown some prior evidence of atopy (197,198). Among American personnel who develop "asthma" in Japanese urban areas, a large bronchitic component is present, and an unremitting course requiring evacuation has occurred in up to 10% of the cases (199).

With its feature of periodic airway obstruction, byssinosis has been regarded as a possible allergic response to crude cotton, flax, and hemp fibers or to associated contaminants. Low levels of antibody to extracts of crude cotton are present in many textile workers and may be especially prominent in those with this condition (200). However, neither these findings nor the demonstrated histamine-like activity in extracts of unfinished cotton (201) can be definitely considered etiological. Since the fibers are often moldy, direct bronchial constrictors, elaborated by contaminating fungi, might also be sought (202).

The nature of sarcoidosis remains unclear, and its suggested relationship to inhaled pine pollen has never been convincingly established (203).

Similarly, whether the granulomatous tissue reaction and positive patch tests seen in berylliosis are related and a consequence of host sensitivity remains controversial (204). No more can be said definitely, at present, for metal fume fever, silicosis, mill fever and many other puzzling environmental conditions. It is doubtful, however, that the full scope of aeroallergen-related morbidity has already been defined. Immunologic considerations remain a prominent part of any comprehensive and thoughtful approach to the health effects of airborne substances.

References

1. S. M. Feinberg, *Allergy in Practice*, 2nd Ed., Year Book, Chicago, 1946, p. 8.
2. C. Von Pirquet, *Muench. Med. Wochschr.*, **53,** 1457 (1906), also translated as Appendix A in P. G. H. Gell and R. R. A. Coombs, *Clinical Aspects of Immunology,* Davis, Philadelphia, 1963.
3. W. C. Boyd, *Fundamentals of Immunology,* 4th ed., Interscience, New York, 1966, p. 414.
4. R. S. Farr, *Arch. Environ. Health,* **3,** 379 (1961).
5. M. Arthus, *Compt. Rend. Soc. Biol.,* **55,** 817 (1903).
6. C. G. Cochrane, W. O. Weigle, and F. J. Dixon, *J. Exptl. Med.,* **110,** 481 (1959).
7. J. P. Lewis and K. F. Austen, *New Engl. J. Med.,* **270,** 597 (1964).
8. W. Feldberg, "Distribution of Histamine in the Body," in *Histamine,* G. E. W. Wolstenholme and C. O'Connor, Eds., Little, Brown, Boston, 1956
9. J. Lecomte, *J. Allergy,* **28,** 102 (1957).
10. G. Katz and S. Cohen, *J. Am. Med. Assoc.,* **117,** 1782 (1941).
11. H. O. Schild, D. F. Hawkins, J. L. Mongar, and H. Herxheimer, *Lancet,* **1951-II,** 376.
12. W. E. Brocklehurst, *Progr. Allergy.* **6,** 539 (1962).
13. H. O. Schild, "Histamine Release and Anaphylaxis," in *Histamine,* G. E. W. Wolstenholme and C. O'Connor, Eds., Little, Brown, Boston, 1956.
14. H. Herxheimer and E. Stresemann, *J. Physiol. (London),* **165,** 78P (1963).
15. H. Herxhemier and E. Stresemann, *J. Physiol. (London),* **158,** 38 (1961).
16. M. E. Webster and J. V. Pierce, *Ann. N.Y. Acad. Sci.,* **104,** 91 (1963).
17. P. J. Lachmann, "Complement," in *Clinical Aspects of Immunology,* P. G. H. Gell and R. R. A. Coombs, Eds, Davis, Philadelphia, 1963.
18. E. Middleton and W. B. Sherman, *J. Allergy,* **31,** 441 (1960).
19. A. H. Sehon and L. Gyenes, "Antibodies in Nontreated Patients and Antibodies Developed During Treatment," in *Immunological Diseases,* M. Samter and H. L. Alexander, Eds., Little, Brown, Boston, 1965.
20. K. F. Austen and K. J. Bloch, "Differentiation *In Vitro* of Antigen-Induced Histamine Release from Complement-Dependent Immune Injury," in *Complement,* G. E. W. Wolstenholme and J. Knight, Eds., Little, Brown, Boston, 1965.
21. G. A. Settipane, J. T. Connell, and W. B. Sherman, *J. Allergy,* **36,** 92, (1965).
22. M. Samter and E. L. Becker, *Proc. Soc. Exptl. Biol. Med.,* **65,** 140 (1947).
23. K. Ishizaka, E. G. Dennis, and M. Hornbrook, *J. Allergy,* **35,** 143 (1964).
24. M. A. Ramirez, *J. Am. Med. Assoc.,* **73,** 984 (1919).
25. M. H. Loveless, *J. Immunol.,* **41,** 14 (1941).

26. H. N. Eisen, L. Orris, and S. Belman, *J. Exptl. Med.*, **95**, 473 (1952).
27. S. Epstein, *Ann. Allergy,* **10,** 633 (1952).
28. R. A. Cooke and A. Vander Veer, *J. Immunol.*, **1,** 201 (1916).
29. U. W. Schnyder, *Acta Genet. Statist. Med. Suppl.*, **10,** (1960).
30. D. Leigh, E. Marley, and D. Braithwaite, *Bronchial Asthma. A Genetic, Population and Psychiatric Study,* Pergamon Press, Oxford, 1967.
31. A. F. Coca and R. A. Cooke, *J. Immunol.*, **8,** 163 (1923).
32. S. Leskowitz and F. C. Lowell, *J. Allergy,* **32,** 151 (1961).
33. J. E. Salvaggio and S. Leskowitz, *Intern. Arch. Allergy Appl. Immunol.*, **26,** 264 (1965).
34. J. E. Salvaggio, J. J. A. Cavanaugh, F. C. Lowell, and S. Leskowitz, *J. Allergy,* **35,** 62 (1964).
35. J. E. Salvaggio, H. Kayman, and S. Leskowitz, *J. Allergy,* **38,** 31 (1966).
36. R. A. Cooke, A. Menzel, P. Meyers, J. Skaggs, and H. Zeman, *J. Allergy,* **27,** 324 (1956).
37. W. J. Kuhns and A. M. Pappenheimer, Jr., *J. Exptl. Med.*, **95,** 363 (1952).
38. E. W. Kailin, E. A. Rossbach, and M. Walzer, *J. Allergy,* **21,** 225 (1950).
39. D. B. Sparks, S. M. Feinberg, and R. J. Becker, *J. Allergy,* **33,** 245, (1962).
40. R. A. Cooke, M. H. Loveless, and A. Stull, *J. Exptl. Med.*, **66,** 689 (1937).
41. J. H. Barnard, *J. Allergy,* **40,** 107 (1967).
42. J. Smith, J. E. Johnson, and L. E. Cluff, *New Engl. J. Med.*, **274,** 998 (1966).
43. V. B. Chambers and J. Glaser, *J. Allergy,* **29,** 249 (1958).
44. W. S. Curran and G. Goldman, *Ann. Internal Med.*, **55,** 777 (1961).
45. R. S. B. Pearson, *Quart. J. Med.*, **30,** 165 (1937).
46. A. S. Weiner and I. J. Silverman, *J. Exptl. Med.*, **71,** 21 (1940).
47. M. Schwartz, *Heredity in Bronchial Asthma,* Ejnar Munksgaards Forlag, Copenhagen, 1952.
48. P. F. DeGara, "The Hereditary Predisposition in Man to Develop Hypersensitivity: A Critical Review," in *Mechanisms of Hypersensitivity,* J. H. Shaffer, G. A. LoGrippo, and M. W. Chase, Eds., Little, Brown, Boston, 1959.
49. A. S. Weiner, I. Zieve, and J. H. Fries, *Ann. Eugenics,* **7,** 141 (1936).
50. R. L. Tips, *Am. J. Human Genet.*, **6,** 328 (1954).
51. P. P. Van Arsdel, Jr., and A. G. Moltulsky, *Acta Genet.*, **9,** 101 (1959).
52. B. Ratner and D. E. Silberman, *Ann. Allergy,* **10,** 1 (1952).
53. W. C. Spain and R. A. Cooke, *J. Immunol.*, **9,** 521 (1924).
54. R. Bowen, *J. Allergy,* **24,** 236 (1953).
55. D. Spaich and M. Ostertag, *Z. menschl., Vererb. Konstitutionslehre,* **19,** 731 (1936) through Ref. 48.
56. W. P. Buffam and B. Feinberg, *J. Allergy,* **11,** 604, (1940).
57. G. D. Barkin and J. P. McGovern, *Ann. Alergy,* **24,** 602 (1966).
58. Definition and Classification of Chronic Bronchitis, Asthma and Pulmonary Emphysema, *Am. Rev. Respirat. Diseases,* **85,** 762, (1962).
59. G. Simon and W. S. Jordan, *J. Pediat.*, **70,** 533 (1967).
60. H. J. Witting, N. J. Crawford, J. Glaser, and G. Lanoff, *J. Allergy,* **30,** 19 (1959).
61. W. T. Vaughan and J. H. Black, *Practice of Allergy,* 3rd Ed., Mosby, St. Louis, 1954, Chapter 8.
62. I. Broder, P. P. Barlow, and R. J. M. Horton, *J. Allergy,* **33,** 513 (1962).
63. C. J. Maternowski and K. P. Mathews, *J. Allergy,* **33,** 130 (1962).
64. W. Schepegrell, *Hay fever and Asthma,* Lea and Febiger, Philadelphia, 1922.

65. Commission on Chronic Illness, *Chronic Illness in a Large City, The Baltimore Study*, Harvard Univ. Press, Cambridge, 1957.
66. V. J. Derbes and H. T. Engelhardt, *Am. J. Med. Sci.*, **205,** 675 (1943).
67. A. A. Thommen in *Asthma and Hay Fever in Theory and Practice*, A. F. Coca, M. Walzer and A. A. Thommen, Thomas, Springfield, 1931.
68. I. Broder, P. P. Barlow, and R. J. M. Horton, *J. Allergy*, **33,** 524 (1962).
69. M. M. Peshkin, *Am. J. Diseases Children*, **36,** 89, (1928).
70. H. G. Rapaport, *N. Y. State J. Med.*, **58,** 393 (1958).
71. J. A. Clarke and H. C. Leopold, *J. Allergy*, **11,** 494 (1940).
72. W. R. Solomon, K. P. Mathews, and J. M. Sheldon, unpublished observations.
73. H. A. Allard, *Science*, **98,** 292 (1943).
74. D. A. Williams, in *International Textbook of Allergy*, J. A. Jamar, Ed., Thomas, Springfield, Illinois, 1959, Chapter 4.
75. J. M. Quintero, in *Regional Allergy of the United States, Canada, Mexico and Cuba*, M. Samter and O. C. Durham, Eds., Thomas, Springfield, Illinois, 1955, Chapter 14.
76. H. H. Shilkret and L. C. Lazarowitz, *Ann. Allergy*, **11,** 194 (1953).
77. A. J. Fine and L. E. Abram, *J. Allergy*, **31,** 375 (1960).
78. R. F. Hughes, *Can. Med. Assoc. J.*, **80,** 651 (1959).
79. E. W. Phillips, *J. Allergy*, **11,** 28, (1939).
80. E. W. Phillips, *J. Allergy*, **12,** 24 (1940).
81. H. L. Green and W. R. Lane, *Particle Clouds: Dusts, Smokes and Mists*, 2nd ed., Spon, London, 1964.
82. P. H. Gregory, *The Microbiology of the Atmosphere*, Interscience, New York, 1961.
83. K. Maunsell, *Progr. Allergy*, **4,** 457 (1955).
84. P. H. Gregory, *Nature*, **180,** 330 (1957).
85. H. D. Landahl, *Bull. Math. Biophys.*, **12,** 161 (1950.)
86. T. F. Hatch and P. Gross, *Pulmonary Deposition and Retention of Inhaled Aerosols*, Academic Press, New York, 1964, p. 54.
87. K. D. Figley and R. J. Elrod, *J. Am. Med. Assoc.*, **90,** 79 (1928).
88. D. Ordman, *Intern. Arch. Allergy Appl. Immunol.*, **7,** 10 (1955).
89. L. F. Daws, "Movement of Air Streams Indoors," in *Airborne Microbes*, P. H. Gregory and J. L. Monteith, Eds., Cambridge, Cambridge, England, 1967.
90. C. M. Christensen, *J. Allergy*, **31,** 409 (1950).
91. M. Corn and F. Stein, "Mechanics of Dust Redispersion," in *Surface Contamination*, B. R. Fish, Ed., Pergamon, Oxford, 1966.
92. R. M. Brown, D. A. Larson, and H. C. Bold, *Science*, **143,** 583 (1964).
93. C. T. Ingold, *Dispersal in Fungi*, Clarendon, Oxford, 1953.
94. R. S. Smith, *Trans. Brit. Mycol. Soc.*, **49,** 33 (1966).
95. M. H. Zoberi, *Ann. Botany*, (*London*), **25,** 53 (1961).
96. R. A. Bagnold, *Intern. J. Air. Pollution*, **2,** 357 (1960).
97. J. B. Tyldesley, "Movement of Particles in the Lower Atmosphere," in *Airborne Microbes*, P. H. Gregory and J. L. Monteith, Eds., Cambridge, Cambridge, England, 1967.
98. F. S. Duckworth and J. S. Sandberg, *Bull. Am. Meteorol. Soc.*, **34,** 198 (1954).
99. A. R. Weinhold, Technical Report, Office of Naval Research, ONR Contract N9 onr 82400, 1955, cited by J. M. Hirst and G. W. Hurst, in *Airborne Microbes*, P. H. Gregory and J. L. Monteith, Eds., Cambridge, Cambridge, England, 1967.

100. R. S. Shapiro and R. Rooks, *J. Allergy,* **22,** 450 (1951).
101. N. Polunin and C. D. Kelly, *Nature,* **170,** 314 (1952).
102. G. Erdtman, *An Introduction to Pollen Analysis,* Chronica Botanica, Waltham, Mass., 1954, p. 177.
103. J. M. Hirst and G. W. Hurst, "Long Distance Pollen Transport," in *Airborne Microbes,* P. H. Gregory and J. L. Monteith, Eds., Cambridge, Cambridge, England, 1967.
104. P. H. Gregory, *Pollen Spores,* **4,** 348 (1962).
105. J. E. McDonald, *Science,* **135,** 435 (1962).
106. A. C. Chamberlain, "Deposition of Particles in Natural Surfaces," in *Airborne Microbes,* P. H. Gregory and J. L. Montheith, Eds., Cambridge, Cambridge, England, 1967.
108. F. H. Milner, E. C. Tees, B. Dybas, and P. M. Dean, *Acta Allergol.,* **20,** 379 (1965).
109. M. B. Cohen, T. Nelson, and B. H. Reinarz, *J. Allergy,* **6,** 517 (1935).
110. W. E. Vannier and D. H. Campbell, *J. Allergy,* **32,** 36 (1961).
111. R. D. A. Peterson, P. E. Wicklund, and R. A. Good, *J. Allergy,* **35,** 134 (1964).
112. B. Pernis, E. C. Vigliani, C. Cavagna, and M. Finulli, *Brit. J., Ind. Med.,* **18,** 120 (1961).
113. L. Berrens, J. H. Morris, and R. Versie, *Intern. Arch. Allergy Appl. Immunol.,* **27,** 129 (1965).
114. R. Voorhorst, F. Th. M. Spieksma, H. Varekamp, M. J. Leupen, and A. W. Lyklema, *J. Allergy,* **39,** 325 (1967).
115. L. Berrens, *Intern. Arch. Allergy Appl. Immunol.,* **29,** 575 (1966).
116. H. C. Wagner and F. M. Rackemann, *Ann. Internal Med.,* **11,** 505 (1937).
117. F. M. Newmark and I. H. Itkin, *Ann. Allergy,* **25,** 251 (1967).
118. G. L. T. Wright and H. T. Clifford, *Med. J. Australia,* **2,** 74 (1965).
119. W. Crocker, *Botan. Rev.,* **4,** 235 (1938).
120. J. M. Sheldon and E. W. Hewson, *Atmospheric Pollution by Aeroallergens,* Progress Rept. #4, National Institute of Allergy and Infectious Diseases, Research Grant No. E 1379 (C), University of Michigan Research Institute, Ann Arbor, Michigan, 1960, p. 10.
121. A. D. Groulx, *Can. J. Public Health,* **45,** 329 (1954).
122. M. Walzer and B. B. Siegel, *J. Allergy,* **27,** 113, (1956).
123. F. A. Simon, *J. Exptl. Med.,* **77,** 185 (1943).
124. A. R. Goldfarb, *J. Asthma Res.,* **2,** 7 (1964).
125. R. L. Meyers, *Federation Proc.,* **25,** 729 (1966).
126. G. L. Waldbott and M. S. Ascher, *Ann. Internal Med.,* **14,** 215, (1940).
127. R. A. Bruce, *Intern. Arch. Allergy Appl. Immunol.,* **22,** 294 (1963).
128. H. Herxheimer, H. A. Hyde, and D. A. Williams, *Lancet,* **1966-*I*,** 572.
129. I. L. Bernstein and R. S. Safferman, *J. Allergy,* **38,** 166 (1966).
130. J. P. McGovern, T. J. Haywood, and T. R. McElhenney, *Ann. Allergy,* **24,** 145 (1966).
131. S. M. Pady and C. L. Kramer, *Mycologia,* **59,** 714 (1967).
132. F. Perlman, *Acta Allergol.,* **21,** 241 (1966).
133. H. Osgood, *J. Allergy,* **28,** 113, 292 (1957).
134. K. D. Figley, *Am. J. Med. Sci.,* **178,** 338 (1929).
135. D. D. Stevenson and K. P. Mathews, *J. Allergy,* **39,** 274 (1967).

136. T. Matsumura, K. Tateno, S. Yugami, and K. Toshisada, *J. Asthma Res.*, **4,** 205 (1967).
137. D. W. Cowan, H. J. Thompson, H. J. Paulus, and P. W. Mielke, Jr., *J. Air Pollution Control Assoc.*, **13,** 546 (1963).
138. H. Herxheimer, *Lancet*, **1967-I,** 83.
139. R. A. Kern, *J. Allergy*, **10,** 164 (1939).
140. T. G. Johnston, A. G. Cazort, H. N. Marvin, R. B. Pringle, and J. M. Sheldon, *J. Allergy*, **22,** 494 (1951).
141. S. M. Feinberg and R. M. Watrous, *J. Allergy*, **16,** 209 (1945).
142. K. P. Mathews, *Am. J. Med.*, **44,** 310 (1968).
143. H. H. Gelfand, *J. Allergy*, **34,** 374 (1963).
144. H. G. Brugsch and H. B. Elkins, *New Engl. J. Med.*, **268,** 353 (1963).
145. H. C. Bruckner, S. B. Avery, D. M. Stetson, J. J. Ronayne, and V. N. Dodson, *Arch. Environ. Health*, **16,** 619 (1968).
146. K. Hirschorn, F. Bach, and R. L. Kolodny, *Science*, **142,** 1, 185 (1963).
147. H. A. Hyde and K. F. Adams, *An Atlas of Airborne Pollen Grains*, Macmillan, London, 1958.
148. W. R. Solomon and O. C. Durham, "Aeroallergens. 2. Pollen and the Plants that Produce Them," in *A Manual of Clinical Allergy*, J. M. Sheldon, R. G. Lovell, and K. P. Mathews, Saunders, Philadelphia, 1967.
149. R. P. Wodehouse, *Pollen Grains*, McGraw-Hill, New York, 1935. Reprinted by Hafner, New York, 1965.
150. G. Smith, *An Introduction to Industrial Mycology*, 5th Ed., Arnold, London, 1960.
151. S. Funder, *Practical Mycology—Manual for Identification of Fungi*, 2nd ed., Hafner, New York, 1962.
152. H. L. Barnett, *Illustrated Genera of Imperfect Fungi*, 2nd ed., Burgess, Minneapolis, Minn., 1960.
153. J. C. Gilman, *A Manual of Soil Fungi*, 2nd. ed., Iowa State College Press, Ames, Iowa, 1957.
154. W. R. Solomon, "Aeroallergens. 3. Fungi," in *A Manual of Clinical Allergy*, J. M. Sheldon, R. G. Lovell, and K. P. Mathews, Saunders, Philadelphia, 1967.
155. H. W. Wolf, P. Skalig, L. B. Hall, M. M. Harris, H. M. Decker, L. M. Buchanan, and C. M. Dahlgren, "*Sampling Microbiological Aerosols*," *Public Health Monograph* **60** (1959).
156. *Air Sampling Instruments. For Evaluation of Atmospheric Contaminants*, 2nd Ed., Am. Conference of Governmental Industrial Hygienists, Cincinnati, Ohio, 1962.
157. O. C. Durham, *J. Allergy*. **17,** 79 (1946).
158. Public Health Significance, Distribution and Control of Air-borne Pollens, *Am. J. Public Health*, **39,** (Year Book), 86 (1949).
159. A. N. Dingle, *Bull. Am. Meteorol. Soc.*, **38,** 465 (1957).
160. E. C. Ogden and G. S. Raynor, *J. Allergy*, **3,** 307 (1960).
161. M. M. Albert, *N.Y. State J. Med.*, **66,** 2409 (1966).
162. J. M. Hirst, *Ann. Appl. Biol.*, **39,** 257 (1952).
163. S. Cryst, C. W. Gurney, and E. Hansen, *J. Lab. Clin. Med.*, **46,** 471 (1955).
164. J. B. Harrington, G. C. Gill, and B. R. Warr, *J. Allergy*, **30,** 357 (1959).
165. E. C. Ogden and G. S. Raynor, *J. Allergy*, **50,** 1 (1967).
166. P. H. Gregory, *Ann. Appl. Biol.*, **38,** 357 (1951).

167. W. R. Solomon, "Aeroallergens. 1. Techniques of Air Sampling," in *A Manual of Clinical Allergy,* J. M. Sheldon, R. G. Lovell, and K. P. Mathews, Saunders, Philadelphia, 1967.
168. M. B. Morrow, G. H. Meyer, and H. E. Prince, *Ann. Allergy,* **22,** 575 (1964).
169. R. Rooks and R. S. Shapiro, *J. Allergy,* **29,** 24 (1958).
170. A. A. Andersen, *J. Bacteriol.,* **76,** 471 (1958).
171. K. R. May, *Appl. Microbiol.,* **12,** 37 (1964).
172. O. M. Lidwell and W. C. Noble, *J. Appl. Bacteriol.,* **28,** 280 (1965).
173. C. W. Gurney and S. Cryst, *Ann. Allergy,* **15,** 367 (1957).
174. J. E. Goodwin, J. A. McLean, F. M. Hemphill, and J. M. Sheldon, *Federation Proc.,* **16,** 628 (1957).
175. J. Rankin, W. H. Jaeschke, Q. C. Callies, and H. A. Dickie, *Ann. Internal Med.,* **57,** 505 (1962).
176. J. Rankin, H. A. Dickie, M. Kobayashi, and M. A. Stahmann, *J. Lab. Clin. Med.,* **60,** 1008 (1962).
177. J. Pepys and P. A. Jenkins, *Thorax,* **20,** 21 (1965).
178. J. Rankin, M. Kobayashi, R. A. Barbee, and H. A. Dickie, *Med. Clin. N. Am.,* **51,** 459 (1967).
179. J. V. Williams, *Thorax,* **18,** 182 (1963).
180. P. H. Gregory and M. E. Lacey, *J. Gen. Microbiol.* **30,** 75 (1963).
181. F. H. Staines and J. A. S. Forman, *J. Coll. Gen. Practit.,* **4,** 351 (1961); J. Rankin, M. Kobayashi, R. A. Barbee, and H. A. Dickie, *Med. Clin. N. Am.,* **51,** 459 (1967).
182. Editorial, *Lancet,* **1965-II,** 223.
183. T. Lowry and L. M. Schuman, *J. Am. Med. Assoc.,* **162,** 153 (1956).
184. H. A. Buechner, E. Aucoin, A. J. Vignes, and H. Weill, *J. Occupational Med.,* **6,** 437 (1964).
185. J. E. Salvaggio, H. A. Buechner, J. H. Seabury, and P. Arguembourg, *Ann. Internal Med.,* **64,** 748 (1966).
186. J. W. Towey, H. C. Sweany, and W. H. Huron, *J. Am. Med. Assoc.,* **99,** 453 (1932).
187. D. Emanuel, B. Lawton, and F. Wenzel, *New Engl., J. Med.,* **266,** 33 (1962).
188. C. E. Reed, A. Sosman, and R. A. Barbee, *J. Am. Med. Assoc.,* **193,** 261 (1965).
189. F. E. Hargreave, J. Pepys, J. L. Longbottom, and D. G. Wraith, *Lancet,* **1966-I,** 445.
190. M. M. Braverman, C. Theophile, F. Masciello, and C. Smith, *J. Air Pollution Control Assoc.,* **12,** 570 (1962).
191. L. S. Bringhurst, R. N. Byrne, and J. Gershon-Cohen, *J. Am. Med. Assoc.,* **171,** 15 (1959).
192. P. G. Tufnell and I. Dingwall-Fordyce, *Brit. J. Ind. Med.,* **14,** 250, (1957).
193. J. Pepys, P. A. Jenkins, P. J. Lachmann, and W. E. Mahon, *Clin. Exp. Immunol.,* **1,** 377 (1966).
194. J. L. Fromer and W. S. Burrage, *J. Allergy,* **24,** 425 (1953).
195. R. G. Lovell, K. P. Mathews, and J. M. Sheldon, *J. Allergy,* **26,** 408 (1955).
196. J. B. Howell, *Arch. Dermatol.,* **50,** 306 (1954).
197. M. Sponitz, *Am. Rev. Respirat. Dis.,* **92,** 371 (1965).
198. H. Weill, M. M. Ziskind, V. J. Derbes, R. J. M. Horton, and R. C. Dickerson, *Arch. Environ. Health,* **10,** 148 (1965).

199. H. W. Phelps, *Arch. Environ. Health,* **10,** 143 (1965).
200. A. Massoud and G. Taylor, *Lancet,* **1964-II,** 607.
201. A. Bouhuys and S. E. Lindell, *Experientia,* **17,** 211 (1961).
202. S. Onashi, I. Yamaguchi, I. Yamamoto, and Y. Kobayashi, *Proc. Soc. Exptl. Biol. Med.,* **113,** 977 (1963).
203. M. M. Cummings, *Acta Med. Scand. Suppl.,* **425,** 48 (1964).
204. P. Gross, *Arch. Environ. Health,* **3,** 379 (1961).

Catalytic Removal of Potential Air Pollutants from Auto Exhaust

ROBERT H. EBEL
Stamford Research Laboratories, American Cyanamid Company, Stamford, Connecticut

I. Introduction . . . 237
II. The Niche for Catalytic Devices . . . 239
III. History of Catalytic Devices . . . 241
IV. Catalyst Technology . . . 244
V. Catalytic Device Technology . . . 250
VI. Gasoline Engine Technology Related to Emission Control . . . 254
VII. Engine-Device Systems and Their Future . . . 261
VIII. Fuel Technology Related to Emission Control . . . 263
IX. Auto Exhaust Emission Standards . . . 265
X. Summary . . . 269
XI. Appendix A . . . 269
XII. Appendix B . . . 270
XIII. Appendix C . . . 271
XIV. Appendix D . . . 274
XV. Appendix E . . . 274
XVI. Appendix F . . . 275
XVII. Appendix G . . . 276
XVIII. Appendix H . . . 277
XIX. Appendix I . . . 278
XX. Appendix J . . . 279
XXI. Appendix K . . . 280
References . . . 280

I. INTRODUCTION

Successful widespread use of catalysts to purify gases and vapors in manufacturing processes in the United States is supported by extensive research competence in our academic and industrial laboratories. This fact leads one to wonder why catalysts are not widely used to purify automotive exhaust. The major difficulty with past approaches to catalytic devices as a solution to the automotive exhaust problem has been philosophical. Most proposers and critics of catalytic devices have until recently taken the viewpoint that such devices should be capable of bearing either all or none of the burden of reducing emission levels (13,14,51,82,88). There were a few exceptions (34,39). The earlier objective of a single catalytic device adaptable to any car and capable of purifying the exhaust of virtually any engine while meeting rigorous standards of cost, maintenance, and durability (Appendix B) has faded away. Much has been said (15,24,36,39) of the deactivation of exhaust catalysts by lead and by high temperature. Basically, these can be viewed as economic problems rather than as technical difficulties which must be overcome. Lead deactivation can be made tolerable by periodic catalyst replacement. Protection of devices from high temperatures is not an insuperable engineering problem. Since 1966 in California, and nationwide since 1968, engine modifications have reduced the amount of hydrocarbons emitted by the average new car to about one-third and carbon monoxide to about one-half the level exhausted by new cars in earlier years (108). In a context of well-maintained engines modified for relatively low exhaust-emissions levels, disposable or refillable catalytic devices will perform well. They will not perform well unless designed into specific power systems by car manufacturers.

Gasoline engines have been evolving steadily over the years in terms of performance and economy of operation. Drivers of cars have become conditioned to demanding and expecting good drivability, instantaneous power at their toe-tips, good mileage per gallon of fuel, and minimum maintenance cost. In the majority of car purchasing decisions, cost, performance, and styling continue to be the major deciding factors. The value system of the driving public does not yet generally place any premium on superior exhaust control systems or on a program of regular maintenance of their cars by highly skilled mechanics. Car advertising, heavily directed at styling and performance, may be taken as a sometimes fallible indicator of the driving public's priorities. The public's opinion of car maintenance is indicated by the relatively low estate of skilled garage mechanics and the relatively large amount of do-it-yourself car repair. Therefore, to control and to maintain control of automotive air pollution, enforced legislation is an absolute necessity. Legislated standards for automotive exhaust are a reality (22,27–29). Of extreme importance is the fact that a program has been laid out (Appendix E) for making these standards progressively more stringent over the next five years. Car manufacturers will meet the new standards by making economic choices between feasible alternative methods. There is a high probability that in the early seventies catalytic exhaust devices will be economically preferable to the feasible alternatives then available.

A number of new types of engines with less severe emission problems than present internal combustion spark ignition engines have been publicized and discussed (3,118,121). Among them are electrical motors energized by batteries or fuel cells, external combustion engines using water or some other working fluid, and gas turbine engines (1). As a matter of necessity, car manufacturers have constructed prototypes and made cost–performance studies of such new power systems for at least two reasons. First, car manufacturers need to know if their competitive position might be underminable. Second, they need to know the cost differential available for further improving the internal combustion spark ignition engine. Thus far it appears that automobiles will be powered by present types of gasoline engines for at least the next decade or two (3,17,24). More fundamental is the criticism of the present type of automobile powered by any engine as a costly, hazardous, and inefficient means of transport (21). Development of an alternative to our present car–highway system is likely to require several decades.

II. THE NICHE FOR CATALYTIC DEVICES

The proliferation and evolution of techniques for the control of air pollutants in automotive exhaust has been dramatically rapid over the past ten years and promises to be equally dramatic over the next ten years. There is a mountain of literature recording the data of the past and projecting possible paths of the future. To relate this welter of data and techniques to catalytic devices, a systems approach is needed. A catalytic device for exhaust treatment cannot be designed in isolation because it is only another significant variable in the particular power system of which it is a part. Other parts of the power system must be optimized to include the capabilities and limitations of the device. To be sure, a single overdesigned type of device could be built to handle the exhaust from the unmodified engines of well-maintained compact, sport, or luxury cars and some poorly maintained cars. Because of the large car population, the economic waste of such overdesign has been and will continue to be unacceptable to legislators and to the automobile manufacturers.

In terms of a systems approach, the power train of cars has been subjected to a new constraint. Carbon monoxide (CO), hydrocarbons (HC), and nitrogen oxides (NO_x) in the exhaust must now not exceed certain limiting values. As frequently happens when a new, significant variable is introduced in a system, large blocks of earlier data become irrelevant or must be repeated with measurement of the new variable included. Most of the significant variables affecting exhaust-emission levels of CO, HC, and NO_x have been either identified or studied thoroughly only within the past 15 years. For reasons already cited, car manufacturers cannot summarily sacrifice the performance optimization of engines to achieve minimal exhaust-emission levels. However, the mechanisms controlling emission levels of CO, HC, and NO_x have been sufficiently well clarified that car manufacturers have been able to meet present emission standards without impairing engine performance. In fact, car manufacturers have done so well so far (108,124) that one might well ask what factors will compel the use of catalytic devices as emissions standards approach closer to zero. Two factors come to mind. First, the car manufacturers have had to reduce CO and HC emission levels but not yet NO_x levels. Present standards require only that NO_x emission levels not be increased by more than 15% in the process of reducing CO and HC emission levels (6). The point is that combustion conditions within the cylinders favoring low CO and HC emissions also favor increased NO_x emissions. A catalytic device in the power system will allow considerably greater flexibility in design when lower NO_x emission standards become law in California in 1971 (Appendix E).

The second factor that might compel the use of catalytic devices is somewhat more subtle. It concerns the way in which compliance with legislated emission standards is to be enforced. To comply with federal standards car manufacturers must now obtain certification that the cars they produce for sale will *on the average* not exceed standard emission levels of CO, HC, and NO_x (27). Eventually, each state will have to enforce continued compliance to car emission standards from the amorphous public. Such a situation virtually demands that each and every car periodically must pass an emissions test administered by the state. In California this will happen in 1970 (11). New Jersey is studying exhaust analyzers for use in its state-operated car inspection stations (127). Other states will follow. Consider then the plight of the motorist in the next decade whose car does not pass inspection because its HC emission level is somewhat high. Engine flaws or maladjustments causing relatively small increases in emission levels can be very elusive of diagnosis. Recycling through an inspection station time after time without passing is a frustrating experience. At such a juncture there would be much appeal to a snap-on type of catalytic device or "exhaust filter" similar to present air and oil filters with respect to maintenance and installation. Another aspect of the foregoing concerns quality control in car manufacture. Car manufacture is a mass production operation. Performance of cars produced will fall on the familiar Gaussian bell-shaped probability distribution curve. As emission standards are pushed toward zero, the curve will be compressed. Two recent papers (64,72) on carburetor quality control illustrate the problem very well. The cost of keeping the performance of *all* new engines within a very narrow quality control range may well begin to exceed the cost of designing a catalytic device into the power system.

III. HISTORY OF CATALYTIC DEVICES

There has been a steady stream of patents and publications on catalytic exhaust treatment for the past 50 years. At times the stream has slowed to a trickle but there have been three major surges. The first surge came in the late 1920's and early 1930's (Appendix F). Because the engines of the day were of low compression ratio and were richly carburetted, NO_x emissions were not a real problem. As a matter of fact, because of the low automobile population, there was not a real automotive exhaust problem at all in today's terms. Even so, the exhaust smoke was visible, the smell of unburned gasoline was apparent, the toxicity and presence of CO were well known, and tetraethyl lead was finding use as an antiknock additive for gasoline. Many scientists were genuinely concerned. Those who worked on the problem used the approach described in the first paragraph of this chapter, namely to devise a sort of "black box" to be attached to any engine. Improvements in car manufacture did reduce the visible manifestations of automobile exhaust (1). There was no documentation that the invisible portion of automobile exhaust was directly or indirectly an air pollution problem. Therefore, there was no basis for legislation regulating automotive exhaust.

The second surge in effort expended on catalytic exhaust treatment started slowly in the 1950's, grew to enormous proportions in the early 1960's, and ended abruptly in 1964 (Appendices G–K). Conclusive documentation was then at hand (111,112,120) demonstrating that the HC and NO_x in automotive exhaust were the major cause of photochemical smog in the Los Angeles basin. Much has been written about the triangular Los Angeles bowl with mountains for two of the walls and a very light sea breeze for the third wall. Thermal inversion layers of air put a lid on the bowl. The picture is completed by the bright California sun shining through the lid and irradiating the exhaust of about three million automobiles to make photochemical smog inside the covered bowl. Smog incidents regularly aroused public clamor that something be done.

California's situation was almost desperate in the early 1960's. Most of the nonautomotive sources of air pollution had been or would soon be cleaned up. Photochemical smog had been publicly identified with automotive exhaust. The California Motor Vehicle Pollution Control Board (CMVPCB) had been established to do something about the problem. California had passed a law requiring that the concentration of HC and CO in exhaust gas be substantially reduced (Table I), (24).

TABLE I

Automotive Pollutant Levels vs. Standards in California ca. 1960

	Approximate uncontrolled exhaust emission level per car	Maximum exhaust emission level required by law—if triggered
HC	900 ppm	275 ppm
CO	3.2%	1.5%
NO_x	1500 ppm	350 ppm[a]

[a] The NO_x standard has not been triggered to date.

The law was inoperative until at least two devices that could reduce emission levels as in Table I had met the requirements detailed in Appendix A. In other words, the CMVPCB was faced with public clamor to do something about smog, knew that auto exhaust gas was the primary cause of smog, but had an inoperative law for remedial action (4). Car manufacturers, collectively known as Detroit, were generating much valuable information (14). However, at that time the likelihood that Detroit would trigger the law seemed remote (5,67).

At this juncture, various teams of catalyst manufacturers and muffler manufacturers (35) such as Walker Manufacturing Co. with American Cyanamid Co., W. R. Grace Co. with Norris-Thermador Co., and Arvin Co. with Universal Oil Products (UOP) began to run fleet tests to qualify catalytic mufflers or noncatalytic afterburners for HC and CO removal under the California law. These teams were later to be characterized as "unsung heroes" (5). Again the approach was as described in the first paragraph of this chapter. The test fleets were selected by the CMVPCB to be representative of the California car population. It was surprising and discouraging to discover what poor condition an engine could be in and still provide quite adequate transportation. In initial fleet tests, no catalytic devices were able to cope with the formidable emission levels of these badly worn and/or maladjusted engines. In subsequent fleet trials the teams were allowed an initial tune-up of the engines to manufacturer's specifications. A sufficient number of devices then passed fleet tests to trigger the law. At the same time Detroit announced a capability to meet California's standards with their own devices (2). Detroit's devices relied primarily on lean carburetion with or without injection of air into the exhaust system for further combustion of HC and CO to meet standards (62,68,87). In due course, Detroit's fleets also passed the requirements detailed in Appendix A. In 1965 the law was revised. The new requirements detailed in Appendix B include 50,000-mile device durability and a cost limitation of not more than $65 per car for the exhaust control system. The devices that had triggered the law were decertified for not meeting the revised requirements. In retrospect, this outcome was best for the nation's economy.

The third surge of effort on catalytic devices is now well started. For the first time a systems approach is being used with Detroit and catalyst manufacturers cooperating to optimize power systems including a catalytic device. One group consists of the Ford Motor Company and the American Cyanamid Company. Another group is the Inter Industry Emissions Control Group (IIECG) made up initially of Ford and the Mobil Oil Corp., later joined by American Oil, Atlantic Richfield, Fiat S.p.A., Marathon Oil, Mitsubishi Heavy Industries Ltd., Nissan Motor Company (Datsun), Standard Oil of Ohio, Sun Oil, and Toyo Kogyo Co. Ltd. (8). A third group made up of Esso and Chrysler appears to be focussing on fuel and engine technology without the use of catalytic devices. If any one group's efforts are successful, their technology will become available to the rest of Detroit through the Automobile Manufacturer's Association (AMA) (67). Present litigation between the government and car manufacturers may sharply reduce or eliminate the exchange of technology through the AMA.

The three major centers of technical, financial, and regulatory effort directed at the control of automobile exhaust are the automobile industry, the oil industry, and the government (10,25). The automobile industry is served by the AMA and the oil industry by the American Petroleum Institute (API). Liaison between these two industries is provided by the Coordinating Research Council (CRC). The arm of the federal government concerned with automotive exhaust is the

National Air Pollution Control Administration (NAPCA), part of the Department of Health, Education and Welfare (HEW). At the state level there are or will be boards or commissions such as California's Air Resources Board (ARB), successor to the CMVPCB.

IV. CATALYST TECHNOLOGY

Over the past 50 years engine and fuel technologies have made tremendous strides. Catalyst technology also has made great strides. An enormous catalysis literature has been built up. A large array of sophisticated empirical methods for finding out a great deal about specific catalyst systems has been developed. There is still lacking, however, a quantitative theory explicitly describing the mechanism of heterogeneous catalysis. In the absence of predictive theory, a great deal of screen testing of different catalyst compositions must be performed. Catalyst compositions with excellent activity for exhaust gas purification have been found as a result of screen testing (Appendix G). Whether these catalysts are or are not adequate depends on the way minimum performance standards are set, as discussed earlier.

Catalysts used in devices are composite materials usually in the form of free-flowing granules. Individual granules are either cylindrical or spherical with the diameter of the circular cross-section about $\frac{1}{16}$ to $\frac{1}{4}$ in. Catalyst granules are usually built up from a porous refractory inorganic oxide, such as alumina, silica, silica–alumina mixtures, or calcined clays, called the catalyst support. Covering the surface of the pore walls is a thin layer, hopefully not more than a few molecules deep, of the active catalytic ingredient called the promoter. The promoter is usually one or more metals or metallic oxides. Promoters decrease the original pore volume of the support depending on the amount of promotor added. Porosity of catalysts for use in exhaust treatment ranges from about 0.2 cc/g, a dense catalyst, to over 1.0 cc/g, a light catalyst. For tolerance toward lead from leaded gasoline, high porosity is desirable. On the other hand, low porosity is desirable for good mechanical strength of the catalyst granules (Appendix G—U.S. Pat. 3,282,861). In practice, one compromises on intermediate porosities. The average diameter of the pores in an exhaust catalyst is measured in Angstroms (1×10^{-8} cm) and is usually less than 100 Å. Internal surface area due to these small pores ranges from 100 to 300 m^2/g in fresh catalyst. In more mundane terms, a 10-lb bed of such catalyst has a pore wall surface area of roughly 100–300 acres. Because the thin film of active promoter is spread out over this surface it is of the utmost importance to avoid conditions which cause the pores to become blocked with inert material, cause the surface area to decrease by destruction of small pores, or cause the film of promoter to agglomerate into crystalline masses.

Considering first the thermal degradation of typical catalyst supports, alumina, silica, and clays begin to lose surface area rapidly in dry heat at elevated temperatures (47). The effect of the loss of surface area on catalyst activity is cumulative, irreversible, and adverse. At high temperature there is migration of molecular moieties of silica or alumina in the support. As the smaller pores become filled, the remaining pores become larger. For relatively pure alumina there is not a net decrease in pore volume (47). There is a net decrease in pore volume for catalysts made from silica–alumina, which may also involve the promotors (Appendix C—U.S. Patent 3,094,394) (45,54). A net decrease in pore volume simply means that the catalyst granules shrink in size as sintering progresses. Mobility of the molecular moieties of alumina or silica is enhanced by the presence of water vapor. Automobile exhaust contains about 15% by weight of water vapor. Table II illustrates the surface area decline of gamma alumina as a function of time and temperature (47).

TABLE II[a]

Effect of Thermal Treatment on Alumina Surface Area

	Surface area of gamma alumina after	
Temperature	5 hr in air	5 hr in 3: 1 air–steam
Room temperature	278	
600°C (1112°F)	210	154
700°C (1292°F)	192	131
800°C (1472°F)	155	106
900°C (1652°F)	129	81

[a] Courtesy of Levy, Bauer, and Roth; reprinted from page 218 of reference 47 in part. Copyright September 20, 1966, by the American Chemical Society. Reprinted by permission of the copyright owner.

Large quantities of gamma alumina-based catalysts are used by the petroleum refining industry. Their operations with these catalysts are essentially steady-state continuous hydrogenations or dehydrogenations at temperatures ranging up to about 900°F. A carbon deposit called coke, although it contains some hydrogen, does accumulate in the catalyst pores. It is periodically burned off under carefully controlled conditions. Even so, the catalyst is exposed to transient temperatures of perhaps 1200–1300 °F during coke burnoff. The life of such catalysts will range from one to eight years depending on the temperature history of the particular catalyst. Life of catalyst in exhaust devices can certainly be cut short thermally as evidenced by the initial fleet failures in California. However, in subsequent fleet tests (59,60) the primary cause of catalyst deactivation was deposition of lead on the catalyst rather than catalyst sintering. In these fleet tests there were high-temperature (above 1200 °F) excursions but not prolonged high-temperature situations.

The three components of exhaust gas desired to be catalytically converted are CO, HC, and NO_x. Most catalytic device work has been directed at the oxidation of CO and HC to the innocuous products CO_2 and H_2O. Over the long term, CO_2 may not be innocuous (22). Removal of NO_x will be considered later. Catalysts disclosed in the patent literature for HC and CO oxidation appear to include virtually all the metals of the Periodic Table single or in combination on about all of the known refractory porous inorganic oxide supports. Appendices F, G, and H list some but by no means all of the pertinent U.S. patents. Those listed are intended to illustrate the diversity of catalyst compositions and the diversity of people and organizations active in the area. There have also been publications on HC and CO oxidation catalysts containing chromium (37), cobalt (52), vanadium (41), uranium (32), copper (44), and other metals (33,38,49). Filiform (45) and honeycomb (42) catalyst structures have been described and tested. In sum, catalysts of excellent oxidation activity for HC and CO are known. The problem is rather to design specific catalysts for specific devices to serve specific engines.

Catalyst activity may be considered in terms of the Arrhenius rate equation (eq. 1).

$$r = A \exp - E_{\text{act}}/RT \tag{1}$$

where r = reaction rate, A = preexponential term, E_{act} = energy of activation, R = gas constant, T = absolute temperature, and $\exp = e = 2.718$.

For a reaction normally proceeding in the vapor phase the activation energy is lower when the reactants are adsorbed on a catalyst surface because of the energy of chemisorption. For first-order reactions this may be shown by writing the exponent term as $-(E_{\text{act}} - Q)/RT$ where Q is the energy of chemisorption (48). More generally the numerator of the exponent is called the apparent activation energy. Catalytic oxidation of HC, CO, and H_2 in exhaust gas appears to follow first-order kinetics as long as there is an excess of oxygen present (39,41). A catalyst not only effectively lowers activation energy requirements, it also substantially increases the preexponential term A (58). In noncatalytic vapor phase reactions, A is presumed to be a measure of the number of effective collisions of reactant molecules per unit time. An effective collision is defined as one resulting in reaction. For catalytic reactions A is presumed to be a measure of the number of active sites on the catalyst surface. An active site may be defined as a place on the catalyst capable of inducing effective collisions between reactants, one or more of which is chemisorbed on the catalyst. There is usually an effective increase in concentration when a reactant goes from the vapor phase to the chemisorbed state. Because chemisorption lowers activation energy, a higher percentage of collisions is likely to be effective. Both the increase in concentration and the lower activation energy requirement will increase A and hence increase the reaction rate.

When exhaust gas containing O_2, CO, and HC is put through a catalyst bed at room temperature not much happens. As the exhaust gas temperature is raised a point will be reached, usually in the range of 300–700°F, where the temperature in the catalyst bed begins to exceed the temperature of the inlet exhaust. This is called the lightoff temperature for that particular catalyst.

Starting with a cold engine, the time elapsed until lightoff will depend on the temperature, composition, and rate of flow of the exhaust gas and on the mass and heat capacity of all the hardware that can cool off the exhaust gas. Heat sinks during warmup are the exhaust manifolds, the exhaust pipe, any exhaust pretreating devices such as lead traps, the catalytic device, and the catalyst bed.

The lightoff temperature depends, other things being equal, on the promoters and the number of active sites or surface area of the catalyst. Different promoters form different chemisorbed species with the reactants, resulting in greater or lesser apparent activation energies for reaction to occur. While the number of active sites is obviously important, there are other factors. These can be seen when the sequence of steps in a catalyzed reaction is examined.

Diffusion of reactants into catalyst pores
Chemisorption of reactants
Formation of activated complex
Breakdown of activated complex
Desorption of reaction products
Diffusion of products out of catalyst pores

Promoters have differing rates for the sequence of chemisorption–reaction–desorption (58), i.e., they are able to handle greater or lesser amounts of reactants per unit of time per active site. Platinoid metal promoters usually provide for lower lightoff temperatures than base metals, probably not because of greater site density but more likely through a combination of greater lowering of activation energy requirement and higher rates of processing reactants at active sites.

As the catalyst in a device loses surface area the lightoff temperature increases. For practical purposes, when the surface area of an exhaust catalyst reaches about 40 m^2/g, the lightoff temperature is usually well above 1000°F and the catalyst is dead. Appendix H lists a number of patents directed at solving the lightoff problem. Proposed solutions range from use of an "igniter catalyst" to electrical heaters to injection of raw gasoline into the inlet exhaust.

The importance of early lightoff can be seen in the way HC and CO are determined for device certification. As detailed in Appendix C, vehicles are tested from a cold start on an appropriately loaded chassis dynamometer. The vehicle engine is put through a seven-mode cycle (idle, acceleration, cruise, and deceleration) four times for warmup and five times after it is hot. The HC and CO analyses from the four warmup cycles and from two of the hot cycles are used to calculate emission levels. Weighting factors reduce the influence of warmup more than the engine choke increases it. If a catalytic device does not achieve lightoff until the first hot cycle, the very best HC and CO conversions that could be expected from it would be less than 60%. Note that emission levels formerly expressed in parts per million or per cent are now expressed in terms of grams per mile. The change was made to correlate emission standards with car size. Appendix D approximates emission concentrations equivalent to 1970 weight emission standards for HC and CO at several car sizes.

The problem of lead deactivation of exhaust catalysts is well known (39,59,60). At present, leaded gasoline may not legally contain more than 4 ml of tetraethyl lead per gallon of gasoline and in practice averages about 2.5 ml/gal. Scavengers must also be added, usually ethylene dichloride and dibromide, to prevent large deposits of lead oxides from building up in the engine cylinders. Volatile lead halide salts are formed via the scavengers. Gasoline also contains combined sulfur to the extent of about 0.05% by weight. This sulfur is oxidized to SO_2 when the gasoline is burned. At the high flame-front temperatures in the cylinder (ca. 5000°F) any SO_3 formed will dissociate back to SO_2 and O_2. The equilibrium constant shifts very much in favor of SO_3

formation at the temperature of the exhaust system, usually below 1200°F. At these temperatures the reaction rate of SO_2 and O_2 is so low that very little SO_3 is formed. However, almost any oxidation catalyst effective for CO and HC is also a good catalyst for SO_2 oxidation. The dewpoint of the lead halide salts in undiluted exhaust gas ranges from about 650 to 850°F (60). Diffusion of lead halide salts as liquid or vapor into the catalyst pores followed by reaction with SO_3, either in the vapor phase or just before desorption of SO_3 from the catalyst surface, undoubtedly accounts for the fact that lead deposits on exhaust catalysts are invariably mostly lead sulfate (60).

Nonvolatile lead salts block the pores of the catalyst and lower the diffusion rate of CO and HC to the interior of the catalyst granules. Appendix I lists a number of patents related to lead deactivation of catalysts. Among them are patents on the washing of lead from catalyst and on subjecting catalysts to impact to dislodge the lead salts and restore catalyst activity. Another indication that lead salts act by imposing a diffusion barrier rather than by poisoning the promoter is that CO conversion falls off much less rapidly than HC conversion as lead deposits build up (59). CO is a relatively small and light molecule with a higher diffusion coefficient than the average HC in exhaust gas.

Periodically, questions concerning the elimination of all lead from gasoline are raised. A recent report issued by the API (101) shows that eliminating all lead from gasoline would involve a capital cost to the petroleum industry of about 4.25 billion dollars. Amortization and added operating cost would appear publicly as an increase of several cents per gallon in the selling price of gasoline. The API report did not indicate the cost of lowering the lead content of gasoline to intermediate levels such as 1, 1.5, or 2 ml TEL per gallon. Undoubtedly, it is only a matter of time before this question is raised.

An important characteristic of catalysts is their selectivity in catalyzing some reactions more than others. Selectivity is related to the energy of chemisorption of reactants and to the activation energy needed for reaction to occur. Both of these factors can be measured for a given catalyst–reactant system. There is, however, no known way to predict quantitatively how selectivity will change as the catalyst is changed. Generally, oxidation catalysts for exhaust gas tend to oxidize preferentially the more reactive hydrocarbons such as olefins, isoparaffins, and cycloparaffins (41). Aromatics and acetylenes are oxidized more slowly. As might be expected, methane and ethane are the most difficult to oxidize. As reaction temperature is raised, the selectivity of catalysts in general decreases and ultimately disappears.

The literature on removal of NO_x from exhaust is relatively sparse in comparison with the literature on HC and CO. A few patents on catalysts (Appendix K) are based on the premise that there is not enough O_2 present to interfere with NO_x reduction. It has been shown that CO reacts with oxygen preferentially to NO over catalysts such as copper–cobalt–alumina (31) and chromia (55). In the former study (31) it is stated that sufficient CO must be present to consume all of the O_2 first with at least a stoichiometric amount of CO left for NO_x reduction. Obviously, for this to occur, the air–fuel mixture in the cylinders must contain less oxygen than required for complete combustion. When there was a sufficient excess of CO, NO_x removals of about 90% were obtained (31). In the latter study (55) relatively poor conversions of NO to a mixture of N_2 and N_2O were obtained over chromia on alumina even in the absence of oxygen. It must be pointed out that the objective of the latter study was to develop a hypothesis to explain catalyst selectivity. Other investigators using catalysts of copper on silica (56) or barium promoted copper chromite (30) obtained about 90% conversions of NO_x essentially in the absence of oxygen. The present trend toward lean carburetion of engines is not conducive to the absence of oxygen in exhaust, particularly in acceleration and cruise modes when most of the NO_x is formed. It is likely that significant reductions in NO_x emissions will be made by engine modifications. As was the case with HC and CO, catalytic devices for NO_x removal will probably have to wait a more thorough exploration and exploitation of less costly methods for reducing NO_x emissions.

V. CATALYTIC DEVICE TECHNOLOGY

A catalytic device is also a catalytic reactor. Chemical engineering as well as mechanical engineering skills are required to design an operable device. Such skills have been and are still being applied. Basic requirements for a device are now well known (34,39,41). Two simple designs are shown, a flat catalyst bed in Figure 1 and a radial exhaust flow device in Figure 2.

Neither of the above designs is complete, of course, as drawn. Even distribution of gas flow through the bed is important. Flow of gas into the dead-end center pipe of Figure 2 or into the simulated tidal bore of Figure 1 gives rise to a ram effect. Most of the exhaust will flow through the downstream end of the catalyst bed, focussing the heat load of HC and CO combustion. Lead deposition on catalyst will similarly be focussed. Distribution of lead in the catalyst bed can be

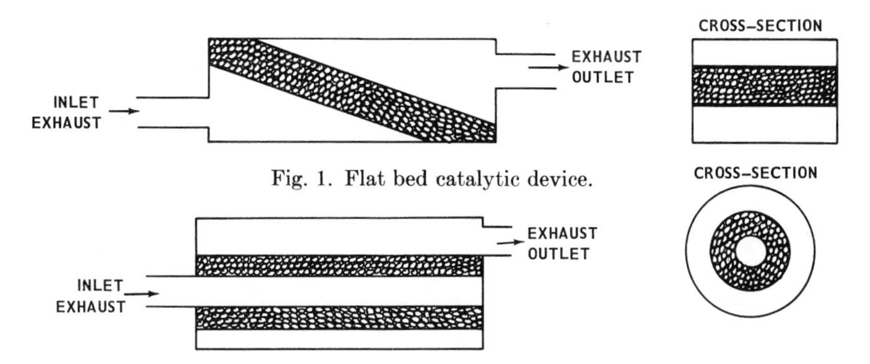

Fig. 1. Flat bed catalytic device.

Fig. 2. Radial flow catalytic device.

used to diagnose the flow pattern of exhaust gas through the bed. The device is detached while warm, plugged at one end, and filled with molten wax. After the wax has hardened, the device can be sectioned as desired with a metal saw and the lead distribution determined by X-ray photographs. The devices of Figures 1 and 2 must be provided with baffles upsteam of the catalyst bed to distribute the gas flow.

Pressure drop through the catalyst bed is theoretically influenced by bed thickness, catalyst granule dimension, and the per cent of hole area in the perforated sheet steel supporting the catalyst bed. High back pressure can be generated by a catalyst that crumbles to small pieces and cakes or by a catalyst bed that has been partially fused. Normally the back pressure across a device is primarily controlled by the per cent hole area in the catalyst support grid. This is because it is difficult to perforate alloy sheet steel with holes of smaller diameter. It is relatively easy to make a narrow-slot louver alloy sheet steel of relatively low per cent hole area. High per cent hole area can be attained with perforated stock but not with louvered stock.

When a car is started up on a cold winter morning, water vapor condenses in the exhaust system. The condensate contains dissolved SO_2, lead halides, plus a little hydrochloric and sulfuric acids. It is called "muffler acid." As the engine and the exhaust system become hot, the muffler acid evaporates or is blown out the tailpipe. When the car is allowed to cool, residual exhaust gases in the exhaust system condense to form muffler acid. Most mufflers corrode from the inside to the outside over a span of 1–4 years because of condensate. Any moving parts in the exhaust system, such as butterfly valve axles and heat riser valves, tend to corrode, rust, and become frozen unless constructed of special materials. Many exhaust catalysts are capable of withstanding regular leaching by muffler acid.

Location of a catalytic device in terms of distance from the exhaust manifold appears to be of marginal importance (58). The cooling rate of exhaust gas in regular single-walled exhaust pipe has been measured. It varies from 30 to 150°F per linear foot as a function of exhaust gas flow rate and temperature. For double-walled or ceramic-coated exhaust pipe, the cooling rate can be close to nil.

Differential thermal expansion is another device-design problem. If the catalyst support grids of Figures 1 or 2 are at 1200°F at the same time that the shell is at 600°F, the resulting stress will either crack the end plate weld seams or cause the grid supports to warp. Solutions to some of the foregoing problems are advanced in patents listed in Appendices I and J.

Device design modifications to provide protection to the catalyst against lead deactivation involve some sort of lead trap. Some of the lead traps described in Appendix I are: (*1*) a bed of metal promoter or lead reactant on a porous support to scavenge lead from the exhaust gas upstream to the main catalyst bed, (*2*) the same but with the lead-entrapping material comprising the upstream portion of the catalyst bed, (*3*) a cyclone separator or filter to remove particulate lead upstream to the catalyst. Mechanical lead filters and separators are generally ineffective because of the relatively low dewpoint of the lead halide salts in the exhaust. Some of the chemical lead traps, e.g., phosphoric acid and chromate, undoubtedly have stoichiometric limitations but that is not their major disadvantage. Any lead trap is a thermal luxury with respect to warmup of the catalyst bed. Should a lead trap be effective, the warmup problem would get worse with time. Assume a vehicle that attains 20 miles per gallon using a premium gasoline containing 3 ml of TEL per gallon. Three milliliters of TEL is roughly equivalent to 3 g of elemental lead. In 12,000 miles of travel, about 4 lb of lead are converted to about 6 lb of lead halides. About 25% of this is either entrapped as exhaust system deposits or finds its way to the engine lubricating oils (114). A portion of the remaining 75% or a maximum of about 4 lb as lead sulfate or phosphate could accumulate in the lead trap. This accumulation plus the mass of the trap itself acts as a heat sink. Essentially the same thing happens in a catalyst bed. Catalyst at the inlet side becomes leaded first. As the more distant parts of the catalyst bed begin to carry the conversion load, the leaded part of the bed extends the warmup time.

Very high temperatures can easily be reached in a catalyst bed because of its poor heat-rejection characteristics. Catalysts consisting of metal promoter on a porous refractory support are extremely poor conductors of heat. It is quite possible to measure a catalyst bed hot spot of 1500°F within millimeters of catalyst at 700°F. Heat rejection by radiation is insignificant in the desired temperature range, below about 1200°F. Convection is the only heat-rejection mechanism remaining. Essentially all of the heat generated by combustion of HC and CO in a catalyst bed is removed as sensible heat by the exhaust gas.

The temperature potential of an exhaust gas stream is defined as the temperature increase resulting from complete combustion of the HC and CO content of the gas stream. Equation 2 is a means of calculating temperature potential (39,57). It includes several factors not readily apparent. One hidden approximation is that the $CO–H_2$ ratio in exhaust averages about 3:1. Another is that the parts per million of hexane as measured by hexane-sensitized, nondispersive infrared (HSNDIR) is only about two-thirds of the true value as measured by flame ionization detector (FID) (24). The heat capacity of exhaust gas is approximately 0.02 Btu/°F standard ft^3.

$$\text{Temp. potential} = [(0.0062)\ (\text{ppm hexane}) + (4.1)\ (\%\text{CO})]/0.02 \qquad (2)$$

Cars adjusted for rich idle, or with maladjusted spark timing or with a sticking choke, or worn distributor bearings, or any of a dozen other flaws, can have extraordinarily high emission levels of CO and HC. A rich idle may give rise to as much as 2000 ppm HC and 10% CO in the exhaust gas. By eq. 2, these emission levels could cause a temperature rise in a catalyst bed of 2670°F. If a single cylinder is misfiring, the HC level in the exhaust may go to 10,000 ppm, equivalent to a temperature potential of about 3000°F.

Before engines were modified to reduce HC and CO emissions, catalytic devices had to be designed to handle, at least in part, the high emissions from rich idle, generously power-jetted acceleration and, worst of all ,the high raw gasoline load from poor combustion during deceleration. Otherwise the device would not pass the California test. Penalties were provided for the time that exhaust was bypassed around the device in the fleet tests of 1964. To illustrate the economics of the 1964 high-temperature problem, the absolutely essential combination of temperature sensor, bypass piping, bypass valve, and actuator for the bypass valve designed to reroute exhaust gas around the device when the catalyst bed exceeded about 1400°F would have cost about $30 in 1964 marketed through parts distribution channels. At present new car sales of about 10,000,000 per year, the cost to the nation would now be well over $300,000,000 per year. This is an example of the economic waste of over-design referred to earlier.

Now that engines have been modified for lower HC and CO emissions, catalytic devices do not have to be designed to handle any high emission levels. The device still needs protection against high temperature but this is a different and easier problem. High temperatures now occur only in unusual operating modes such as very high-speed cruise or rapid, prolonged acceleration. In normal operating modes, high temperature in a catalytic device would be a symptom of relatively severe engine malfunction. Malfunction is indeed the proper word because certification requirements are that engine modifications be capable of keeping emission levels at a relatively low and steady level for 50,000 miles.

A catalytic device loaded with fresh catalyst operating as part of a relatively simple power system design is capable of initially converting at least 70% of the HC and 80% of the CO in the engine exhaust to CO_2 and H_2O (59). This is without added air using today's modified engines. Conversions are based on the seven-mode cycle cold-start test (Appendix C). As the vehicle is driven, the conversion percentages will drop, partly as a function of mileage but mainly as a function of the cumulative amount of tetraethyl lead fed to the engine in the gasoline (59,60). In one particular reference cited (59) there was some loss in conversion attributable to loss of catalyst by attrition.

VI. GASOLINE ENGINE TECHNOLOGY RELATED TO EMISSION CONTROL

Perhaps the best way to illustrate some of the optimization choices and remedial measures taken by or available to car manufacturers to minimize exhaust emissions is to trace the course of gasoline from carburetor bowl to tailpipe. The schematic diagram of Figure 3 shows how filtered air flowing through a venturi throat creates a suction that inducts liquid gasoline from the carburetor bowl through the carburetor main jet into the intake manifold. Air flow is controlled by the throttle plate which in turn is controlled by the accelerator pedal. The rate of flow of gasoline into the airstream depends on the size of the metering orifice and the rate of flow of air through the venturi. As the air flow increases, the venturi suction also increases, thus pulling fuel through the carburetor jet at a greater rate (90).

The physical state of the air–fuel (A/F) mixture as it proceeds from the carburetor has been particularly well described by investigators at Ford (78) who installed a window in the full length of the intake

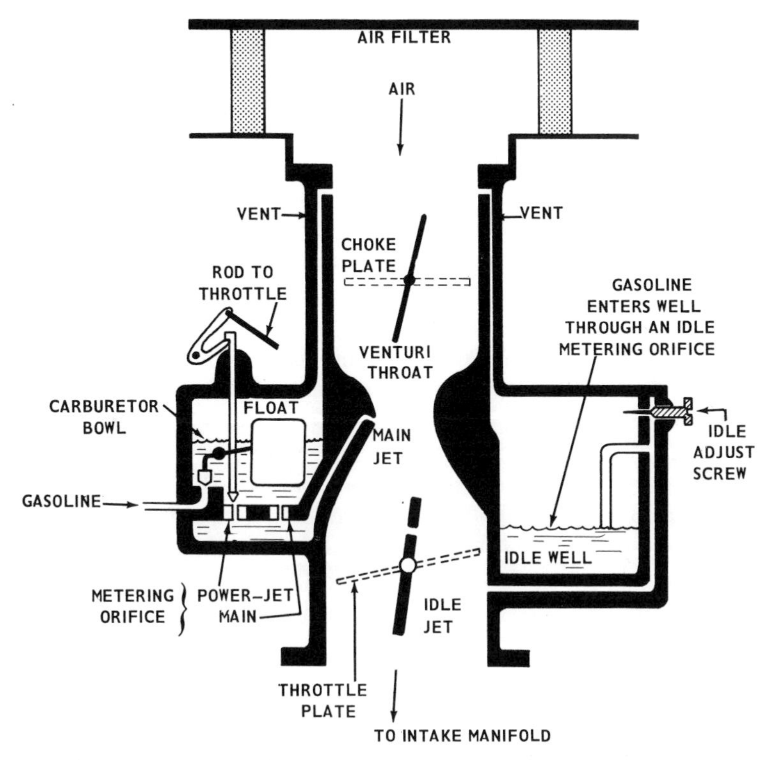

Fig. 3. Schematic drawing of a carburetor.

manifold of a six-cylinder engine. At idle, the A/F mix appeared to be a mist or fog. As the throttle was opened, droplets and globules of gasoline appeared. The interior surface of the intake manifold became wet with gasoline and actual puddles of gasoline appeared. When the throttle plate was closed and the engine began to pull a vacuum on the intake manifold the puddles of gasoline began to boil and the wetted surfaces became dry. Obviously, the A/F mixture going to the cylinders at this point was very rich. The actual driving condition simulated was a deceleration. During steady-state operation in idle or cruise modes the intake manifold equilibrates at a certain level of wetness. We have seen that a change of mode involving the closing of the throttle plate decreases the intake manifold wetness and enriches the A/F delivered to the cylinders. Conversely, acceleration which involves opening the throttle plate increases the intake manifold wetness and leans out the A/F mixture delivered to the cylinders. For cylinders farthest from the carburetor, the situation is worsened by the lower transport rate of liquid phase versus gas phase in the intake manifold. If the carburetor is designed to provide lean mixtures in the first place, then the further leaning in the initial phase of an acceleration will cause misfires in the cylinders. The resulting lack of power is called tip-in stumble, or lean-limit misfire.

The function of the power jet of Figure 3 is to permit the flow of sufficient gasoline both for power and for wetting the intake manifold when the throttle is opened. Emission control has necessitated reduction in the rate at which power jets formerly supplied gasoline to the intake manifold.

There are three commonly used measures of the air–fuel mixture. One is the A/F ratio which is the pounds of air per pound of fuel. The inverse of this, the F/A ratio, is also used. For the average gasoline the stoichiometric A/F ratio for theoretically complete combustion is approximately 14.6 or an equivalence ratio of 1.0, the third measure. An engine operating at an A/F ratio of 13.1 or (14.6 − 1.46) has an A/F equivalence ratio of 0.9 or is said to be 10% rich.

Two other terms of interest need also be defined. One is the mean break effective pressure (mbep) which is a measure of the useful power that can be gotten from a given engine. The other is the brake specific fuel consumption (bsfc) which for a given engine puts the constraint of fuel economy on power optimization. Other things being equal, the mbep is optimum at an equivalence ratio of about 0.9 or a 10% rich carburetion. The bsfc is optimum at an equivalence ratio of about 1.1 or about a 10% lean carburetion. Until the advent of emission-control legislation, automotive carburetors were uniformly designed to deliver A/F mixtures on the rich side of the stoichiometric ratio thus maximizing power. Figure 4 shows roughly how HC, CO, and NO emissions can vary as a function of the A/F ratio (61,73,77). Other factors are, of course, involved and will be discussed later.

Most interesting is the way in which various car manufacturers have coped with the problems of optimizing the induction system (carburetor plus auxiliaries) for emissions control without sacrificing car drivability. The automotive industry, associated industries, and universities have generated a great deal of data in a relatively short period of time. References 62, 68, 73, 76, 78, 80, 81, 87, 92, and 94 represent a selection of their work. The bulk of the references deal with much more than the induction system. Only one manufacturer (Volkswagen) has gone to a fuel injection system (66).

An essential change in carburetor appurtenances was a remodelled idle adjust screw. With the earlier type idle screw, the idling jet

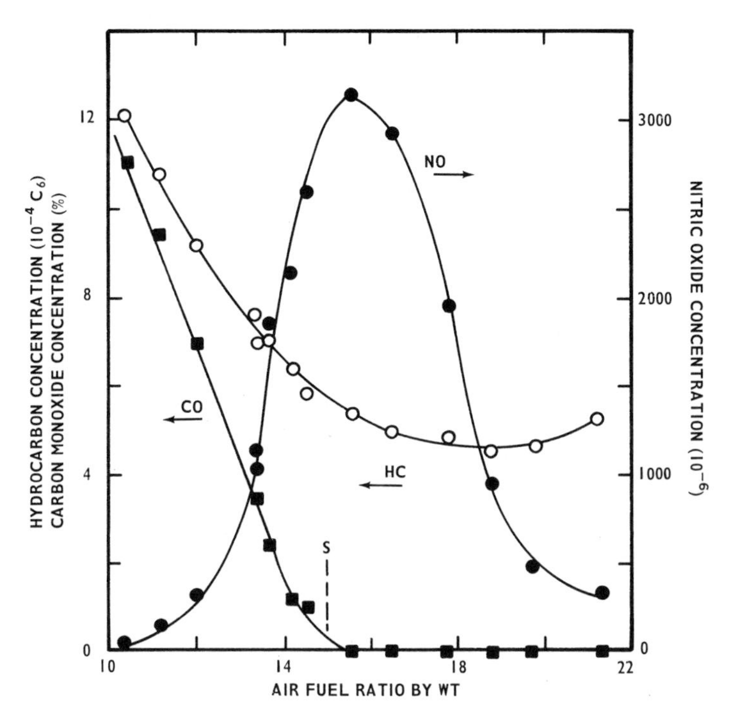

Fig. 4. The effects of air-fuel ratio on hydrocarbon, carbon monoxide, and nitric oxide exhaust emissions. Courtesy W. Agnew, General Motors Research Laboratories (1).

could be easily opened wide, thus undoing the efforts of the car manufacturer toward emission control. One remedy (Toyota) is to put a stop on the idle screw limiting the adjustment to a range of about ± 1 on the A/F ratio (92). Another way to achieve the same objective is to put an elongated tapered end on the idle screw (Chrysler). The tapered end fits into the idle jet and even when screwed out to the maximum does not provide adverse enrichment of the A/F mixture (indicated in Fig. 3.)

At idle, the air flow rate is low because the throttle is closed and the intake manifold vacuum is low. Linear velocity of the A/F mix is low, enhancing undesirable cylinder-to-cylinder variation in A/F ratio. To remedy this, most engines are now set for higher idling speeds, about 550–650 rpm versus the former 500–600 rpm. In some cars the size of the intake manifold has been reduced to provide for more turbulence and better mixing of air and fuel. One manufacturer (Volvo) uses two throttle plates. At idle and low engine load, one throttle plate is closed and forces the A/F mixture coming through the first throttle plate to go through a heated labyrinth where the air and fuel become better mixed. At higher engine loads, the second throttle plate opens and bypasses the labyrinth (80).

Other changes to the carburetor that have helped reduce exhaust emissions include some sort of air bleed into the intake manifold during deceleration. This can be done by drilling a hole in the throttle plate (indicated in Fig. 3) or by installing a throttle positioner to keep the throttle from closing completely. An engine is necessarily an effective air-pump because it must pump its air requirement in through the air filter and carburetor. When the throttle plate is completely closed, the engine is pumping from a closed system. A vacuum of up to 22–26 in. of mercury in the intake manifold can occur during a deceleration. This vacuum provides engine braking power. Air bleeds can only be used to reduce the intake manifold vacuum to about 22 in. of mercury. Even this reduction is of help in slowing the evaporation of liquid gasoline in the intake manifold. If the vacuum is further reduced, the reduction in engine braking power becomes noticeable (80) as does increased brake wear.

The choke plate in Figure 3 reduces the flow of air through the carburetor and permits the flow of gasoline through the jets to be increased by the intake manifold vacuum. Prolonged and tight closure of the choke plate can result in A/F ratios as low as 9:1 with exhaust emissions correspondingly rich in CO and HC. Duration and richness of choke have both been reduced in emission-controlled engines.

The form of spark ignition internal combustion engine most used operates on the Otto cycle. The four strokes of the Otto cycle are the intake, compression, power, and exhaust strokes. Position of a particular piston is given in terms of crank angle. This angle is between the axis of the cylinder and the specific crank in the crankshaft to which the piston is attached. To indicate ascent or descent of the piston, the angle is described as degrees before top dead center (btdc) or degrees after top dead center (atdc). Similar terms involve the bottom dead center.

As a piston at top dead center starts downward on the intake stroke, the exhaust valve is still open and stays open until the crank angle is about 15–30° atdc. In the combustion chamber above the piston there is residual gas from the previous exhaust stroke. Additional exhaust gas is sucked back through the exhaust valve until it closes. The intake valve opened earlier at about 15–30° btdc. Valve overlap is the period when both valves are open. At about 15–30° abdc the intake valve closes. For an engine running at 3000 rpm, the elapsed time for the intake stroke is 10 msec. Elapsed time for each of the other strokes is of course the same. At an engine idling speed of 600 rpm, the time for the intake stroke is 50 msec. Because the piston is in harmonic motion, it has maximum speed at 90° before or after tdc. Most of the actual flow of A/F into the cylinder therefore effectively occurs in less than the time of the full intake stroke. Turbulence of flow of A/F into the cylinder varies considerably as a function of engine speed, manifold vacuum, and the size of the intake valve port (89).

After the piston passes the bdc position, the compression stroke has begun. Timing the spark to ignite the A/F mixture is crucial both to engine power and to emissions control. For best power the combustion should be complete by about 15–30° atdc. Delay past this point encounters the possibility of the flame being quenched in the expansion or power stroke. The speed of propagation of the flame front from the spark plug out to the walls of the combustion chamber is obviously a key factor. Remarkably and fortunately the flame front travels faster and faster as engine revolutions per minute increase (90). Combustion is completed in the 10 msec available at 3000 rpm as well as in the 50 msec available at 550 rpm. The correspondence is not exact and may be compensated for by relatively small but important changes in spark timing (90). The increased turbulence of the A/F mixture at higher engine speeds is the reason for the increased flame speeds. Eddies and vortices small in size relative to the flame front enhance the propagation of the flame (90).

A number of interesting phenomena occur in the flame front. Bulk temperature in the combustion chamber reaches about 1500°F as measured at the nose of the spark plug. Flame front temperatures have been calculated at 4000–5000°F. At these temperatures the equilibrium constant and the k_1 rate of eq. 3 are favorable for NO formation (84).

$$N_2 + O_2 \underset{k_2}{\overset{k_1}{\rightleftharpoons}} 2NO \tag{3}$$

As soon as the flame front has passed, the burned gases cool by expanding against the unburned gases (99). At the bulk temperature of the burned gases k_2 is sufficiently low that the NO concentration is frozen at its highest level. There is some disagreement whether the highest level is an equilibrium NO concentration (75,84,129). One investigator has reported that addition of NO to the A/F charge of a cylinder did not increase the NO output of the cylinder after firing (84).

While a number of engine variables affect NO emission level (77), it is apparent that major reduction in NO must involve a lowering of the flame-front temperature. This can be done by adding inert material to act as a heat sink. In an earlier day, excess gasoline, i.e., rich mixtures, served this purpose. Exhaust gas recycled to the intake manifold can also serve (79). Experimental work indicates reduction of about 80% of NO emission levels by recycling of exhaust up to about 25% of the cylinder charge (71,83). Water injection into the intake manifold also lowers flame-front temperature (85). At an injection rate of 1.25 lb of water per pound of fuel, NO emissions were reduced by more than 90% and with some improvement in power and fuel consumption. Neither water injection nor exhaust recycle has been developed into a practical system for the reduction of NO emission levels. Recent development work on exhaust recycle encountered problems of fuel economy and drivability of the test vehicles (63).

The level of NO_x in exhaust depends primarily on cylinder flame-front temperature. CO level is primarily a function of A/F ratio. HC emissions have a more complex dependency. Only during misfires and poor combustion during deceleration does the composition of exhaust HC's closely correspond to fuel composition (100). A considerable amount of analytical work (100,115,126) is summed up in the statement that about 50% of the photochemically reactive hydrocarbons in exhaust are due to low molecular weight combustion products not present in gasoline (126). An additional conclusion is that exhaust hydrocarbons are made up of about 60% combustion products and 40% fuel components (126). Theoretically, one would expect all the HC to burn up in the cylinder if there were enough air present. A theory to explain the presence of low molecular weight combustion products and unburned fuel was developed and substantiated in a series of remarkable studies conducted at General Motors (70,93) over the past 18 years. One elegant experimental study (69) illustrates the technique. A single-cylinder engine using propane as fuel was equipped with an ingenious gas-sampling valve. During engine operation samples were taken from the cylinder. Sampling occurred immediately before and after flame-front passage. The sampling valve was adjustable to withdraw gas out to varying distances from the cylinder wall. Analysis was by chromatography. The results showed clearly that right at the cylinder wall the propane concentration was very high. Within about 0.001–0.005 in. of the wall were cracked propane products such as methane, ethane, ethylene, and propylene. Further out from the cylinder walls, combustion was relatively complete. The zone of cracking is called the quench zone. It has effects on NO and CO as well as on HC (86). Post-flame reaction continues to oxidize both CO and HC at a rate depending on the time–temperature exposure and oxygen concentration of the gases during and after the expansion and exhaust strokes (61,69). Continuance of this homogeneous nonflame oxidation of CO and HC in the exhaust manifold is the basis for the Ford Thermactor and the GM AIR (Air Injection Reactor) devices. In both devices the exhaust manifold is enlarged to provide residence time and insulated to keep the temperature up. Air is injected at the exhaust ports to supply an excess of oxygen. The devices operate at about 1100°F and are capable of oxidizing about half the HC and a lesser proportion of the CO. As with catalytic devices, overtemperature protection must be provided, although their conversion capability does not decline with lead deposits and cumulative time at high temperature. Manifold air injection became feasible only when a suitable air pump became available. Development of such a pump by GM was a considerable engineering achievement (91).

During the work on quench zones, additional factors were observed to affect HC emissions. The combustion flame front does not penetrate crevices with one small dimension. For example, the crevice representing the clearance between cylinder and piston down to the piston rings is not penetrated by the flame front. Significant amounts of HC are trapped there and in various other crevices (93) during the compression stroke. They escape during the expansion and exhaust strokes. Perhaps best advantage of this finding was taken by American Motors (76). Figure 5*b* shows most of the crevices eliminated by their redesign of combustion chamber. The initial design, Figure 5*a*, with center-well in the piston, called a squish piston, formed numerous crevices with the cylinder head. In Figure 5*b*, the present design, the surface-to-volume ratio of the combustion chamber has been decreased. The equivalent of crevice volume is apparently supplied by porous lead deposits in cylinders because such lead deposits have been shown to cause an increase in HC emission (97). There is some disagreement on the size of the increase with one study (Dupont) indicating an HC emission increase of about 5% (105) and another study (Ford) indicating about 20% (98). Both studies agree that engine deposits do not affect CO or NO emission levels.

VII. ENGINE-DEVICE SYSTEMS AND THEIR FUTURE

As stated earlier the problem of exhaust emission control will be solved by a series of economic choices between feasible alternatives. Table III emanated from discussions by representatives of government, industry, and universities as described in reference 24. Two columns

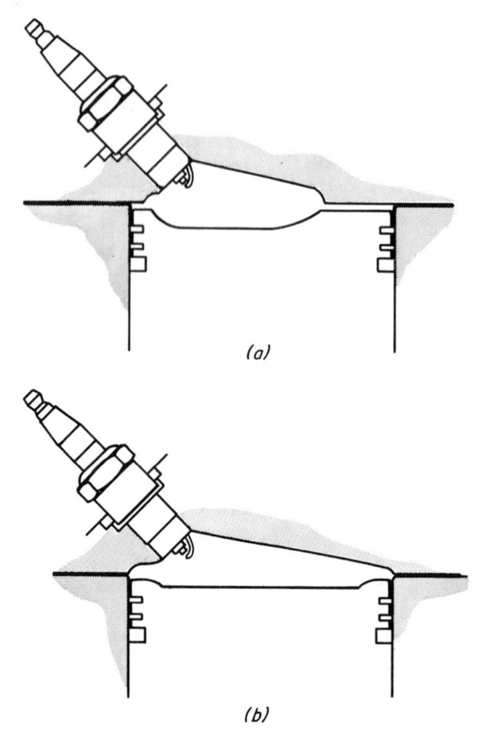

Fig. 5. Cross section of 232 CID combustion chamber; (*a*) quench and (*b*) low quench. Courtesy of D. L. Hittler and L. R. Hamkins, Society of Automotive Engineers, Inc. (76).

of numbers under a date represent a choice between either lowering HC and CO or lowering NO_x

Because of the lead time, about 20 months, needed in the automobile industry for manufacture of new car models, the choices between alternative emission control methods for the 1970 model year have surely already been made. Corresponding choices for the 1975 model year have almost certainly not been made. There is undoubtedly room for improving manifold air injection devices (65,74). Speculating further, perhaps carburetion problems will disappear with adoption of fuel injection (89). On the other hand, considerable progress has been made in carburetor and intake manifold design. Hence it may be economically preferable to continue with an improved carburetor and treat the somewhat higher emissions with an improved manifold reactor. Since manifold reactors are less effective in converting CO and HC than catalytic devices, a catalytic device might be preferrable. Over the life of a

TABLE III

Estimated Emissions of Future Control Systems[a,b]

	Demonstrated technical feasibility, 1967		Commercial feasibility				Ultimate feasibility
			1967	1970		1975	
	Either	Or		Either	Or		
Hydrocarbons, ppm (as hexane)	50	275	275	180–200	275	50	<25
Carbon Monoxide, %	0.7	1.5	1.5	1.0–1.2	1.5	0.5	<0.25
Nitrogen Oxides, ppm	500	350	1500	600–800	350	250	<100

[a] Spark-ignition gasoline engines tested according to California Standard Cycle Test. Procedures for hydrocarbons and carbon monoxide (bag sample for nitrogen oxides) and weight-averaged according to vehicle sales volume.

[b] Table based on simultaneous attainment of three (HC, CO, NO_x) emission levels.

car the cost of a manifold reactor may probably be lower than the cost of a catalytic device. Counterbalancing this the catalytic device might provide a relatively easy and reliable way for motorists to pass emissions inspection tests. Perhaps two catalytic devices, one to convert NO_x and the other to convert HC and CO, will provide the ultimate feasibility levels of Table III (50).

Catalytic devices were thoroughly evaluated and dismissed by a member (Ford) of the automoble industry in 1962 (53). In retrospect, this conclusion was economically inevitable considering the early stage of development of emission control by motor modification. At this time catalytic devices are being reevaluated by Ford utilizing the advances in engine technology made since 1962 in a systems approach. It is not presently possible to state that catalytic devices will or will not appear and survive in the marketplace. One thing is certain; within the next decade the gasoline engine will be pushed to the wall with respect to emissions control. Car manufacturers will need to use all the resources at their disposal to meet this challenge.

VIII. FUEL TECHNOLOGY RELATED TO EMISSION CONTROL

The oil industry has a major stake in the fate of the gasoline engine. It has been succinctly stated that the energy to run our economy is supplied in the ratio of about 40% by oil, 40% by gas, 20% by coal and 0.2% by food (16). The oil industry will be expected to and will desire to supply directly or indirectly the energy to power the vehicles of the future whatever the nature of their engines. A handicap common to both the oil and automotive industries is that their scale of operation necessitates a large investment, i.e., a large capital mass of inertia. Changes in their direction or rate of motion require the application of correspondingly large forces. For gasoline specifically, any significant change in present oil refinery configuration (the combination of processing equipment to produce a particular product spectrum) is virtually certain to entail severe economic penalties.

As described earlier, gasoline and exhaust gas do have quite different although not entirely unrelated compositions. However, for evaporative emissions of gasoline vapor from carburetor bowl and gas tank, the composition of the gasoline is obviously important. California has limited the olefin content of gasoline. Significant reductions in evaporative emissions could be achieved by widespread use of gasoline of lower volatility (104). To change refinery configurations to reduce gasoline volatility, an API task force estimate (102) would entail new refinery investments of one to three billion dollars and add one to 2.3 cents per gallon to the cost of gasoline. Devices containing activated charcoal for redirecting evaporative emissions to the intake manifold have been described and tested (43,71). However, the economic choice does not appear to be between devices and gasoline volatility reduction. An API-sponsored study (96) clearly shows severe starting and warmup problems for carbureted engines operating on gasoline of low volatility. Probably an engine with fuel injection and throttled intake valves could successfully use a fuel of low volatility (89). Hence the economic choice really appears to be either devices or gasoline volatility reduction plus engine modification. The odds in such a choice favor devices. A more recent study by GM showed satisfactory startup and drivability in the equitable climate of Los Angles for a group of 1800 cars using a low volatility gasoline (104). Evaporative emissions could be reduced an estimated 50% in the Los Angeles basin by changing the Reid vapor pressure of gasoline supplied to the area from the present 10 psi average to about 7 psi. Since the nation as a whole does not enjoy Los Angeles average temperature ranges there still appears to be a niche for evaporative emission control devices.

Alternate fuels to gasoline would undoubtedly involve more capital expense than gasoline volatility reduction. A recent study (103) shows that the use of methane or propane instead of gasoline reduces exhaust emission levels of HC, CO, and NO_x significantly. The cost to produce and market either methane or propane instead of gasoline is dismaying to contemplate.

Historically the sequence has been to develop or improve fuels after a new or improved engine has been developed. The invention of TEL antiknock additive might be considered an exception. Without TEL the introduction of high compression engines would undoubtedly have been delayed. At this juncture it appears probable that fuel technology will eventually provide important contributions to the solution of the automotive exhaust emissions problem. However, leadership in the attack on the problem is supplied by other technologies.

IX. AUTO EXHAUST EMISSION STANDARDS

From the point of view of those working on remedial measures for air pollution, the quantitative standards and the test methods are the immediate concern. Standards are, however, but one element in a mesh of continuing operations for air pollution control. The major operations for most pollution control situations are:

1. Data gathering—air samples are periodically taken and analyzed to define distribution, concentrations, sources, and fate of pollutants.

2. Criteria writing—supporting data are assembled to show adverse time-concentration effects of pollutants.

3. Control method review—feasibility and effectiveness of available control technology are summarized.

4. Standards setting—Air quality standards and emission standards are set using considered judgment, preferably in the light of information from the above three operations.

5. Standards enforcement—a policing apparatus is needed to enforce compliance with established standards. Part of its work includes the gathering of some data.

6. Legislation enactment—funding and the force of law are essential for pollution control.

7. Research and development—new knowledge is required to pave the way for continued purification of our atmosphere.

8. Public education—the citizenry must be informed and concerned for acceptance of enforcement and to preclude unrealistic expectations.

The above listing is not intended to be complete nor is there any implication that the real-life application of these stages necessarily follows a logical or inviolable sequence.

The magnitude of the automotive exhaust problem in relation to our total air pollution problem can be seen in Table IV (24). These figures are estimates based on a vehicle population of 90,000,000 with uncontrolled evaporative, crankcase, and exhaust emissions. The indicated multipliers in the reference 24 are the pounds of pollutants discharged per vehicle per year. These may be approximately calculated from the appropriate data of Table I, the conversion formula of Appendix C, and some interesting average numbers. Emission of hydrocarbons from cars is roughly 55% from the exhaust, 25% from the crankcase, and 20% from evaporative losses. The average car is presently driven about 10,000 miles per year and consumes about 700 gallons of gasoline (24). In 1962 the mileage was about the same but about 800 gallons of gasoline were consumed (78). Gasoline averages about 0.05% sulfur and 2.5 ml of TEL per gallon. As stated earlier, 1 ml of TEL contains about 1 g of Pb which ends up as lead salts.

TABLE IV

Estimated Total U.S. Emissions of Various Pollutants in Millions of Tons per Year

	Auto exhaust	Nonautomotive	Total
CO	66	5	77
HC	12	7	19
NO_x	6	7	13
SO_2	1	25	26
Lead	0.2	—	0.2
Particulates	1	11	

Nonautomotive sources include industrial operations, heat and power generation, and waste incineration. They comprise the source of about one-third the nation's air pollutants. The two-thirds share of automobiles has already been significantly reduced. Installation of positive crankcase ventilation (PVC) has eliminated about 25% of the HC emitted by cars since the 1966 model year. Exhaust emissions have been reduced as indicated in Table I starting with the 1968 model year. Evaporative emissions are scheduled to disappear in 1971. Offsetting these gains is the increase in car population. Car sales are now at a rate of about 10,000,000 per year. Life of a car is about 10 years. It is easy to see that by the mid-seventies the U.S. car population will probably exceed 100,000,000. California accounts for the purchase of about 10% of new cars sold and has had crankcase emissions controlled since 1963 and exhaust emissions controlled since 1966.

The size of the figures in Table IV becomes even more alarming when one realizes that these tonnages are not evenly spread out but are concentrated in the cities and suburbs (24,119). Our populous areas vary widely in the amount of ventilating winds and cleansing rains they receive. Some correspondences must be established between pollutant levels in the atmosphere and adverse effects such as vegetation damage, eye irritation, increase of lead and combined CO levels in the blood of people, and impairment of visibility. The need for data gathering, research, and criteria writing is immediately obvious. An enormous amount of work has been done and is represented or cited in part by references 7,20,24, and 122. The authoritative summaries of the information supporting the establishment of standards for ambient air quality will issue in the form of criteria from the Secretary of HEW in accordance with the U.S. Air Quality Act of 1967. Air quality standards involve dosage, i.e., the time duration of a concentration of a pollutant. As an example, the California Department of Public Health has set CO standards calling for less than 30 ppm CO for an eight-hour exposure or 120 ppm CO for one-hour exposure. Colorado's Air Pollution Control Act of 1967 declares unacceptable an NO_x level of 0.1 ppm for one hour and a total oxidant level of 0.1 ppm for one hour.

Emission standards for specific pollutant sources could be set by summing average emission times and rates of the pollutant from all sources in an area, subtracting the amount of pollutant removed assuming an average rate of removal, and comparing the result with the air quality standard for the volume of affected air over the geographical area in question. Emission standards so set do not preclude aggravated localized situations or incidents of poor air quality caused by nonaverage weather affecting pollutant removal rates. The first automotive emission standards were set in California (see Table I) with the objective of attaining by 1970 an air quality equivalent to air quality in 1940 (123,125). The records apparently indicate that smog was not a serious problem in California prior to 1940.

For purposes of data gathering and standards enforcement, adequate analytical methods are available for CO and NO_x (7,110,128,130,131) although not without international complications (117). Hydrocarbons in exhaust are currently analyzed by HSNDIR (hexane sensitized nondispersive infrared). Results are expressed as parts per million of HC as hexane. Because the infrared spectrophotometer is sensitized to the C—H group in paraffins it does not register in full other exhaust components such as olefins, acetylenes, and aromatics. Other instruments such as a Flame Ionization Detector (FID) are capable of registering all the carbon-containing components of exhaust. Seven mode cycle HC test results by HSNDIR must be multiplied by a factor of about 1.5 (26) to arrive at the real (FID) HC emissions. The multiplying factor varies from mode to mode and is larger during the acceleration and cruise modes when the olefin content of exhaust is higher. Olefinic HC has been shown to be a primary raw material in the formation of photochemical smog in atmospheric inversion situations similar to Los Angeles (113,120). On the basis of photochemical reactivity, smog indices have been developed for individual hydrocarbons (116,128). The work on indices was performed by mixing the pure individual HC with NO and air followed by irradiation in a smog chamber. There is general recognition that the automotive HC emission standard should eventually be based on reactive HC (26). However, there is still considerable disagreement in the scientific community on where to draw the line between reactive and unreactive HC. Virtually all agree that methane is unreactive. Most agree that ethane too is unreactive. Many would also include propane and benzene in the unreactive category. Considering the rest of the HC in exhaust gas as reactive, the sticking point is the assignment of numerical values to smog indices or molar reactivity constants. There is some evidence that changes in the molar ratio of HC to NO_x may significantly increase the reactivity constant of HC formerly considered unreactive (106,107). While studies continue (109) the fact is that all legislated HC emission standards are in terms of total HC without regard to reactivity considerations.

During the catalytic oxidation of HC in exhaust, olefins are more completely oxidized than methane, ethane, propane, or benzene (39,59, 40). Obviously, analytical methods directed at measurement of reactive HC in exhaust would more accurately reflect the contribution of a catalytic device to smog reduction.

The research effort on air pollution generally and on automotive air pollution in particular is at an all-time high and increasing (1,12,17,18, 25). There is little question but that our pollution control technology and knowledge of air pollution will be greatly augmented by research and development already in progress or projected. Of more concern is the area of public education. The automobile of a driver who prefers the acceleration–deceleration modes may have up to ten times the pollutant emission levels of the same automobile operated by a driver who prefers steady cruising (9). Regulation of traffic flow patterns to induce steady cruise driving is obviously desirable (9). Education of drivers toward a steady cruise preference is likely to be difficult, particularly since it is counter to Detroit's historic advertising emphasis.

X. SUMMARY

The new factor in the long history of catalytic devices for exhaust purification is the systems approach. This approach treats devices as a component to be designed into an automotive power system. There is a considerable reservoir of catalyst and device technology to draw on in designing specific devices for specific power systems. Protection of catalytic devices from deactivation by lead and high temperature are economic rather than technical problems. The contribution that catalytic devices can make toward the reduction of automotive air pollution is additive to reductions achievable by engine modifications.

In the broader view, our present systems for transporting people cannot indefinitely cope with continued growth of both population and affluence. An innovative revolution is needed and it will come at some time decades away in the future. However, we live in today's world where air pollution is an increasing problem and legislated emission standards for automobiles are becoming increasingly stringent. Catalytic devices are likely to be relevant as long as the spark ignition internal combustion engine is our prime source of automotive power.

Acknowledgments

The advice of W. E. Bambrick in the sections concerned with Catalytic Device Technology and Engine Technology is gratefully acknowledged.

XI. APPENDIX A

A. California Requirements for Device Certification in 1964 (19)

Criteria for exhaust control devices are:

a. The purchase or cost of installation of such device shall not constitute an undue cost burden to the motorist.

b. Such device shall operate on a designated classification of motor vehicle is specified in Section 2104 without replacement of the device or any major component thereof so that for 12,000 miles its emissions are within the limits established by the State standards.

c. Such device shall operate in a safe manner and so that the device will not result in any unsafe condition resulting from excessive heat applied to the floorboards, hydraulic brake cylinders, brake lines, gasoline tank, fuel pump, fuel lines, transmission, or other components of the motor vehicle or otherwise result in an unsafe motor vehicle.

d. Malfunction or failure of the device shall not endanger life or property.

e. Such device shall not malfunction or fail under the stress of backfire in the exhaust system.

f. Such device shall not allow exhaust products of the motor vehicle to enter the passenger compartment in a volume beyond the volume characteristic of the motor vehicle with a standard exhaust system.

g. Heat emanating from an operating device shall not create a hazard to persons or property who are in close proximity to the motor vehicle.

h. Such device shall not cause an increase in back pressure in the exhaust system greater than 25% beyond the back pressure characteristic of the motor vehicle equipped with a standard exhaust system.

i. Such device shall not be permanently impaired by the variety of severe motor vehicle operating conditions frequently encountered in California including heavy rains, mountain and desert driving.

j. Such device shall operate in a manner so as not to create excessive noise or odor beyond the standard characteristics of the motor vehicle equipped with a standard exhaust system; nor should the installation of such device create a noxious or toxic effect in the ambient air.

XII. APPENDIX B

A. California Requirements for Device Certification as of 1965 (28)

Criteria for device certification (article III, part 4):

To issue certificates of approval for any motor vehicle pollution control device which after being tested by the Board or tested and recommended by a laboratory designated by the Board as an authorized vehicle pollution control testing laboratory, the Board finds that the device operates within the standards set by the Board and meets the criteria adopted; provided that no certificate of approval shall be issued for any device required by subdivision (*d*) of paragraph 39090 of this code if:

a. The cost of such device, including installation, is more than $65.

b. The annual maintenance cost of the device, including any adjustment necessary for its proper operation in order to meet prescribed standards set is likely to exceed $15 a year; or

c. The device does not equal or exceed the performance criteria established by the Board for devices for new motor vehicles or, in the alternative, have an expected useful life of 50,000 miles of operation.

XIII. APPENDIX C

A. Determination of CO and HC Levels in Engine Exhaust (27)

85.75 *Dynamometer Operation Cycle*

a. The following seven-mode cycle shall be followed in dynamometer operation tests.

Sequence no.	Mode	Acceleration mph/sec	Time in mode seconds		Cumulative time seconds	Weighting factor
1	Idle	—	20		20	0.042
2	0–25		11.5		31.5	0.244
		2.2		14	—	—
3	25–30		2.5		34	—[a]
4	30	—	15		49	0.118
5	30–15	−1.4	11		60	0.062
6	15	—	15		75	0.050
7	15–30		12.5		87.5	0.455
		1.2		29	—	—
8	30–50		16.5		104	—[a]
9	50–20	−1.2	25		129	0.029
10	20–0	−2.5	8		137	—[a]

[a] Data not read.

b. The following equipment will be used for dynamometer tests:

1. Chassis dynamometer—equipped with power absorption unit and flywheels.

2. Cooling fan—A fixed-speed fan will be used. It will have sufficient capacity to maintain engine cooling during sustained operation on the dynamometer and its air moving capacity shall not exceed 5,300 cfm.

85.76 *Dynamometer Procedure*

a. The vehicle shall be tested from a cold start. Four warmup cycles and five hot cycles make a complete dynamometer run. Exhaust emission measurements for hydrocarbons and carbon monoxide will be performed during the four warmup cycles and during the sixth and seventh (hot) cycles.

85.87 *Calculations (Exhaust Emissions)*

The final reported test results shall be derived through the following steps:

a. Exhaust gas concentrations shall be adjusted to a dry exhaust volume containing 14.5 mole percent carbon atoms by applying the following dilution factor to the individual mode data.

$$\frac{14.5}{\%CO_2 + (0.5)\ \%CO + (1.8 \times 6)\ \%HC}$$

Since hydrocarbons, carbon monoxide, and carbon dioxide all are measured with the same moisture content, no moisture correction is required to convert the results to a dry basis.

b. The adjusted mode data of paragraph *a* of this section shall be weighted in proportion to the time spent in each mode during a typical metropolitan trip by applying the appropriate "weighting factor" shown in Table I of the calculation example.

c. Composite hydrocarbon and carbon monoxide concentrations are determined for each cycle by summing up the respective weighted mode data of paragraph *b* of this section.

d. The composite cycle data of paragraph *c* of this section shall be weighted in proportion to the time spent in each cycle, classified according to "warmup cycle" and "hot cycle," during a typical metropolitan trip by applying the appropriate "weight factor" shown in Table II of the calculation example.

e. Composite hydrocarbon and carbon monoxide trip concentrations are determined by summing up the respective weighted cycle data of paragraph (*d*) of this section.

f. The overall composite concentration values of paragraph *e* of this section shall be converted into mass emission values by substituting in the appropriate formula given in paragraph *g* of this section.

g. Formulas for converting concentration into mass:

1. For light duty vehicles, excluding off-road utility vehicles.

(*i*) For hydrocarbons:

$$HC_{mass} = \frac{HC^*_{conc}}{1{,}000{,}000} \times (1.8 \times 6) \times \frac{\text{Exhaust volume}}{\text{mile}} \times \text{Density}_{HC}$$

(*ii*) For carbon monoxide:

$$CO_{mass} = \frac{CO_{conc}}{100} \times \frac{\text{Exhaust volume}}{\text{mile}} \times \text{Density}_{CO}$$

2. Example calculation of composite cycle concentrations.

*Author's note: For vehicles without evaporative emission control. Those with, receive a credit of 20 ppm HC as shown in Section g-3.

TABLE I

Mode	Concentration as measured HC	CO	CO_2	Dilution factor	Adjusted HC	CO	Weighting factor	Weighted HC	CO
Idle	99	1.2	14.0	14.5/14.8	97	1.2	0.042	4.1	0.05
0–25	125	0.6	13.5	14.5/13.9	130	0.6	0.244	31.7	0.15
30	159	0.6	13.6	14.5/14.1	164	0.6	0.118	19.4	0.07
30–15	93	1.2	14.0	14.5/14.7	92	1.2	0.062	5.7	0.07
15	90	1.0	14.1	14.5/14.7	89	1.0	0.050	4.4	0.05
15–30	116	0.3	13.5	14.5/13.8	122	0.3	0.455	55.5	0.14
50–20	210	1.3	13.6	14.5/14.5	210	1.3	0.029	6.1	0.04
Sum				(cycle composite)				127	0.57

TABLE II

Cycle	Concentration as determined HC	CO	Weighting factor	Weighted HC	CO
1	319	4.68	0.35/4	27.9	0.41
2	136	0.66	0.35/4	11.9	0.06
3	127	0.57	0.35/4	11.1	0.05
4	128	0.48	0.35/4	11.2	0.04
5	Not read				
6	122	0.46	0.65/2	39.6	0.15
7	133	0.58	0.65/2	43.2	0.19
Sum		(trip composite)		145	0.90

3. Example Calculation of Mass Emissions Values

Assuming: 1970 model light-duty vehicle with automatic transmission, equipped with evaporative emission control system and tested with a 3,500-lb dynamometer inertia wheel ($W = 3{,}500$):

$$HC_{mass} = \frac{(145\text{-}20)}{1{,}000{,}000} (1.8 \times 6)\ (-6.69 + 0.0277 \times 3{,}500 - 0.00000201 \times 3{,}500^2) \times (16.33) = 1.45 \text{ g/mile}$$

Assuming: 1971 model off-road utility vehicle with manual transmission and tested with a 4,500-pound dynamometer inertia wheel ($W = 4{,}500$):

$$CO_{mass} = \frac{(0.90 \times 0.85)}{100} (-6.00 + 0.0249 \times 4{,}500 - 0.00000181 \times 4{,}500^2) \times (33.11) = 17.6 \text{ g/mile}$$

XIV. APPENDIX D

Using the calculation procedure of Appendix C, the U.S. standards for 1970 of 2.2 g of HC and 23 g of CO per vehicle mile can be compared to the U.S. 1968 standards

Inertia wheel weight,[a] lb	Calculated 1970 standards		1968 Standards	
	ppm HC	% CO	ppm HC	% CO
2000	340	1.90	410	2.3
3000	225	1.25	275	1.5
4000	180	1.00	275	1.5
5000	155	0.85	275	1.5

[a] The inertia wheel loading corresponds roughly to the weight of the loaded vehicle; e.g., a Volkswagen would require a 2,000-lb inertia wheel and a Cadillac, Lincoln, or Imperial a 5000-lb inertia wheel.

XV. APPENDIX E

Present, Projected, and Target Engine Exhaust Emissions

U.S. standards	HC	CO	NO_x
1968	275 ppm	1.5%	—
1970	2.2 g/mile	23 g/mile	—
California standards[a]			
Uncontrolled 1959 cars	900 ppm	3.2%	1500 ppm
1968	275 ppm	1.5%	
1970	2.2 g/mile	23 g/mile	—
1971	—	—	4 g/mile
1972	1.5 g/mile	—	3 g/mile
1974			1.3 g/mile
Seeking low emission car	0.5 g/mile	11 g/mile	0.75 g/mile
IIECG			
Objective	65 ppm	0.5%	175 ppm

[a] California must have a waiver from the federal government to permit the state of California to have standards more stringent than U.S. standards. The waiver has been requested and apparently will be granted. Terms and conditions of the waiver are not known at this time but could be of critical importance.

XVI. APPENDIX F

Selected Pre-1950 U.S. Patents Concerned with Catalytic Devices for Oxidizing HC and CO in Exhaust Gas

U.S. patent no., date of issue, inventor and assignee	Catalysts disclosed	Comment
1,522,111; Jan. 6, 1925; A. Franck-Phillipson	MnO_2	Catalyst in removable cartridge protected from particles by an upstream bed of light, porous material
1,605,484; Nov. 2, 1926; C. E. Thompson	Pt or Pd on asbestos, CuO, zirconia	Filter and recycle portion of exhaust back through engine. All exhaust catalytically treated before discharge
1,789,812; Jan. 20, 1931; J. C. W. Frazer	Chromite of a metal in Groups VI, VII, or VIII	Directed primarily at CO removal
1,867,325; July 12, 1932; H. A. Neville	Oxide of Fe, Co, or Ni plus one or more of another metallic oxide	Directed at catalyzing the water–gas shift reaction between CO and H_2O in exhaust gas
1,902,160; Mar. 21, 1933; J. C. W. Frazer et al	Copper chromite prepared by a specified method	Add additional air to exhaust. Omit air initially to reduce catalyst to lower its lightoff temperature
1,875,024; Aug. 30, 1932; Mine Safety Appliances	Oxides of Co, Ni, Mn, or Fe on pumice or asbestos	Two venturis to add air to exhaust. Larger venturi opened by thermostat to cool catalyst bed
2,071,119; Feb. 16, 1937; J. Harger	Oxides of Fe, Mn, Cr, Cu, or Ti singly or in combination on alumina	Remarkable grasp of problem. Radial flow reactor with venturi for 2° air, overtemperature bypass, spot heater for lightoff

XVII. APPENDIX G

Selected Post-1950 U.S. Patents Primarily Directed at Catalyst Compositions for Oxidizing CO and HC in Exhaust Gas

U.S. patent no.; date of issue; inventor and assignee	Catalysts disclosed	Comment
3,220,794; Nov. 30, 1965; Stiles to DuPont	Mangano–chromia–manganite, ± metal oxide or chromate, ± refractory interspersant	High Mn:Cr ratio lends selectivity for CO oxidation, low Mn:Cr for HC oxidation
3,228,746; Jan. 11, 1966; Howk et al. to DuPont	Manganese chromite plus a chromite of Cu, Ni, Fe, Zn, Cd, Co, Sn, or Bi	Considerable detail on process of preparation
3,230,034; Jan. 18, 1966; Stiles to DuPont	List 108 catalytic materials which can be mixed with 35 refractory dispersants	High temperature stability obtained by building up catalyst from 1500-Å crystallites of promoters and dispersants
3,282,861; Nov. 1, 1966; Innes to American Cyanamid	Vanadia and cupria on alumina ± small amount of Pd. SiO_2 may be used to stabilize Al_2O_3 to heat	Discusses lead tolerance and crush strength versus catalyst pore volume and particle size
3,271,324; Sept. 6, 1966; Stephens et al. to Ethyl	Alumina impregnated with Cu, Cu + Fe, Cu + Fe + Mn, Cu + Pd, Cu + Co, Cu + Co + V, or Cu + rare earth	Catalysts said to be heat stable and of low lightoff temperature. Road tested
3,346,328; Oct. 10, 1967; F. J. Sergeys et al.	Copper exchanged crystalline zeolite. Part of Cu^{2+} converted to CuO by NaOH	Ratio of Cu^{2+} to CuO said to govern selectivity for CO or HC oxidation
3,362,783; Jan. 9, 1968; Leak to Texaco	Metal wool coated with Al_2O_3. Promoter then put on Al_2O_3. Refractory wools are disclosed	Better attrition resistance and lead tolerance is claimed
3,288,558; Nov. 29, 1966; Briggs et al. to W. R. Grace	10% CuO, 5% Cr_2O_3, and 0.02% Pd on gamma alumina	Road tested and said to have performed satisfactorily
3,249,558; May 3, 1966; Kearbey to Esso	80% Al_2O_3, 10% V_2O_5, and 10% of either CuO, Ag_2O, Cr_2O_3, or MnO	Catalysts said to have excellent activity and lead tolerance
3,304,150; Feb. 14, 1967; Stover et al. to W. R. Grace	0–4% CoO, 0–4% CuO, 12% MnO_2, 0.02% Pd + 2–10% Fe_2O_3 on gamma Al_2O_3	The iron oxide is said to confer Pb tolerance to catalyst
3,409,390; Nov. 5, 1968; Hoekstra to UOP	Stabilize with Ca, Ba, or Sr, metals of Groups VB, VIB or VIII singly or in combination on a support	Alkaline earth said to increase thermal stability and Pb tolerance

XVIII. APPENDIX H

Selected U.S. Patents Directed at Device and Catalyst Improvements to Cause Lightoff at Lower Temperature

U.S. patent no.; Date of issue; inventor and assignee	Catalysts disclosed	Comment
2,898,202; Aug. 4, 1959; Houdry et al. to Oxy-Catalyst	Pt, Cu, or Pd on alumina, beryllia, thoria, or magnesia	Inject fuel into exhaust upstream of device as needed to maintain catalyst above lightoff temperature
2,937,490; May 24, 1960; Calvert to Oxy-Catalyst	Pt, Pd, CuO, or CuO-CrO_3 on alumina	Interrupt spark and cause misfires so that unburned fuel will cause catalyst bed to heat up rapidly
3,222,140; Dec. 7, 1965; Scivally et al. to UOP	Metals of Groups I, V, VI, and VII singly or in combination $\pm$ refractory support	Direct flow of exhaust gas with dampers to save part of catalyst bed for lightoff only
3,224,831; Dec. 21, 1965; Stephens to Ethyl Corp.	Metals Groups IB and VIII, 5th 6th periods on alumina plus 1–15% H_3PO_4	Phosphoric acid said to lower lightoff temperature
3,253,883; May 31, 1966; Jaffe to Monsanto	Hydroxides of Cu, Co or Fe, Cr, and Zr and/or Hf, dried, tabletted, and calcined	Low lightoff and high strength claimed
3,254,966; June 7, 1966; Block et al. to UOP	Pt, Fe, Cu and/or Cr on alumina plus Pt or Pd plated on alloy steel wire	Main bed of metal wool catalyst with pockets of alumina catalyst for lightoff
3,259,454; July 5, 1966; Michalko to UOP	Pt group metal concentrated at periphery of alumina beads	Life and lightoff of catalyst said to be improved
3,291,564; Dec. 13, 1966; Kearby to Esso	Metal oxidation promoter on support of alumina stabilized with barium chloride	Barium salt said to enhance thermal stability of alumina
3,310,366; Mar. 21, 1967; Koepernik to Kali Chemie	Metal oxidation promoter on alumina prepared by a special process	Claim high thermal stability and activity
3,375,059; Mar. 26, 1968; Gerhold to UOP	Igniter catalyst of smaller particle size and higher promoter level	Igniter catalyst is minor component in device catalyst bed
3,378,334; Apr. 16, 1968; Bloch to UOP	Igniter catalyst of higher promoter level but same size as rest of bed	Igniter catalyst said to facilitate lightoff

XIX. APPENDIX I

Selected U.S. Patents Related to Lead Deactivation of Catalysts

U.S. patent no.; date of issue; inventor and assignee	Technique disclosed	Comment
2,772,147; Nov. 27, 1956; Bowen et al. to Oxy-Catalyst	External reservoir of catalyst to maintain catalyst level in catalyst bed	Use of partly filled bed leaves room for catalyst to move and abrade away fouling deposits
2,867,497; Jan. 6, 1959; Houdry et al. to Oxy-Catalyst	Leach Pt catalysts with dilute HNO_3 and Cu catalysts with ammonium acetate	Periodic leaching of lead salts from catalyst to restore activity
3,247,665; April 26, 1966; Behrens to Texaco	Porous alumina coating containing phosphate or chromate on steel wool	Pretreat exhaust to remove lead as phosphate or chromate before entry to catalyst bed
3,259,453; July 5, 1966; Stiles to Dupont	Chambered device with Pb scavenger upstream to catalyst	Scavengers are sulfates of Na, K, Li, NH_3 and Mg
3,024,593; Mar. 13, 1962; Houdry to Oxy-Catalyst	Catalyst geometric surface of 115–250 $in.^2$ per CID of engine	Geometric surface, bed thickness and bed cross-section ranges said to optimize lead tolerance of catalyst
3,025,133; Mar. 13, 1962; Robinson to M&C Phillips	Granules of aluminosilicate activated with H_3PO_4	Pretreat exhaust to remove Pb as phosphate upstream of catalyst bed
3,053,612; Sept. 11, 1962; DeRossit to UOP	Pellets containing metal above Pb in electromotive series and an alkali	Upstream half of catalyst bed replaced with pellets containing, e.g., Al and CaO to trap lead
3,072,457; Jan. 8, 1963; Bloch to UOP	Regenerate Pb deactivated catalyst by heating in a reducing atmosphere	Presumably operable because Pb metal occupies less volume than equivalent amount of Pb salt
3,117,936; Jan. 14, 1964; Michalko to UOP	Wash Pb deactivated catalyst with detergents to remove Pb salts	Extent of reactivation said to be unrelated to amount of Pb removal
3,162,518; Dec. 22, 1964; Briggs et al. to W. R. Grace	Combination of cyclone separator and Pb absorbent bed	Cyclone for particulate lead salts and bed for vaporized lead salts
3,168,368; Feb. 2, 1965; Mills to Air Products	Add $R = Mn(CO)_3$ to gasoline where R is methylcyclopentadienyl	Catalyst said to be protected from lead by an upstream cyclone separator
3,295,919; Jan. 3, 1967; Henderson et al to W. R. Grace	Alumina as Pb absorbent. Catalyst 10% CuO, 4% Cr_2O_3, and 0.02% Pd on alumina	Cooling loop in exhaust pipe to protect catalyst from overheating

XX. APPENDIX J

Selected U.S. Patents Relating to Device Designs to Protect Catalyst

U.S. patent no; date of issue; inventor and assignee	Design feature	Comment
3,215,507; Nov. 2, 1965; Horstmann et al.	Multiple tubes for catalyst. Joints designed for differential thermal expansion	
3,297,400; Jan. 10, 1967; Eastwood to Mobile Oil	Temperature activated valves to direct exhaust to either of two catalyst beds	If temperature too high, then bypass to atmosphere. Save first bed for lightoff
3,303,003; Feb. 7, 1967; Zimmer to W. R. Grace	Temperature sensor in catalyst bed to control speed of pump adding air to exhaust	Add air to cool catalyst bed particularly during rich idle
3,223,491; Dec. 14, 1965; Maillie et al. to Firestone	Improved design for bypass of bed and for withstanding heat warping	Compact design with provision for catalyst replacement
3,380,810; Apr. 30, 1968; Hamblin to UOP	Layer of inert refractory next to catalyst support grid to prevent warping	Designs of radial and flat bed devices described
3,413,096; Nov. 26, 1968; Britt to UOP		Prevent channeling of exhaust through bed in case catalyst volume lost by attrition or settling
3,094,394; June 18, 1963; Innes et al. to American Cyanamid	Spring-loaded movable plate to prevent voids as catalyst settles and shrinks	Fusible cap for over-temperature protection. Voids permit catalyst attrition by movement

XXI. APPENDIX K

Patents Related to NO Removal

U.S. patent; date issued; inventor and assignee	Catalysts disclosed	Comment
3,316,057; Apr. 25, 1967; Houk et al. to DuPont	Mangano-chromia-manganites in a refractory matrix claimed to remove HC and CO with added air and NO when no air added	Use tandem reactors with air added in between. First reactor removes NO, second HC and CO
3,370,914; Feb. 27, 1968; Gross et al. to Esso	20–50% nickel and 3–20% barium on alumina. Catalyst must be in reduced state to remove NO. Disclose many metallic gasoline additives	Claim equilibration of CO, HC, H_2O, and NO through water–gas shift, steam reforming, and reduction to end up with N_2, CO_2, and H_2O

References

A. *General and Historical References*

1. Agnew, W. G., "Automotive Pollution Research," in *Proc. Roy. Soc., Ser. A*, **307,** 153–181 (1968).
2. Applied Science and Technology Index 1964, p. 27.
3. Bolt, J. A., "Air Pollution and Future Automotive Powerplants," SAE paper No. 680191, Meeting of the Detroit Section, November 6, 1967.
4. California Motor Vehicle Pollution Control Board J. T. Middleton, Chairman, "Report to Governor E. G. Brown and the Legislature," Jan. 12, 1961.
5. *California Motor Vehicle Pollution Control Board Bull.*, "Executive Officer's Personal Corner," D. A. Jensen, Volume 3, No. 8 (1964).
6. California State Department of Public Health, Bureau of Sanitation, "A Report on the Effects of Hydrocarbon Control on Oxides of Nitrogen Emissions and Air Pollution," prepared in compliance with House Resolution 567 of the 1965 Regular Session of the California Legislature, Jan. 31, 1966.
7. "Carbon Monoxide—A Bibliography with Abstracts," Public Health Service Publication, No. 1503 (1966).
8. Clewell, D. H., "The Search for the Clean Car," in *Technol. Rev.*, 17–23 (1968).
9. Duckstein, L., M. Tom and L. L. Beard, "Human and Traffic Control Factors in Automotive Exhaust Emission," SAE 680398, May 20, 1968.
10. *Environmental Currents, Environmental Sci. Technol.*, **2,** No. 2, 87 (1968).
11. *Environmental Currents, Environmental Sci. Technol.*, **2,** No. 9, 659 (1968).
12. "Guide to Research in Air Pollution—1966," Public Health Service publication No. 981, U.S. Govt. Printing Office, Washington, D.C.

13. Haagen-Smit, A. J., "The Control of Air Pollution," in *Sci. Am.*, **21,** No. 1, 25–31 (1964).
14. Heinen, C. M., "The Car and Air Control," National Pollution Control Exposition and Conference, Houston, Texas, April 3–5, 1968.
15. Hurn, R. W., "Mobile Combustion Sources," *Air Pollution*, 2nd ed. Volume 3 A. C. Stern, Ed., pp. 55–95, Academic Press, 1968.
16. Larsen, R. I., "Air Pollution from Motor Vehicles," in *Annals N.Y. Acad. Sci.* **136,** Art. 12, 275–301 (1966).
17. Ludwig, J. H., "Progress in Control of Vehicle Emissions," *J. Proc. Am. Soc. Civil Engrs.*, Sanitary Engineering Division, 73–79 (1967).
18. Ludwig, J. H. and B. J. Steigerwald, "Research in Air Pollution—Current Trends," *Am. J. Public. Health,* **55,** No. 7, 1082 (1965).
19. Middleton, J. T., "Criteria for Certification of Motor Vehicle PollutionDevices in California," California Motor Vehicle Pollution Control Board, 55th Annual Meeting, Chicago, Ill., May 20–24, 1962, Paper No. 62–57.
20. "Motor Vehicles, Air Pollution and Health," a report of the Surgeon General to the U.S. Congress in compliance with public law 86–493, the Schenck Act, June 1962.
21. Ross, H. R., "The Future of the Automobile," *Sci. Technol.*, 14–24 (1968).
22. Stanford Research Institute, "Sources, Abundance and Fate of Gaseous Atmospheric Pollutants," prepared for the American Petroleum Institute, Feb. 1968.
23. Sweet, A. H., B. J. Steigerwald, and J. H. Ludwig, "The Need for a Pollution-Free Vehicle, *JAPCA*, **18,** No. 2 (1968). *J. of the Air Pollution Control Assoc.*, 111–113.
24. U.S. Department of Commerce, "The Automobile and Air Pollution: A Program for Progress," Part II, Dec., 1967, pp. 1, 2, 8, 11, 23, 40, 45, and 46.
25. U.S. Department of HEW, "First Report of the Secretary of HEW Pursuant to Public Law 90–148—The Air Quality Act of 1967," U.S. Govt. Printing Office, 1968, pp. 15–28.
26. U.S. Department of HEW, PHS, Proceedings: The Third National Conference on Air Pollution, Washington, D.C., Dec. 12–14, 1966.
27. U.S. Department of HEW, Federal Register, Vol. 33, No. 108, June 4, 1968, Part II.
28. U.S. Department of HEW, "A Digest of State Air Pollution Laws," Public Health Service publication no. 711, 1967.
29. U.S. 90th Congress, "Air Quality Act of 1967," Public Law 90–148, S. 780, Nov. 21, 1967.

B. *Catalyst and Exhaust Device Technology*

30. Ayen, R. J., and Y. S. Ng, "Catalytic Reduction of Nitric Oxide by CO," *Intern. J. Air Water Pollution,* **10,** No. 1, 1–13 (1966).
31. Baker, Sr., R. A. and R. C. Doerr, "Catalytic Reduction of Nitrogen Oxides in Automobile Exhaust," *J. of the Air Pollution Control Assoc.*, **14,** No. 10, 409–414 (1964).
32. Bienstock, D., E. R. Bauer, T. H. Field, and R. C. Kurtzrock, "Removal of Hydrocarbons and Carbon Monoxide from Automotive Exhaust Using a Promoted Uranium Catalyst," U.S. Bureau of Mines Report Investigation No. 6323 (1963).

33. Cannon, W. A. and C. E. Welling, "Catalytic Oxidation of Automotive Exhausts," Industrial and Engineering Chemistry, Product Research and Development, **1**, No. 3, 152–156 (1962).
34. De Rycke, D., "Catalytic Purification of Car Exhausts—2nd Paper," *Proc. Inst. Mech. Engrs.* (A.D.), No. 7, 220–227, (1962–1963).
35. Faith, W. L., "Automobile Exhaust Control Devices," *JAPCA*, **13**, No. 1, 33–35 (1963).
36. Faith, W. L., "Status of Motor Vehicle Exhaust Afterburners," *Proc. Am. Petroleum Inst.*, **40** (III) 358 (1960).
37. Feenan, J. F., R. B. Anderson, H. W. Swan, and L. J. E. Hofer, "Chromium Catalysts for Oxidizing Automotive Exhaust," *J. of the Air Pollution Control Assoc.*, **14**, No. 4, 113–117 (1964).
38. Hoffer, L. J. E., P. Gussey, and R. G. Anderson, "Specificity of Catalysts for Oxidation of CO-Ethylene Mixtures," *J. Catalysis*, **3**, No. 1, 451–460 (1964).
39. Innes, W. B., and K. Tsu, "Automobile Exhaust Control," *Encyclopedia of Chemical Technology*, 2nd ed., Vol. 2, Interscience, New York, pp. 814–839.
40. Innes, W. B., and A. J. Andreatch, "Characterization of Petroleum Waste Products by Selective Combustion Thermal Effects," Symposium of Developments in Petroleum Environmental Chem., Div. of Petroleum Chem., American Chemical Society, San Francisco, April 2–5, 1960, Page E-11 in Pre-prints.
41. Innes, W. B., and R. Duffy, "Exhaust Gas Oxidation on Vanadia-Alumina Catalysts," *J. of the Air Pollution Control Assoc.*, **11**, No. 8, 369–373 (1961).
42. Johnson, L. L., W. C. Johnson, and D. L. O'Brien, "The Use of Structural Ceramics in Automobile Exhaust Converters," Chemical Engineering Progress Symposium Series, Pollution and Environmental Health, **57**, No. 35, 55–67 (1961).
43. Joyce, R. S., P. D. Langston, G. R. Stoneburner, C. B. Stunkard, and G. S. Tobias, "Activated Carbon for Effective Control of Evaporative Loss," SAE Paper No. 690086, International Automotive Engineering Congress, Detroit, Michigan, Jan. 13–17, 1969.
44. Kontsoukos, E., and K. Nobe, "Catalytic Combustion of Hydrocarbons IV Effect of Preparation Method on Catalytic Activity," Report No. 64–12, Feb., 1964, Univ. of California, Los Angeles, Office of Technical Services, U.S. Department of Commerce (AD 601026 $4.60).
45. Leak, R. J., J. T. Brandenburg, and M. D. Behrens, "Use of Alumina-coated Filaments in Catalytic Mufflers. Testing with Multicylinder Engine and Vehicles", Environmental Science and Technology, **2**, No. 10, 790–794 (1968).
46. Levy, R. M., and D. G. Bauer, "The Effect of Foreign Ions on the Stability of Activated Alumina," *J. Catalysis*, **9**, 76–86 (1967).
47. Levy, R. M., D. G. Bauer, and J. F. Roth, "Effect of Thermal Aging on the Physical Properties of Activated Alumina," Industrial and Engineering Chemistry, Product Research and Development, Vol. 7, No. 3 Sept. 1968 pp. 217–220.
48. Prettre, M., "Catalysis and Catalysts," Dover Publications Inc., New York, 1963.
49. Reddi, M. M. and R. A. Baker, "Catalytic Burner for TE-Generator," 18th Annual Proceedings, Power Sources Conference, May 19–21, 1964 (Franklin Institute, Phila., Pa.).
50. Researchers Try a Catalytic Cure for Auto-Emission Smog, Product Engineering, Dec. 16, 1968, 0. 40.

51. Ridgway, S. J. and J. C. Lair, "Automotive Air Pollution—A Systems Approach," *J. Air Pollution Control Assoc.* **10,** No. 4, 336–340 (1960).
52. Schachner, H., "Cobalt Oxides as Catalysts for the Complete Combustion of Automobile Exhaust Gases," Communication, Cobalt Information Center, Columbus, Ohio, Dec., 1960.
53. Schaldenbrand, H., and J. H. Struck, "Development and Evaluation of Automobile Exhaust Catalytic Converter Systems", SAE Technical Progress Series, Vol. 6, pp. 274–298, March 1962, Ford.
54. Schlaffer, W. G., C. R. Adams, and J. N. Wilson, "Aging of Silica and Alumina Gels," *J. Phys. Chem.*, **69,** No. 5, 1530–1536 (1965).
55. Shelef, M., K. Otto, and H. Gandhi, "The Oxidation of CO by O_2 and by NO on Supported Chromium Oxide and other Metal Oxide Catalysts," *J. of Catalysis,* **12,** p. 361–375 (1968).
56. Sourirajan, S. and J. L. Blumenthal, "The Application of a Copper-Silica Catalyst for the Removal of Nitrogen Oxides Present in Low Concentration by Chemical Reduction with Carbon Monoxide on Hydrogen," *Intern. J. Air Water Pollution,* **5,** No. 1, 24–33 (1961).
57. Sweeney, M. P., "Standardized Testing of Smog Control Devices for Vehicular Exhaust," revision of paper presented at the Annual Meeting of the American Institute of Chemical Engineers Chicago, Ill., Dec. 4, 1962.
58. Thomas, J. M. and W. J. Thomas, *Heterogeneous Catalysis*, Academic Press, New York, 1967, pp. 60, 300.
59. Weaver, E. E., "Effects of Tetraethyl Lead on Catalyst Life and Efficiency in Customer Type Vehicle Operation," SAE Paper No. 690016 International Automotive Engineering Congress and Exposition, Detroit, Mich., Jan. 13–17, 1969.
60. Yarrington, R. M. and W. E. Bambrick, "Deactivation of Automobile Exhaust Control Catalyst," in press.

C. *Automotive Technology*

61. Agnew, J. T., "Unburned Hydrocarbons in Closed Vessel Explosions, Theory vs. Experiment Applications to Spark Ignition Engine Exhaust," SAE No. 670125, Jan. 9, 1967.
62. Beckman, E. W., W. S. Fagley, and J. O. Sarto, "Exhaust Emission Control by Chrysler—the Cleaner Air Package," SAE Progress in Technology, Vol. 12, pp. 178–192, 1967.
63. Benson, J. D., "Reduction of Nitrogen Oxide in Automobile Exhaust," SAE Paper No. 690019, Intern. Automotive Eng. Congr., Detroit, Mich., Jan. 13–17, 1969.
64. Bier, K. C., J. J. Frankowski, and D. K. Gonyon, "A Study of Factors Affecting Carburetor Performance at Low Air Flows," SAE Paper No. 690137, Intern. Automotive Eng. Congr., Detroit, Mich., Jan. 13–17, 1969.
65. Brownson, D. A. and R. F. Stebar, "Factors Influencing the Effectiveness of Air Injection in Reducing Exhaust Emissions," SAE Paper No. 650526, May 17, 1965.
66. Buttgereit, W., C. H. Voges, and C. Schilter, "Exhaust Emission Control by Fuel Injection: The VW 1600 with Electronically Controlled Fuel Injection System," SAE Paper No. 680192, Southern California Section, Oct. 9, 1967.

67. Caplan, J. D., "The Automobile Manufacturers' Vehicle Emissions Research Program," Annual Meeting of the Air Pollution Control Association, May 20–24, 1962, Chicago, Ill., Paper 62–76.
68. Chandler, J. M., J. H. Struck, and W. J. Voorheis, "The Ford Approach to Exhaust Emission Control." SAE Progress in Technology Vol. 12 pp. 161–177.
69. Daniel, W. A., "Engine Variable Effects on Exhaust Hydrocarbon Composition, SAE Paper No. 670124, Jan. 9, 1967.
70. Daniel, W. A. and J. T. Wentworth, "Exhaust Gas Hydrocarbons-Genesis and Exodus, SAE Tech. Prog. Series Vol. 6, p. 192, Mar. 1962.
71. Deeter, W. F., H. D. Daigh, and O. W. Wallin, Jr., "An Approach for Controlling Vehicle Emissions," Society of Automotive Engineers Paper No. 680400, Mid-year Meeting, Detroit, Mich., May 20–24, 1968.
72. Eberhardt, J. E. and J. A. Beck, "Process and Control for Producing Anti-Emission Carburetors," SAE Paper No. 690138, Int'l. Automotive Engrg. Congress, Detroit, Mich., Jan. 13–17, 1969.
73. Eltinge, L., "Fuel-Air Ratio and Distribution from Exhaust Gas Composition," SAE Paper No. 680114, Automotive Engineering Congress, Detroit, Mich. Jan. 8–12, 1968.
74. Eltinge, Marsee, F. J., and A. J. Warren, "Potentialities of Further Emissions Reduction by Engine Modifications", SAE Paper No. 680123 Jan. 8, 1968.
75. Eyzat, P. and J. C. Guibet, "A New Look at Nitrogen Oxides Formation in Internal Combustion Engines," SAE Paper No. 680124, Jan. 8, 1968.
76. Hittler, D. L. and L. R. Hamkins, "Emission Control by Engine Design and Development," SAE Paper No. 680110 Automotive Engineering Congress, Detroit, Mich., Jan. 8–12, 1968.
77. Huls, T. A. and H. A. Nichol, "Influence of Engine Variables on Exhaust Oxides of Nitrogen Concentrations from a Multicylinder Engine," SAE Paper No. 670482, Mid-year Meeting, Chicago, Ill., May 15–19, 1967.
78. Jones, J. H. and J. C. Gagliardi, "Vehicle Exhaust Emission Experiments Using a Pre-mixed and Pre-heated Air Fuel Charge," SAE Paper No. 670485, Mid-year Meeting, Chicago, Ill., May 15–19, 1967.
79. Kopa, R. D., "Control of Automotive Exhaust Emission by Modifications of the Carburetion System," *Vehicle Emissions—Part II, SAE Progr. Technol. Ser.*, **12,** 212–229 (1967).
80. Larborn, A. O. S. and F. E. S. Zackrisson, "Dual Manifold as Exhaust Emission Control in Volvo Cars," SAE Paper No. 680108, Automotive Engineering Congress, Detroit, Mich., Jan. 8–12, 1968.
81. Lawrence, G., J. Buttivant, and C. G. O'Neill, "Mixture Pretreatment for Clean Exhaust—The Zenith 'Duplex' Carburetion System," SAE Paper No. 670484, Mid-year Meeting, Chicago, Ill., May 15–19, 1967.
82. London, A. L., Stanford, Univ., Palo Alto, Calif., "The Application of Research to Motor Vehicle Emission Control—Can We Avoid Afterburners," presented at an Air Pollution Research Conference on Dec. 7, 1961, at USC, Los Angeles.
83. Newhall, H. K., "Control of Nitrogen Oxides by Exhaust Recirculation. A Preliminary Theoretical Study", SAE Paper No. 670495, May 15, 1967.
84. Newhall, H. K. and E. S. Starkman, "Direct Spectroscopic Determination of Nitric Oxide in Reciprocating Engine Cylinders," SAE Paper No. 670122, Jan. 9–13, 1967.

85. Nicholls, J. E., I. A. El-Messiri, and H. K. Newhall, "Inlet Manifold Water Injection for Control of Nitrogen Oxides—Theory and Experiments," SAE Paper No. 690018, Intern. Automotive Eng. Congr., Detroit, Mich., Jan., 13–17, 1969.
86. Starkman, E. S., H. E. Stewart, and V. A. Zvonow, "An Investigation into the Formation and Modification of Emission Precursors," SAE Paper No. 690020, Intern. Automotive Eng. Congr., Detroit, Mich., Jan. 13–17, 1969.
87. Steinhagen, W. K., G. W. Niepoth, and S. H. Mick, "Design and Development of the General Motors Air Injection Reactor System," SAE Progr. Technol., **12,** 146–160, 1967.
88. Stern, A. C., "Prospects for Exhaust Control by Engine Modification," American Pollution Control Association, 55th Annual Meeting, May 20–24, 1962, Chicago, Illinois, Paper No. 62–52.
89. Stivender, D. L., "Intake Valve Throttling—A Sonic Throttling Intake Valve Engine," SAE Paper No. 680399, Mid-year Meeting, Detroit, Mich., May 20–24, 1968.
90. Taylor, C. F. and E. S. Taylor, *The Internal Combustion Engine,* 2nd ed., International Textbook Co., Scranton, Pa., 1961.
91. Thompson, W. B., "Design and Development of Air Pump for GM Air Injector Reactor System," *GM Eng.* J., **13,** No. 3, 30–35 (1966).
92. Toyoda, S., K. Nakajima, and T. Toda, "Development of the Exhaust Emission Control Device for Toyota Vehicles," SAE Paper No. 670687, West Coast Meeting, Portland, Oregon, Aug. 14–17, 1967.
93. Wentworth, J. T., "Piston and Ring Variables Affect Exhaust Hydrocarbon Emissions," SAE Paper No. 680109, Automotive Eng. Congr., Detroit, Jan. 8–12, 1968.
94. Winkler, G. L., S. S. Lestz, and W. E. Meyer, "Exhaust Gas Sampling Technique for Relating Emissions and Cycle Characteristics," SAE Paper No. 680770, National Fuels and Lubricants Meeting, Tulsa, Oklahoma, Oct. 29–31, 1968.

D. *Fuel Technology*

95. Echols, L. S., V. E. Yust, and J. L. Bame, "Review of Research on Abnormal Combustion Phenomena in Internal Combustion Engines," 1959, Fifth World Petroleum Congress, Section VI, Paper 11.
96. Esso-API, "The Effect of Fuel Volatility Variations on the Performance of Automobiles over a Range of Temperatures," April 2, 1968.
97. Gagliardi, J. C., "The Effect of Fuel Anti-knock Compounds and Deposits on Exhaust Emissions," SAE Paper No. 670128, Jan. 9, 1967.
98. Gagliardi, J. C. and F. E. Ghannam, "Effects of Tetraethyl Lead Concentration on Exhaust Emissions in Customer Type Vehicle Operation," SAE Paper No. 690015 Intern. Automotive Eng. Congr., Detroit, Mich., Jan. 13–17, 1969.
99. Graiff, L. B., "Mode of Action of Tetraethyl Lead and Supplemental Antiknock Agents," SAE Paper No. 660780, Nov. 1, 1966, Fuels and Lubricants Meeting, Houston, Texas.
100. Hurn, R. W., T. C. Davis, and P. E. Tribble, "Do Automotive Emissions Inherit Fuel Characteristics," *Proc. API,* **40** (III), 352 (1960).

101. Lawson, S. D., J. F. Moor, and J. B. Rather, Jr., "A Look at the Economics of Manufacturing Unleaded Motor Gasoline," 32nd Midyear Meeting, Division of Refining API, Los Angeles, Calif., May 16, 1967.
102. Lawson, S. D., "Report of the API Lead/Volatility Economics Task Force, AIChE, Los Angeles, Dec. 1–5, 1968.
103. Lee, R. C. and D. G. Wimmer, "Exhaust Emission Abatement by Fuel Variations to Produce Lean Combustion," SAE paper No. 680769, Nat'l. Fuels & Lubricants Meeting, Tulsa, Okla., Oct. 29–31, 1968.
104. Nelson, E. E., "Hydrocarbon Control for Los Angeles by Reducing Gasoline Volatility," SAE Paper No. 690087, Intern. Automotive Eng. Congr., Detroit, Mich., Jan. 13–17, 1969.
105. Pahnke, A. J. and J. F. Conte, "Effect of Combustion Chamber Deposits and Driving Conditions on Vehicle Exhaust Emissions," SAE Paper No. 690017, Intern. Automotive Engr. Congr., Detroit, Mich., Jan. 13–17, 1969.

E. *Analyses of Exhaust Gas and the Atmosphere*

106. Altshuller, A. P., S. L. Kopczynski, W. Lonneman, and D. Wilson, "Photochemical Reactivities of Exhausts from 1966 Model Automobiles Equipped to Reduce Hydrocarbon Emissions," *J. Air Pollution Control Assoc.*, **17**, No., 11 734 (1967).
107. Altshuller, A. P., S. L. Kopczynski, W. Lonneman, and F. D. Sutterfield, "Photochemical Reactivities of *n*-Butane and Other Low-Molecular Weight Hydrocarbons," Paper 68-154, June 23–27, 1968, 61st Annual Meeting, APCA.
108. Brubacher, M. L. and E. P. Grant, "Do Exhaust Controls Really Work—Second Report," Soc. Automotive Engrs., Paper No. 670689, West Coast Meeting, Portland, Oregon, Aug. 14–17, 1967.
109. Chipman, J. C., A. J. Hocker, and J. Chao, "Measuring and Evaluating Automobile Exhaust Hydrocarbon Emissions by Interrelated Techniques," Preprint 53E, AIChE, Los Angeles, Calif., Dec. 1–5, 1968.
110. Cobb, R. M. and F. Perna, Jr., "Development of an On-Line Exhaust Gas Data Reduction System," SAE paper No. 670164, Jan. 9, 1967.
111. Faith, W. L., J. T. Goodwin, Jr., F. V. Moriss, and C. Bolze, "Automobile Exhaust and Smog Formation," *J. Air Pollution Control Assoc.*, **7**, No. 1, 9–12 (1957).
112. Faith, W. L., "The Role of Motor Vehicle Exhaust in Smog Formation," *J. Air Pollution Control Assoc.*, **7**, No. 3, 219–221 (1957).
113. Haagen-Smit, A. J., "Chemistry and Physiology of Los Angeles Smog," *Ind. Eng. Chem.*, **44**, 1342–1346 (1952).
114. D. A. Hirschler, L. F. Gilbert, F. W. Lamb, and L. M. Niebylski, "Particulate Lead Compounds in Automobile Exhaust Gas." *Ind. Eng. Chem.*, **49**, No. 7 1131–1142 (1957).
115. Hurn, R. S., C. L. Dozois, J. O. Chase, C. F. Ellis, and P. E. Ferrin, "The Potpourri that is Exhaust Gas," API, 27th Midyear Meeting, Refining Div., May 17, 1962.
116. "Individual Hydrocarbon Reactivity Measurements: State of the Art," June 1966, Coordinating Research Council Inc., 30 Rockefeller Plaza, N.Y. 20, N.Y., CRC Report No. 398.
117. Klinksiek, K. E., "Particular Problems with Exhaust Gas Analyses at Volkswagenwerk AG", SAE paper No. 680122, Jan. 8, 1968.

118. Korth, M. V., and A. H. Rose, Jr., "Emissions From a Gas Turbine Automobile," SAE paper No. 680402, May 20–24, 1968, Detroit, Mich.
119. Larsen, R. I., E. E. Zimmer, D. A. Lynn, and K. G. Blemel, "Analyzing Air Pollutant Concentration and Dosage Data," *APCA J.*, **17**, No. 2, 85 (1967).
120. Leighton, P. A., *Photochemistry of Air Pollution*, Academic Press, New York, 1961.
121. Lienesch, J. H. and R. W. Wade, "Sterling Engine Progress Report: Smoke, Odor, Noise and Exhaust Emissions," SAE paper No. 680081, Automotive Eng. Congr., Detroit, Mich., Jan. 8–12, 1968.
122. Lonneman, W. A., T. A. Bellas, and A. P. Altshuller, "Aromatic Hydrocarbons in the Atmosphere of the Los Angeles Basin," *Envir. Sci. Technol.* 1017–1020 (1968).
123. Maga, J. A. and J. R. Kinosian, "Motor Vehicle Emission Standards—Present and Future," *SAE Progr. Technol. Ser.*, **12**, 297–306, 1967.
124. McMichael, W. R., R. E. Kruse, and D. M. Hill, "Performance of Exhaust Control Devices on 1966 Model Passenger Cars," *J. Air Pollution Control Assoc.*, **18**, No. 4., 246 (1968).
125. Nissen, W. E., "Automotive Air Pollution Control—Where Are We Now and Where Are We Going," Air Pollution Control Assoc., West Coast Section, Proceedings Third Technical Meeting, Sept. 26–27, 1963, Monterey,Calif., pp. 106–110.
126. Papa, L. J., "Gas Chromatography—Measuring Exhaust Hydrocarbons Down to Parts per Billion," SAE paper No. 670494, May 15, 1967.
127. Pattison, J. N., C. Fegraus, and J. C. Elston, "New Jersey's Rapid Inspection Procedures for Vehicular Emissions," SAE paper No. 680111, Automotive Eng. Congr., Detroit, Mich., Jan. 8–12, 1968.
128. Reckner, L. R. and W. E. Scott, "Photochemical Reactivity of Hydrocarbons in the Presence of Nitric Oxide in Air," Div. Petroleum Chemistry ACS, San Francisco, April 2–5, 1968, p. E-5 in preprints.
129. Sawyer, R. F. and E. S. Starkman, "Gas Turbine Exhaust Emissions," SAE paper No. 680462, May 20, 1968.
130. Stern, A. C., *Air Pollution Volume II, Analysis, Monitoring and Surveying*, 2nd ed., Academic Press, New York, 1968.
131. Westveer, J. A., "Correlation of Exhaust Emission Test Facilities," SAE paper No. 670165, Jan. 1967.

Photochemical Air Pollution: Singlet Molecular Oxygen as an Environmental Oxidant*

JAMES N. PITTS, JR.
Department of Chemistry, University of California, Riverside, California

I. Introduction 290
II. Review of Effects of Photochemical Air Pollution 291
A. Aerosol Formation—Visibility 291
B. Photochemical Oxidant 293
C. Plant Damage 294
D. Primary and Secondary Air Pollutants 297
E. Laboratory Studies 302
F. Current Mechanisms 307
III. Singlet Molecular Oxygen 309
A. Spectroscopy 309
1. Electronic Structure 309
2. Absorption and Emission of Radiation 311
3. Physical Quenching 312
4. Energy Transfer from Singlet Oxygen 313
B. Laboratory Methods for the Generation of Singlet Oxygen 314
1. Chemical 314
2. Physical; Electrical Discharge 315
3. Energy Transfer; Photosensitization 317
C. Detection of $O_2(^1\Delta)$ and $(^1\Sigma)$ 319
D. Chemical Reactions of $^1O_2$ with Organic Compounds 321
E. Reactions of O, O_2, and O_3 Producing $^1O_2$ 322
IV. Singlet Molecular Oxygen and the Chemistry of the Lower Atmosphere 322
A. Sources of $^1O_2$ in Urban Atmospheres 322
1. Direct Absorption of Sunlight 322
2. Production of Singlet Molecular Oxygen by an Energy Transfer Mechanism 325
3. Singlet Molecular Oxygen from Ozone Photolysis 329
4. Singlet Oxygen from Exothermic Chemical Reactions 330
B. Possible Role of $^1O_2$ in the Photoconversion of NO to NO_2 331
V. Other Environmental Implications of Singlet Oxygen 332
References 334

* Supported in part by the National Air Pollution Control Administration grants AP 00109 and AP 00771.

I. INTRODUCTION

The characteristic symptoms of photochemical air pollution were first encountered in the mid-1940's in Los Angeles, California. Several years later, the early 1950's researchers (primarily Haagen-Smit, Middleton, Blacet, and their colleagues) established that it was indeed a new kind of air pollution primarily caused by the action of sunlight on the exhaust emitted by the millions of motor vehicles in the Los Angeles Basin. Over the next two decades qualitative and subsequently quantitative laboratory and field studies, together with Leighton's definitive monograph *Photochemistry of Air Pollution* published in 1961, provided the basis for much of our present understanding of the chemical and mechanistic aspects of photochemical smog whether produced in the laboratory or in the atmosphere (1).

However, despite this excellent past work, current researchers are well aware that today we are still woefully ignorant of many detailed aspects of the physical, chemical, biological, and medical effects of photochemical smog on man and his environment. Thus, while the current implementation of the United States Air Quality Act of 1967 by the National Air Pollution Control Administration is effectively leading to more widespread and stringent control measures than hitherto existed, it would be tragic if a sizeable fraction of the scientific community and the public at large were to believe that we now fully "understand" smog. Thus, while the application of new controls and the development of new control technology are of the highest priority today, there is a continuing need for fundamental research on all aspects of air pollution.

The following specific example illustrates one aspect of the problem facing air pollution scientists today. A key process in the formation of photochemical "smog" is the photoconversion in the atmosphere of the nitric oxide released in automobile exhaust gases to nitrogen dioxide with the concurrent disappearance of the hydrocarbons and the buildup of oxidants, principally ozone and peroxyacetyl nitrate (PAN), and other oxidized organic material such as aldehydes and alkyl nitrates. The reaction system is extremely complex under field conditions and it is also complex when studies are conducted in large photochemical reaction chambers which are used to simulate atmospheric conditions. While a great deal of useful information has been derived from studies in test chambers, current mechanisms advanced to explain such phenomena as this photochemical conversion of nitric oxide into nitrogen dioxide in the presence of hydrocarbons can account for only approximately 50% of the hydrocarbons consumed in the process!*

It is the purpose of this chapter to (*a*) review briefly the characteristic effects of photochemical air pollution, (*b*) stress several important aspects generally recognized by photochemists but which may not be familiar to researchers in other disciplines, and (*c*) discuss in some detail our belief that a highly reactive transient species, singlet molecular oxygen, $O_2(^1\Delta)$ [and possibly $O_2(^1\Sigma)$], which to date has not been generally considered, may play an important role in the chemistry of urban atmospheres including the oxides of the nitrogen-hydrocarbon system cited above. We further believe that singlet oxygen also may play a significant role in the biological and health effects of such atmospheres on man and his environment.

II. REVIEW OF EFFECTS OF PHOTOCHEMICAL AIR POLLUTION

A. Aerosol Formation—Visibility

We shall begin with an overview of a typical area plagued by photochemical smog, the Los Angeles Basin. This is best seen in Figure 1, a photograph of the Southern California area taken from an altitude of approximately 185 miles by the crew of Gemini V. The pall of grey-white smog hanging over central Los Angeles is clearly visible even from that altitude. For a frame of reference, Figure 2 is a relief map of that area with freeways superimposed. This figure shows the "line sources" of automotive pollution. Both figures show the mountain ranges which provide physical barriers to the rapid dissipation of smog and the Pacific Ocean which moderates the climate and is a key factor in producing strong atmospheric inversions which place a meteorological "lid" over the Los Angeles Basin at an altitude of several thousand feet.

Figure 3 shows this inversion and the "Los Angeles smog" when one flies over the Los Angeles Basin at about 4,000 feet from Riverside to the Los Angeles Airport on a warm, bright day on which *no field or incinerator burning has occurred* (*this is illegal*), *and there are no forest fires in the area.* The dense haze is due chiefly to aerosols and particulates formed photochemically by the action of sunlight on the urban atmosphere.

The sequence of photographs shown in Figures 4, 5, and 6 dramatically illustrates how geographical and meteorological conditions can combine

* For details see the earlier chapters in this volume by Schuck and Stephens (2) and Stephens (3) and the general references 1, 4, and 5.

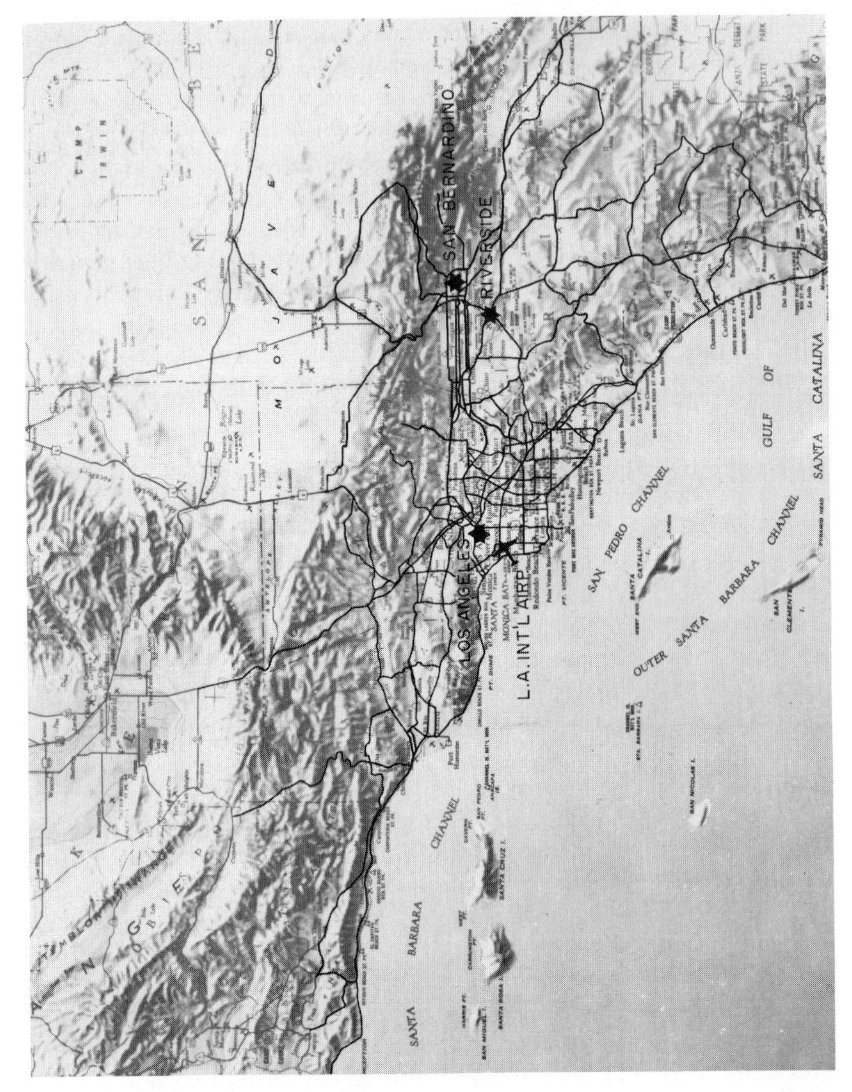

Fig. 2. Photograph of a relief map of the Los Angeles Basin with the freeway system superimposed on it.

to produce moving clouds of photochemical smog, largely generated in cities to the west of us, that literally constitute "gas attacks," in this case on the author's home at Riverside, California (for orientation see map in Fig. 2).

Fig. 1. View of the Los Angeles Basin taken at approximately 185 miles altitude from Gemini V on August 25, 1965. Note the opaque haze characteristic of smog over the central Los Angeles area (for a frame of reference, see Fig. 2). The oxidant level was 0.25 ppm that day. We wish to thank NASA for the use of this photograph.

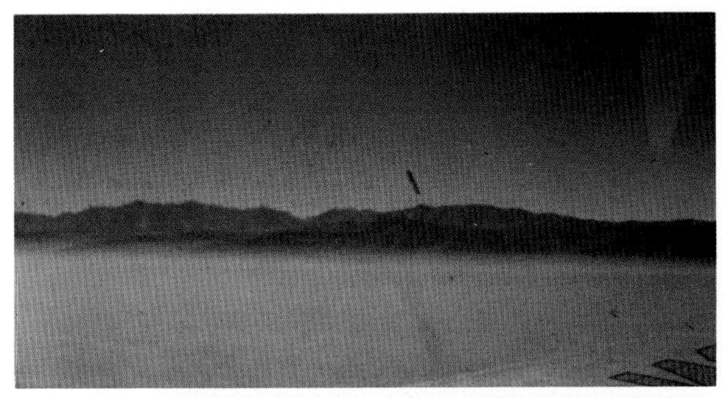

Fig. 3. Photograph of the inversion level of smog taken from an airplane flying over the Los Angeles Basin from Riverside to Los Angeles.

Fig. 4. Photograph in color infrared taken from the UCR campus looking northwest to Mt. Cucamonga. Note smog creeping in at the foot of Mt. Cucamonga. Time: 4:13 P.M., January 20, 1968. We are indebted to Professor Bowden for furnishing us with these color infrared photographs.

Fig. 5. UCR, looking northwest, 5:15 P.M., same day.

Fig. 6. UCR, looking northwest, 5:29 P.M., same day. Note how in only 14 min the smog cloud has broken out of the pass and has literally descended on downtown Riverside.

These photographs were taken January 20, 1968, by Professor Leonard Bowden and associates of the Geography Department at the University of California, Riverside, who used a special film and infrared filters in order to achieve maximum clarity in resolving the smog cloud. The combination gives color infrared prints in which all green plants appear red and in which the obscuring effects of fog droplets are minimized. The UCR campus is in the foreground and in the distance to the northwest about ten miles away is Mount Cucamonga. On the top left of Figure 4, taken at 4:13 P.M., we see the smog cloud approaching from the west but still held back from Riverside by a slight breeze down from the mountains in the east (not seen in the photographs). As darkness begins to fall and the restraining breeze dies out, we see the smog cloud start to spread out (Fig. 5, 5:15 P.M.) and only 14 min later at 5:29 P.M. (Fig. 6) it appears over downtown Riverside and has almost reached the UCR campus.

These photographs illustrate an occurrence which has, unfortunately, become all too common—serious smog occurring in the *winter* as well as the summer. They also show that while smog symptoms usually are at their maximum around noon in downtown Los Angeles and Pasadena, they may not maximize or even appear until late in the day in other cities well to the east. A more detailed analysis of the relationships between certain meteorological factors and photochemical smog has been given by Schuck, Pitts, and Wan (6).

B. Photochemical Oxidant

Accompanying this photochemically produced haze is a complex mixture of gaseous pollutants including the oxidizing species nitrogen dioxide, NO_2; ozone, O_3; and peroxyacetyl nitrate, (PAN), $CH_3C(=O)OONO_2$ (1–5). These are often grouped together and called *oxidant*. Thus, in addition to visual observation, one can also trace the buildup and movement of a smog cloud by use of strategically placed, continuously recording oxidant analyzers. Figure 7 (taken from reference 1) shows a typical change in oxidant concentration as a function of time of day for a single day (September 13, 1955) and averaged over three months (September–November, 1955) at one sampling station in Pasadena (near downtown Los Angeles). The maximum in oxidant concentration coincides with the maximum solar intensity at noon. This can be

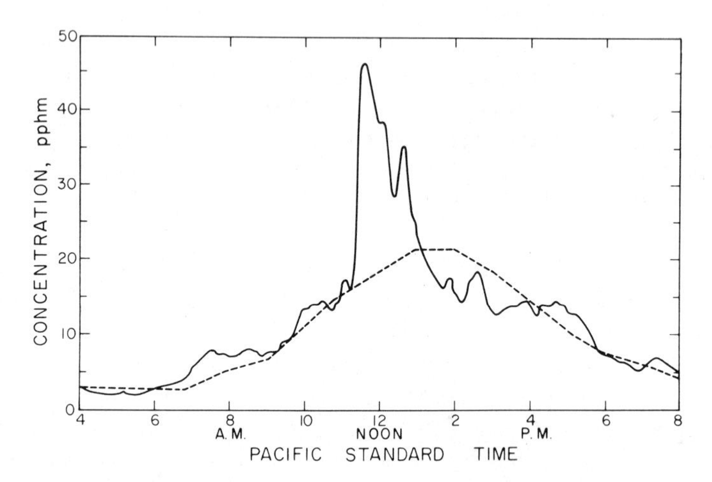

Fig. 7. Curves for oxidant as a function of time of day taken in Pasadena. The dash curve represents a three-month average of oxidant and the solid line represents the concentration on a single day. This figure is taken from reference 1.

contrasted to the graph of oxidant versus time of day taken with an oxidant analyzer on the UCR campus, Figure 8. The oxidant was minimal at about 1–2 pphm until the cloud arrived and in a five-minute interval, starting at 5:00 P.M., the oxidant jumped an increment of 16 pphm and *remained* at or above the 15 pphm level until 8:00 P.M. This dramatic effect parallels the visible arrival of the smog cloud shown in Figures 5 and 6. Incidentally, not only can one see and instrumentally measure the arrival of the smog, on bad attacks one can literally smell it, taste it, and weep from its eye-irritating effects. Furthermore, if a person is strenuously exercising, the smog can dramatically cut down one's "wind" and vitality.

C. Plant Damage

The first definite plant damage due to photochemical air pollution was noted in the mid-1940's by Middleton, Darley, and co-workers. In 1951, for example, they showed that the loss of spinach crops in Los Angeles County was due to photochemical smog and estimated that the total crop damage was as high as $500,000 in Los Angeles County alone. Only a year later plant damage had spread to the surrounding counties of Riverside, San Bernardino, and Orange. The insidious spread of smog throughout the next decade using plant damage as a biological indicator is shown in Figure 9 taken from data collected by Middleton and co-workers.

Plant damage from air pollutants and, in particular, photochemical smog is not confined to California. The spread of photochemical oxidants and crop losses of millions of dollars on the Atlantic Seaboard paralleled, on a year-to-year basis, the spread in Southern California, although occurring somewhat later. For example, in 1953 oxidant from photochemical smog was discovered in New York and in 1957 it was identified in Pennsylvania.

Today in New Jersey, the "Garden State," pollution damage to at least 36 commercial crops has been reported and has been observed in every county (7). By now, damage by photochemical smog has been identified in most states of the U.S. and many foreign countries. Thus, in 1966 the total damage to agriculture in the United States was estimated at *one-half billion dollars* (8,9).

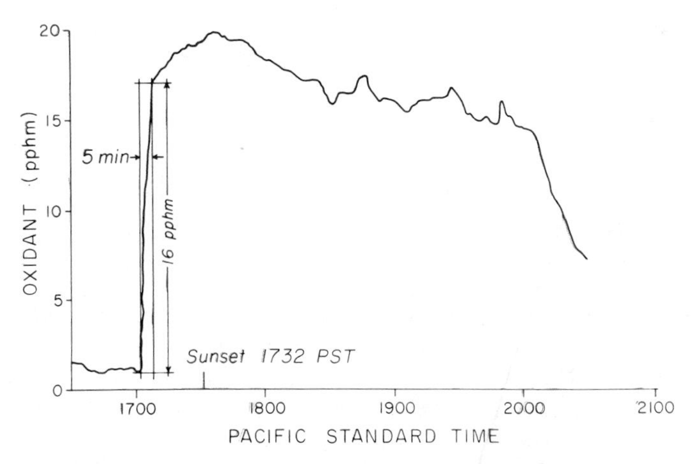

Fig. 8. A plot of oxidant concentration as a function of time of day with an oxidant meter on the University of California, Riverside campus, October 2, 1961. Note the very low level of oxidant until 5:00 P.M. and then the rapid rise as the smog cloud hits Riverside. We are indebted to Dr. E. R. Stephens for this figure taken from reference 106.

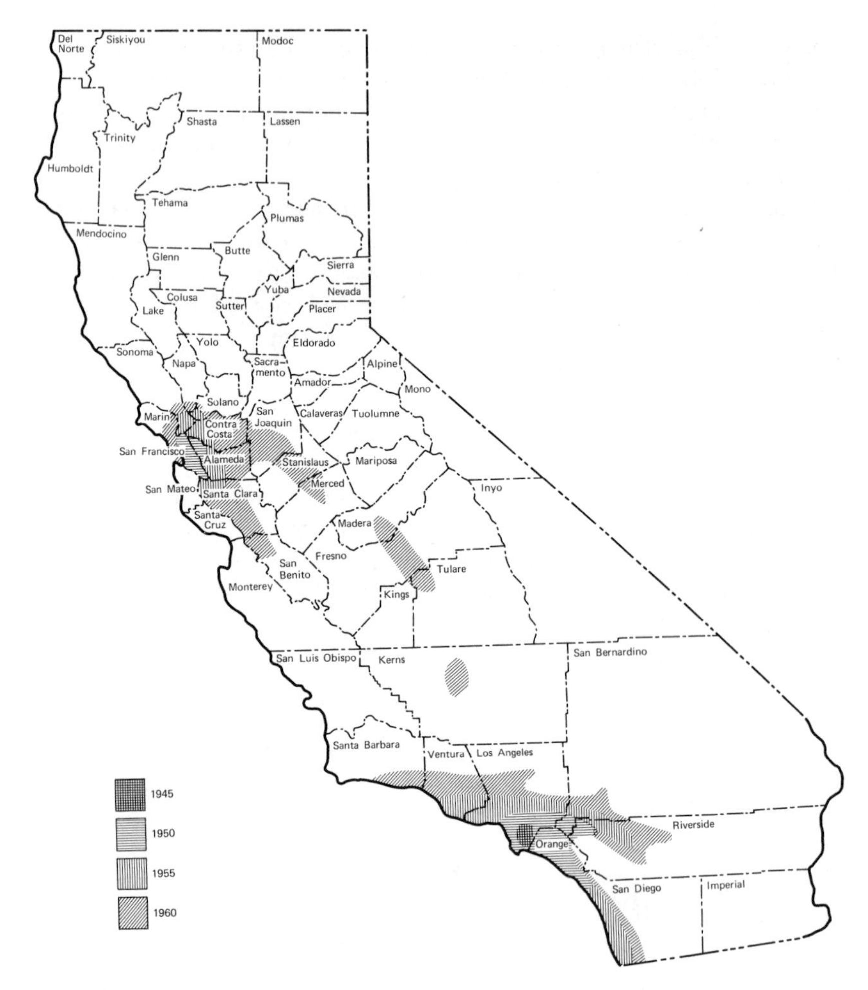

Fig. 9. Spread of plant damage in California. Figure from Middleton and co-workers.

Among the oxidizing gases causing plant damage, two are particularly obvious and characteristic of photochemical smog: ozone and peroxyacetyl nitrate (PAN). PAN is highly toxic to many important field crops including romaine lettuce, alfalfa, and spinach, and ornamental plants such as petunias, snapdragons, and asters (7). Characteristically, with romaine lettuce the damage appears as a silvering on the *bottom* of the leaf.

In contrast, ozone kills cells in the *top* surfaces of leaves and the damage appears as flecking or stippling of these upper surfaces. For example, a three-hour fumigation with air containing only 0.5 ppm of ozone produces lesions in grape leaves. In a recent review, Middleton notes that 57 different species of plants are cited as being susceptible to ozone damage (7).

Nitrogen dioxide (NO_2) has an interesting and more subtle effect on plants. Apparently in fairly low concentrations it appears to affect the growth of young plants without showing other common signs of injury!

Ethylene, the simplest olefin, $H_2C{=}CH_2$, is released by refineries and from auto exhausts and specifically is highly toxic to orchids in concentrations approaching the parts per billion (ppb) range. The presence of this smog constituent has made it impossible for commercial growers to raise orchids in metropolitan areas of California.

Other air pollutants such as sulfur dioxide and hydrogen fluoride are also serious phytotoxicants, but since they are not commonly associated with *photochemical* air pollution, we shall not discuss them here.

D. Primary and Secondary Air Pollutants

We have now cited two general species of pollutants present in photochemical smog, aerosols producing the dense haze and noxious gases such as the oxides of nitrogen, ozone, and PAN. Before going on, the point should be stressed that both the London and Los Angeles types of smog are *mixtures* of particulate matter and toxic gases. It is important to realize that in such a situation possible synergistic effects having important chemical, physical, and/or biological implications can occur. That is, the sum of effects due to the gases alone plus effects due to the particles alone may well be exceeded by the effects of the mixture. This phenomenon is illustrated by the casualties caused by the London type of smog which contains *both* gaseous sulfur dioxide and solid particles of soot and other material. Similar considerations apply to the "excess deaths" reported in an acute smog episode in New York City in 1962 (10,11). Obviously, one must be well aware of such synergistic effects when proposing chemical reaction mechanisms, explaining health effects, or developing models for air pollution control.

Primary pollutants are those emitted directly to the atmosphere. *Secondary pollutants* are formed by chemical or photochemical reactions of the primary pollutants after the latter have been emitted into the atmosphere and have been exposed to the sunlight.

Data for daily emissions of primary pollutants in 1966 compiled by the Los Angeles County Air Pollution Control District (LAAPCD) are shown in Figure 10. It gives average values in the *tons per day* and the

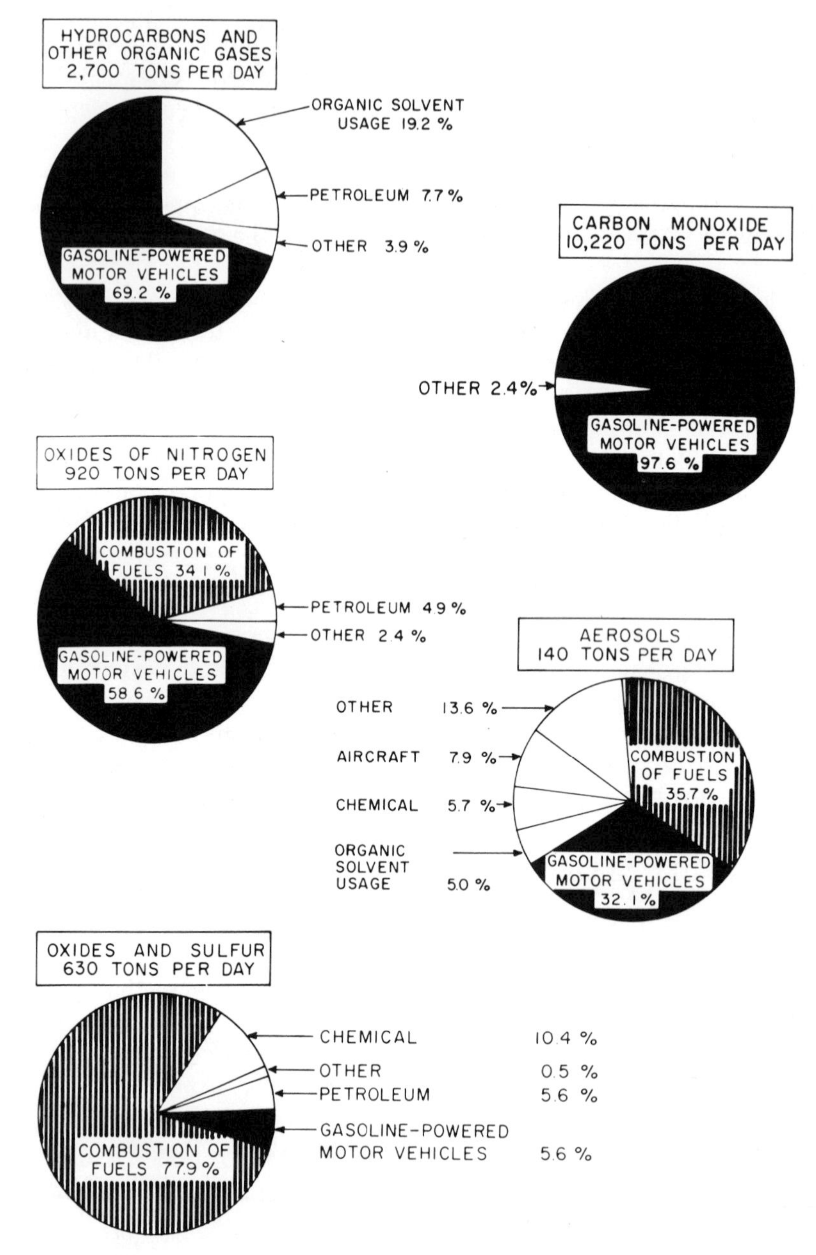

Fig. 10. Percentage contributions of air contaminants from major sources in Los Angeles County. Data of the Los Angeles County Air Pollution Control District, January 1967. Figure taken from reference 12a.

percentage contribution of the important contaminants from the major sources within Los Angeles County (12a). All of the compounds cited are primary pollutants, that is, the organic gases including primarily hydrocarbons and aldehydes, the oxides of nitrogen (NO_x), sulfur dioxide (SO_2), and carbon monoxide.* It is alarming to realize that about *14,000 tons per day* of pollutants are introduced into the atmosphere over Los Angeles County. On the other hand, it is even more sobering to consider what might have been the case if the Air Pollution Control District of Los Angeles County had not been taking stringent measures to prevent the emission of pollutants from stationary sources and the State of California had not initiated controls on automotive emissions. The current LAAPCD program of pollution control prevents the emission of approximately 5,500 tons of pollution daily in Los Angeles County while automobile controls reduce emissions an additional 1700 tons per day (12b).

In terms of the per cent contribution to air pollution from various major sources, in 1966 gasoline-powered motor vehicles in Los Angeles County accounted for approximately 70% of the total hydrocarbons, 60% of the oxides of nitrogen, 6% of the sulfur dioxide, and about 97% of the carbon monoxide. Thus, gasoline-powered motor vehicles accounted for approximately 85% of the total air pollution. Also significant is the fact that 32% of the aerosols released into the atmosphere came from automobiles and trucks (12a).

When these *primary pollutants* emerge from the auto exhaust, they are diluted with approximately 1,000 volumes of air per volume of exhaust gas, producing actual atmospheric concentrations of the order of 0.5 ppm NO_x, 1 ppm hydrocarbons, and 5–20 ppm CO. Actually, these are "order of magnitude" typical concentration ranges; actual maxima in the Los Angeles Basin may be much higher. For example, *maximum* air contaminant concentrations that have been measured in the Los Angeles Basin since 1955 as reported by the LAAPCD are carbon monoxide, 72 ppm; hydrocarbons, 40 ppm (flame ionization method); oxides of nitrogen, 3.9 ppm; ozone, 0.9 ppm and sulfur dioxide, 2.5 ppm (12).

In addition to the gaseous pollutants, both organic and inorganic solids are also emitted to the atmosphere. Of particular interest are polynuclear aromatic hydrocarbons and lead compounds. The latter

* Recall that NO_x actually stands for a mixture of NO and NO_2 plus other oxides of nitrogen present in smaller amounts in an unspecified ratio. As "it" emerges from the auto exhaust, the NO_x is mostly nitric oxide (see reference 2 for a discussion of this important point).

have reached maximum atmospheric concentrations of the order of 50 $\mu g\ M^{-3}$ in Los Angeles traffic, and average concentrations of perhaps 20–30 $\mu g\ M^{-3}$ (13). We shall consider certain aspects of these later in the discussion.

When an urban atmosphere containing pollutants of this type is irradiated by the sun, a series of highly complicated photochemical reactions occur. Products of these photochemical reactions are the *secondary pollutants* referred to earlier. They are responsible for eye irritation and plant damage and include nitrogen dioxide, ozone, the peroxyacyl nitrates, formaldehyde, higher saturated aldhydes such as acetaldehyde, unsaturated aldehydes such as acrolein (a powerful eye irritant), alkyl nitrates, alkyl nitrites, and a variety of other compounds (1,4,5). For mechanistic or atmospheric "modeling" purposes, one should note that the concentrations of these secondary pollutants vary in a complex way with atmospheric conditions, time of the day, etc. One of the earliest reported examples is shown in Figure 11 for downtown Los Angeles. The typical rapid rise in hydrocarbons and nitric oxide occurs around 8:00 A.M. at the peak of the morning traffic. Formation of oxidant and nitrogen dioxide is delayed until the sunlight is intense enough that the photochemical reactions will proceed at rapid rates. These data were taken over a decade ago by the LAAPCD before the analytical techniques for the ozone, NO, and NO_2 were refined and improved, so they are not precise values and must be used with caution. However, the trends are generally correct and it is interesting that air quality data on photochemical air pollution were being taken in the Los Angeles Basin over 15 years ago. In recent years cooperative studies by the Los Angeles County Air Pollution Control District, the California Air Resources Laboratory and the National Air Pollution Control Administration have greatly expanded the amount of quantitative "field data" available on air quality in the Los Angeles Basin (see recent papers and reports by Altshuller et al. that deal with these studies).

Carbon monoxide is of particular interest not only because of its health effects, some of which now appear to occur at substantially lower levels than previously recognized, but also because it is believed to be quite unreactive in the atmosphere (this *may* not be strictly true), and excellent, sensitive instrumentation is available for its monitoring. Thus it serves as a useful "tracer" to indicate the activities of automobiles because, as seen in Figure 10, almost 98% of the carbon monoxide released into Los Angeles atmosphere is from gasoline-powered motor

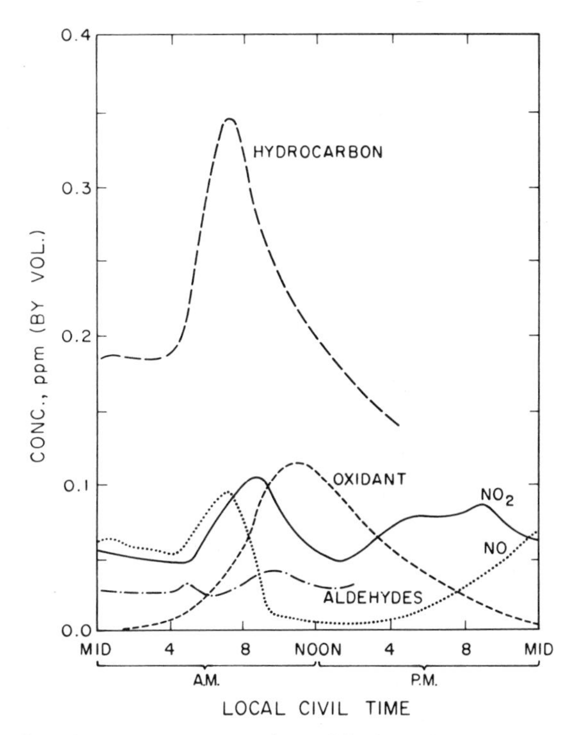

Fig. 11. A plot of average concentrations of hydrocarbons, oxidant, nitrogen dioxide, nitric oxide, and aldehydes in the atmosphere as a function of time of day in downtown Los Angeles. The data were generated by the Los Angeles County Air Pollution Control District for days when eye irritation was present. Data for hydrocarbons, aldehydes and ozone (1953–1954); NO and NO_2 (1958). Figure taken from reference 1.

vehicles. This is not surprising when one recognizes that on the average, 29 lb of carbon monoxide are emitted per 10 gal of gasoline consumed!

Figure 12 shows typical plots of the concentration of carbon monoxide in parts per million as a function of time of day at four stations in the Los Angeles Basin as determined by Colucci and Begeman (14). The concentration drops to a minimum at approximately 5:00 A.M., then rises rapidly to a maximum about 8:00 A.M., the time of peak activity on the freeways. Also at this time there is generally a stable atmospheric inversion which traps the carbon monoxide. A much broader and lower peak occurs between about 4:00 and 8:00 P.M. when the traffic again increases. Here the effect is not as noticeable because the atmospheric inversion is not as intense and the carbon monoxide is

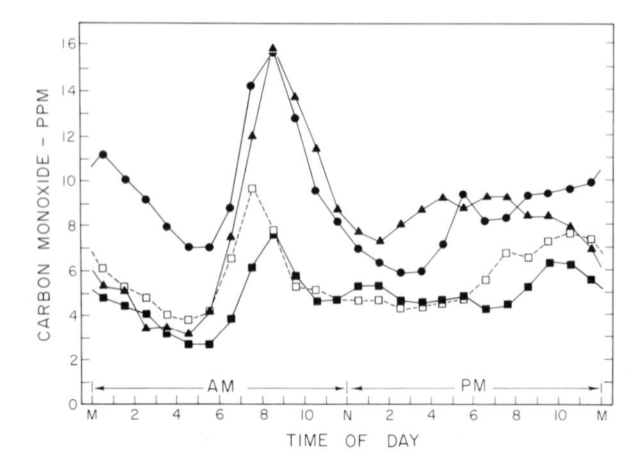

Fig. 12. A plot of the concentration of atmospheric CO as a function of the time of day in several locations in the Los Angeles Basin. (●) Pico Boulevard; (▲) Harbor–Santa Monica Freeway Interchange; (■) Santa Monica; (□) Monrovia. Data of Colucci and Begeman taken from reference 14. We thank General Motors for permission to reproduce this figure.

generally dispersed by the evening breeze. Data of this type are most important to the development of mathematical models for diffusion of pollutants in an urban atmosphere. An interesting paper by Slade on "Modeling Air Pollution in the Washington D.C. to Boston Megalopolis" is one of many on the subject now beginning to appear (15).

E. Laboratory Studies

Briefly, it is well known that if one starts out in the laboratory with a reaction chamber containing nitric oxide, a trace of nitrogen dioxide, and air, and irradiates this mixture with ultraviolet light, the following reactions occur (1,2,4,5,16–19)

$$NO_2 + h\nu(3000\ \text{Å} < \lambda \leq 4200\ \text{Å}) \rightarrow NO + O(^3P) \quad \text{(A)}$$

$$O + O_2 + M \rightarrow O_3 + M \quad \text{(B)}$$

$$O_3 + NO \rightarrow O_2 + NO_2 \quad \text{(C)}$$

The net effect of irradiation on this inorganic system is to set up a dynamic equilibrium, namely

$$NO_2 + O_2 \overset{UV}{\leftrightarrows} NO + O_3 \quad \text{(D)}$$

However, if a hydrocarbon, particularly an olefin or an alkylated benzene (both of which are common constituents of gasoline) is added, the dynamic equilibrium is unbalanced and the following events take place:

1. The hydrocarbons are oxidized and disappear.

2. Reaction products such as aldehydes, nitrates, PAN, etc., are formed.

3. Nitric oxide is converted into nitrogen dioxide.

4. When all of the nitric oxide has been used up, ozone starts to appear. On the other hand, PAN and the aldehydes are formed from the beginning of the reaction.

One can reproduce the essential features of photochemical air pollution simply by irradiating dilute automobile exhaust with light in the wavelength region 3000–4200 Å in a suitable reaction chamber (see references 1,2,4, and 5 for original references). The auto exhaust disappears and various compounds are formed identical to those found in actual urban atmospheres. Figure 13 shows the variation with time of some typical products of ultraviolet irradiation of auto exhaust in a laboratory reaction chamber (1,20). Note that the chemical behavior in this system, as a function of time, resembles that seen in Figure 11 which gives the actual diurnal variations of pollutants in the atmosphere.

The curves of Schuck et al. shown in Figure 13 were generated from data taken in a reaction system, the key feature of which is a long path infrared cell. We are currently using a similar apparatus at UCR. A diagram of our reaction cell is shown in Figure 14. The basic idea is that one employs a conventional infrared spectrophotometer for

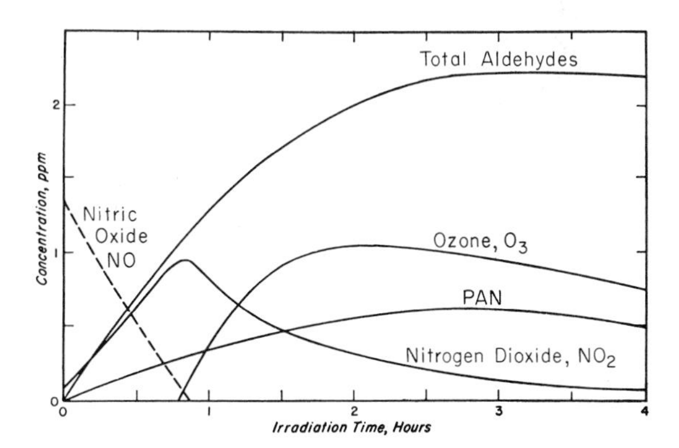

Fig. 13. Variation of concentration of various components in autó exhaust irradiated with ultraviolet light. (Figure from reference 1, data from reference 20.)

analysis of the gases but instead of using small gas cells of about 0.1-liter capacity, one employs two large cylindrical tanks, each of approximately 60-liter volume. One acts as a blank cell and the other as a photochemical reaction chamber. Figure 15 is a photograph of our actual apparatus.

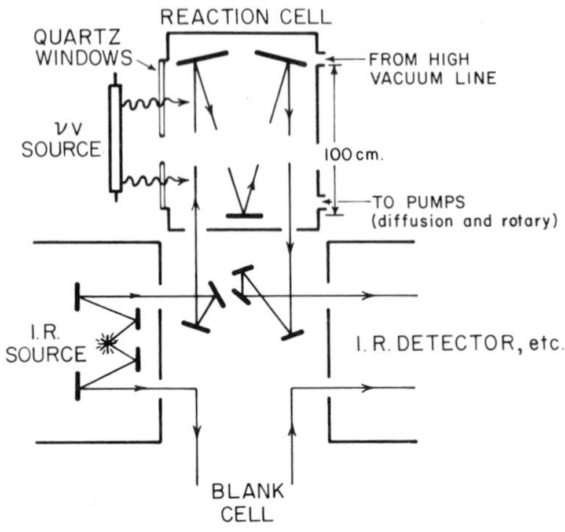

Fig. 14. Diagram of optical path for long path length infrared spectrophotometer used at UCR.

Fig. 15. Photograph of long path length infrared instrument in our UCR laboratories.

Basically, two beams emerge from the infrared source. One traverses the blank cell, which is filled with the nonirradiated material, and then goes into the infrared detector; the other beam goes into the reactor cell which is fitted with quartz or borosilicate windows so that the reaction mixture in the cell can be irradiated with an external ultraviolet light source. The mirrors in the reaction cell and the dummy cell can be aligned so that the infrared analyzer beams traverse a path of up to 40 m before leaving the reaction cell. Compared to the 0.1-m path in a conventional gas cell, this provides a tremendous sensitivity and, with this instrument, one can identify and quantitatively determine various pollutants such as olefins, ozone, aldehydes, PAN, etc., in the fraction of a part-per-million range of concentration.

One does not require the complex mixture of gases and particulates, i.e., auto exhaust, to generate the products and the concentration versus times of irradiation curves characteristic of photochemical air pollution. Thus, for example, irradiation of air containing a pure olefin plus nitric oxide (and a trace of NO_2 to act as the primary light absorber; see below) at wavelengths shorter than about 4200 Å will produce products such as ozone, aldehydes, PAN, NO_2, etc. This is shown in Figure 16 taken from data of Schuck and Doyle (21).

A highly important conclusion arises from experiments conducted in long path length infrared cells of this type, as well as more conventional systems, by researchers such as Altshuller, Cvetanovic, Hamming, Schuck, Stephens, and Tuesday and their colleagues (see references 1,2,4,5, and 22 for original references). The various hydrocarbons emitted in auto exhaust have a *wide range of reactivities toward oxygen atoms and ozone as well as in forming photochemical smog.* That is, hydrocarbon A introduced into a mixture of nitric oxide and air and irradiated might react very much more slowly or, conversely, very much more rapidly than hydrocarbon B under similar conditions. This is not unexpected since the different hydrocarbons have different "chemistries" and one would expect different rates of reactions for different types of hydrocarbons. Various hydrocarbon reactivity indexes for use in air pollution control have been prepared. One of these, developed by Mr. John Maga, Executive Secretary of the Air Resources Board of California, is shown in Table I (23). Critiques on this subject, which is so important to control of air pollutants from automobiles, are given in references 23–25 while reference 26 cites typical laboratory results on hydrocarbon reactivities by Glasson and Tuesday who used long path infrared techniques with a path length of 120 m.

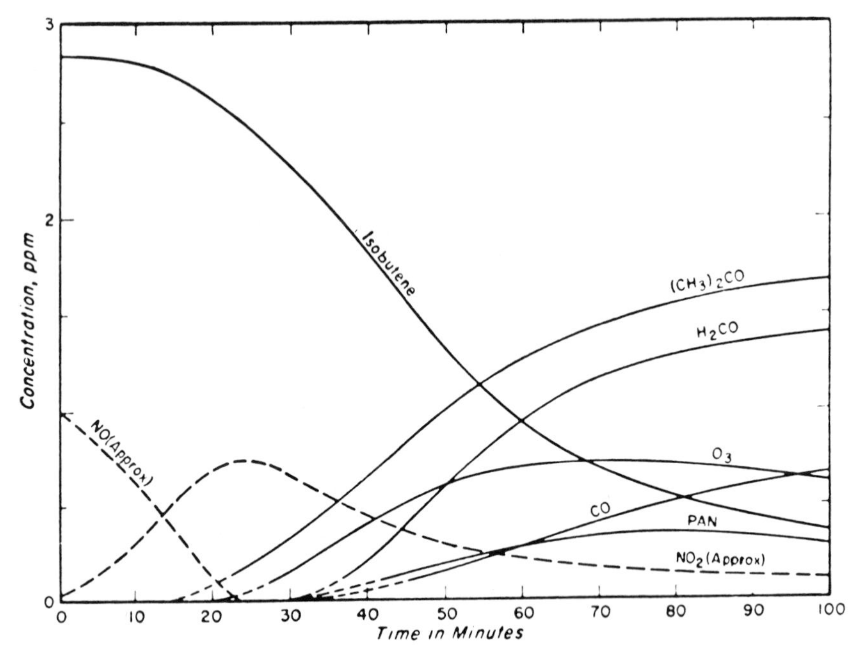

Fig. 16. Products and their variation with time for the irradiation with ultraviolet light of a mixture of isobutene and nitric oxide (plus some nitrogen dioxide) in air. (Figure from reference 1, data of Schuck and Doyle, reference 21.)

TABLE I

Relative Hydrocarbon "Photochemical Reactivities" in the System $HC-NO-NO_2$-Air[a]

Substances	Molar reactivity
C_1-C_3 Paraffins	0
C_4 and higher paraffins	1
Ethylene	4
1-Alkenes	7
Internally double bonded olefins	8
Acetylenes	0
Benzene	0
Toluene and other monoalkyl benzenes	3
Dialkyl and trialkyl benzenes	6

[a] Data of John Maga, California Department of Public Health, September 1966.

While up to 1968, tables of relative hydrocarbon reactivities relegated the saturated hydrocarbons to the lowest end of the scale (least reactive), recent laboratory studies by Altshuller and co-workers have shown that under certain conditions *n*-butane in irradiated *n*-butane–NO_x–air mixtures reacts (27). Thus at relatively high values of the ratio *n*-butane/NO_x, substantial amounts of ozone are formed! This observation has been confirmed by Bonamassa et al., who also note that recent field studies by the California Air Resources Laboratory suggest that similar conditions, i.e., high ratios of saturated hydrocarbons to oxides of nitrogen, can occur in certain parts of the Los Angeles Basin (28). These are particularly significant observations and bear serious consideration when evaluating the multifaceted aspects of photochemical air pollution, particularly its control.

F. Current Mechanisms

As we have seen, the Los Angeles type of smog is formed through a very complex set of processes initiated by ultraviolet light in the range 3000–4200 Å. In order to understand the detailed aspects of smog formation, one must understand the various *primary photochemical* and *photophysical* processes, and the subsequent secondary reactions. Vastly complicating the matter is that in actual urban atmospheres the reactions include not only homogeneous gas-phase reactions but also, and in this author's opinion possibly even more important, hetergeneous gas–solid, gas–liquid, and liquid–solid processes. To date some understanding of the homogeneous systems has been achieved, but far too little quantitative or even qualitative research has been conducted on the chemical, physical, biological, and medical aspects of heterogeneous systems approximating actual "field conditions." Such research, much of which is multidisciplinary, is difficult indeed but it is timely and essential.

Let us now consider several aspects of the homogeneous system which are reasonably clear and which have been set forth in detail elsewhere (1,2,4,5). First, the primary acts are photochemical in nature so that one must consider in detail those species which can absorb solar radiation in the region 3000–4200 Å in order to initiate the reactions. Several classes of compounds known to be primary pollutants in smog can absorb radiation more or less efficiently in the near ultraviolet (1,28). These include nitrogen dioxide, aldehydes, sulfur dioxide, and polynuclear aromatic hydrocarbons. A wide variety of *secondary* pollutants absorbs in that region including ozone, PAN, alkyl nitrates, and alkyl nitrites. Leighton considered in detail all these primary and secondary pollutants in terms of their absorption spectra, atmospheric concentrations, and the intensity of solar radiation as a function of wavelength at the earth's surface and concluded that nitrogen dioxide photolysis was of major importance (1). This is generally recognized to be the case and much attention has been given to the inorganic system $NO–NO_2–O_2$ and the more complex mixture $NO–NO_2–O_2$–hydrocarbons (1,2,4,5). In both cases the primary photochemical process is considered to be reactions (A), the photodissociation of NO_2 into NO and oxygen atoms (2,19,27).

The thermal and photochemical reactions in the *purely inorganic* system, although complicated, appear to raise no fundamental chemical problems (see references 2 and 18). However, this is not true when a reactive hydrocarbon is added. A particularly perplexing aspect of the photochemical production of smog in such systems is the remarkably rapid and highly efficient conversion of NO to NO_2 which has been observed by all investigators. The thermal reaction $2NO + O_2 \rightarrow 2NO_2$ is too slow at the relatively low concentrations of NO in the atmosphere to account for this rapid conversion. Furthermore, the ozonosphere restricts the radiation reaching the surface of the earth to wavelengths greater than about 2900 Å. Hence direct photochemical decomposition of nitric oxide or simple olefins is not possible. Thus the sole absorber of light in chamber studies is NO_2 and the photoconversion of NO to NO_2 is therefore particularly remarkable because it is a rare example of a system in which, as Leighton has pointed out, the compound which absorbs light and photodissociates is produced on an *overall* basis more rapidly than it is destroyed by photolysis.

Clearly, a unique and efficient overall photochemical process or set of processes must exist in chambers and in the atmosphere. Altshuller and Bufalini wrote in mid-1964, "A satisfactory mechanism has not been developed to fully explain the process by which the oxidation of nitric oxide occurs" (4).

While a great deal of additional mechanistic information has been acquired since then and the role of the automobile and the oxides of nitrogen in photochemical air pollution documented in detail (2,30,31), the above quote is unfortunately still valid. Furthermore, as implied earlier in the paper, this same remark can be phrased in more general terms, that is, we do not today have a satisfactory set of mechanisms that can explain *many* key physical, chemical, and biological aspects of photochemical air pollution.

While this observation is in one sense depressing, it is also an exciting challenge for researchers who are interested in applying basic science to relevant problems of man. Thus, in the following sections we shall discuss a specific case in which fundamental research on the spectroscopy, photochemistry, and mechanistic organic chemistry of singlet molecular oxygen recently has been applied to specific problem areas of photochemical smog. Most information presently available comes from laboratory or upper-atmosphere experiments; however, there is good reason to believe that extrapolation to urban atmospheres is not only reasonable but should be demonstrable.

III. SINGLET MOLECULAR OXYGEN

Recently in our laboratory we have come to the conclusion that the lowest electronically excited states of O_2, symbolized as $O_2(a^1\Delta_g)$ and $O_2(b^1\Sigma_g^+)$, may be hitherto unrecognized oxidants present in test chambers as well as in actual polluted urban atmospheres (32,33). Similar conclusions have been reached by Bayes (49) and Kummler et al. (85). We further believe, as do an increasing number of researchers in several disciplines, that certain unexplained chemical and biological effects observed in man, animals, and plants, including some relative to certain observations on photochemical air pollution cited earlier in this paper, might be attributed to these species (for example, see the review by Foote, reference 34). Therefore, in the following sections we shall briefly review the spectroscopic and chemical properties of the two states of singlet oxygen and consider in some detail their possible role in the chemistry of urban atmospheres. For detailed discussions and extensive references to the original literature on photooxidations and singlet oxygen, one should consult the excellent recent reviews by Foote (34,35), Gollnick (36a,b), Gollnick and Schenck (37), Ogryzlo et al. (38), and Wayne (39). The electronic structure and spectra of singlet oxygen are considered in references 29 and 40–42.

A. Spectroscopy

1. *Electronic Structure*

Simplified potential energy diagrams for the ground state and three electronically excited states of the oxygen molecule are shown in Figure 17. The two lowest states, $a\ ^1\Delta_g$ and $b\ ^1\Sigma_g^+$ are singlets located at about 22.5 kcal (0.98 eV) and 37.5 kcal (1.63 eV), respectively, above the ground state, $O_2(X^3\Sigma_g^-)$.* The electronic structures of the two singlet

* We shall abbreviate the spectroscopic notations to $^1\Delta$ or $^1\Sigma$ and refer to these two excited states as singlet oxygen, $^1O_2$.

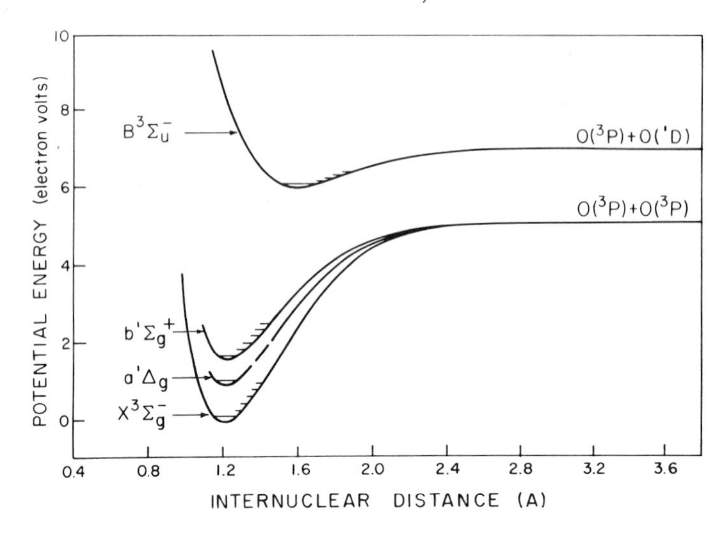

Fig. 17. Franck-Condon curves for several states of the oxygen molecule. $a\,^1\Delta_g \leftarrow X\,^3\Sigma_g^-$ Infrared atmospheric system; $b\,^1\Sigma_g^+ \leftarrow X\,^3\Sigma_g^-$ Atmospheric system; $B\,^3\Sigma_u^- \leftarrow X\,^3\Sigma_g^-$ Schumann-Runge system.

Electronic states of the oxygen molecule	Occupancy of highest orbitals		Energy above ground state	
Second excited state ($^1\Sigma_g^+$)	↑	↓	37.5 kcal	1.63 eV
First excited state ($^1\Delta_g$)	↑ ↓	—	22.5 kcal	0.98 eV
Ground state ($^3\Sigma_g^-$)	↑	↑		

Fig. 18. Occupancy of highest molecular orbitals for ground state and the two lowest singlet states of molecular oxygen according to Foote. (Taken from references 35 and 42.)

species differ significantly as seen in the simplified orbital diagram, Figure 18 (42). The $^1\Sigma$ state resembles ground-state oxygen, $^3\Sigma$, but in the $^1\Delta$ state both electrons are paired in a single orbital, leaving the other vacant (35). Foote points out that $O_2(^1\Delta)$ resembles ethylene electronically, but should be more electrophilic. Thus it would be expected to undergo *two*-electron reactions whereas the $^1\Sigma$ species "might be expected to undergo one-electron free radical reactions" (35).*

* In a recent review, Ogryzlo examines the electronic structures of $O_2(^1\Delta)$ and $O_2(^1\Sigma)$ and comes to the opposite conclusion, that is, that "since the other $^1\Delta_g$ component appears to have its electrons in separate orbitals, this state should display a 'diradical' character similar to the triplet ground state. On the other hand the $^1\Sigma_g^+$ state does not display this 'diradical' configuration, and as we have seen above displays a repulsive interaction with all species." (109)

2. *Absorption and Emission of Radiation*

The "forbidden" transitions

$$O_2(^3\Sigma_g^-) + h\nu(7619\ \text{Å}) \underset{b}{\overset{a}{\leftrightarrows}} O_2(^1\Sigma_g^+) \quad (1)$$

$$O_2(^3\Sigma_g^-) + h\nu(12{,}690\ \text{Å}) \underset{b}{\overset{a}{\leftrightarrows}} O_2(^1\Delta_g) \quad (2)$$

involve magnetic dipole interactions and are very weak relative to the "allowed" electric dipole transition

$$O_2(^3\Sigma_g^-) + h\nu\ (\lambda \leq 2000\ \text{Å}) \leftrightarrow O_2(^3\Sigma_u^-) \quad (3)$$

The latter is responsible for the Schumann-Runge bands which start at about 2000 Å and merge into a strong continuum beginning at 1759 Å (42). The intense absorption by oxygen due to process 3 is a limiting factor in spectroscopic and photochemical studies in air (29). Despite being weak, the radiative transitions 1 and 2 have been well established by spectroscopic studies of both the absorption and emission processes in the laboratory and in the upper atmosphere. The 0–0 bands for the $^1\Sigma$ state fall in the far red region of the visible spectrum, 7619 Å, while the $^1\Delta$ emission at 12,686 Å is in the near infrared. They are often referred to as the "atmospheric" and "infrared atmospheric" bands of oxygen (43,44).

Since transitions to the ground state are forbidden, in an *unperturbed* system both the $^1\Sigma$ and $^1\Delta$ states have long natural radiative lifetimes of the order of 10 sec and 45 min (59), respectively.* However, if an oxygen molecule in the excited $^1\Sigma$ or $^1\Delta$ state collides with a normal ground-state molecule, a short-lived (ca. 10^{-13}sec) *collision complex* is formed and electric dipole transitions become more probable. Thus the collision complex $^1\Delta^3\Sigma$ has a calculated lifetime of four seconds versus about 45 min for the unperturbed O_2 molecule in the $^1\Delta$ state (38).

Under conditions where substantial concentrations of O_2 ($^1\Delta$) are present, collisions between two *excited* O_2 molecules may result and a novel *energy pooling* process may occur. In this, the electronic energy of the two colliding molecules appears as a single photon. Such a case is illustrated by the red glow which is readily observed visually when $O_2(^1\Delta)$ is generated chemically in an alkaline H_2O_2–Cl_2 system or by a microwave discharge (see below). The red light is composed of bands at 6340 and 7030 Å and is called "dimol" emission. It results from an

* For the $^1\Sigma$ state the previously accepted value is 7 sec (reference 45) but recently 12.5 sec has been proposed (46).

energy-pooling process between two $O_2(^1\Delta)$ molecules (see reaction 9). It is not established whether they form a "stabilized dimer" or interact through a collision complex (38,47,48). In any case, the process is sufficiently efficient that the dimol emission at 6340 Å is often used for monitoring the concentration of oxygen molecules in the singlet delta state (see below).

3. *Physical Quenching*

In considering the overall lifetime of the $^1\Delta$ and $^1\Sigma$ states, one must consider several other modes of deactivation besides luminescence and chemical quenching to form products. Particularly important are the radiationless relaxation processes, *physical quenching* and *energy transfer*.

Physical quenching occurs when an excited molecule collides with a normal molecule and its electronic excitation energy is converted to vibrational and rotational energy, i.e., heat.

$$^1O_2 + M \xrightarrow{k_q} O_2 + M \tag{4}$$

It also may occur when the excited molecule strikes the surface of the reaction vessel and is deactivated.

$$^1O_2 \xrightarrow{\text{Wall}} O_2 \tag{5}$$

Both processes may be important modes of deactivation in the laboratory with collisional deactivation being favored at higher pressures and wall reactions efficient at lower pressures.

A particularly important characteristic of the $^1\Delta$ state is that it is remarkably stable to physical quenching. Thus Arnold, Kubo, and Ogryzlo estimate that in most laboratory flow systems it can suffer about 2×10^5 collisions with the walls of the vessel before deactivation. It is even more stable toward collisional deactivation by another molecule; they estimate that at least 10^8 collisions with normal O_2 molecules are required for deactivation of a $O_2(^1\Delta)$ molecule. Collisions of $O_2(^1\Delta)$ with other molecules are even less effective with about 10^9–10^{10} necessary for its relaxation to the ground state (38). Similar conclusions were reached by Winer and Bayes (49a).

In condensed fluid solutions with hydrocarbons or water as quenchers, these quenching results suggest a lifetime of about 10^{-2}–10^{-3} sec for $O_2(^1\Delta)$. In gaseous systems, however, the lifetime of $^1\Delta$ is much longer, of the order of 0.05–0.5 sec at one atmospheric pressure in air (84,85). This is highly relevant to photochemical air pollution. Thus Bayes points out that if $^1O_2$ were generated at the rate of 25 pphm per hour in urban air containing 25 pphm of olefin, "at least 50% of the $^1O_2$ will collide with an olefin molecule before being collisionally deactivated" (49b)!

The $^1\Sigma$ state is much less stable toward deactivation; only about 100 collisions with molecules of water or a hydrocarbon are required for its deactivation, presumably to the $^1\Delta$ state (38).† This corresponds to a lifetime of $^1\Sigma$ of only about 10^{-9} sec in fluid solutions (38). The rate of quenching of $^1\Sigma$ depends on the nature of the quencher, M. Thus Arnold et al. reported quenching rate constants, k_q, ranging from 7×10^5 for M = helium to 1.5×10^6 for N_2, Ar, and CO, to 6×10^8 for H_2O [all in units of liters M^{-1} sec^{-1} (38)]. Izod and Wayne recently have confirmed the value for k_q of N_2 and have shown that normal oxygen is a very inefficient quencher with $k_q = 6 \times 10^4$ liter M^{-1} sec^{-1} (50).

4. *Energy Transfer from Singlet Oxygen*

The transfer of electronic energy from an excited donor molecule, D* to an acceptor molecule A, process 6, requires that the acceptor have an excited electronic state at an energy equal to or less than that of the donor.

$$D^* + A \rightarrow D + A^* \tag{6}$$

We shall discuss this point in detail later when we consider energy transfer from an excited organic molecule to normal oxygen as a means for the production of singlet oxygen in polluted atmospheres, process 7.

$$D^* + O_2(^3\Sigma) \rightarrow D + O_2(^1\Delta \text{ or } ^1\Sigma) \tag{7}$$

Here we shall consider the reverse of process 7, transfer of electronic energy from $O_2(^1\Delta)$ to acceptors.

$$O_2(^1\Delta) + A \rightarrow O_2(^3\Sigma) + A^* \tag{8}$$

Arnold et al. observed energy transfer from $O_2(^1\Delta)$ to (*a*) another $^1\Delta$ molecule, (*b*) dibenzanthrone, (*c*) nitrogen dioxide, and (*d*) iodine atoms (38). The list is limited because the excitation energy of the $^1\Delta$ state, 22 kcal, is so low that few other molecules have excited triplet states at *lower* energies and so are capable of acting as acceptors in this system. [Note: Conversely, electronic energy transfer from the excited triplet states of many organic molecules to normal O_2 to produce $O_2(^1\Delta)$ is a highly favored process in the laboratory and in the atmosphere (see below).]

† It would appear that deactivation of $^1\Sigma$ might also be directly to the ground state, $^3\Sigma$.

The first process, reaction 9, is of particular significance in upper atmosphere chemistry for it is considered as one possible source for the substantial quantities of $O_2(^1\Sigma)$ observed there (44,51).

$$O_2(^1\Delta) + O_2(^1\Delta) \rightarrow O_2(^1\Sigma) + O_2 \tag{9}$$

Recent studies point to a rate constant of about 10^3–10^4 liters M^{-1} sec^{-1} for the process (38,39,50), far below the first reported value of ca. 10^7 liters M^{-1} sec^{-1} of Young and Black (51).

Energy transfer from $O_2(^1\Delta)$ to nitrogen dioxide, while interesting to contemplate from the standpoint of photochemical air pollution, is apparently not particularly efficient. Thus it is probably not important in the atmosphere and will not be considered here (52).

In concluding this section, we should note the much greater ease of quenching of the $^1\Sigma$ state relative to $^1\Delta$ is a particularly significant consideration when one considers the relative contributions of these states to chemical reactions with organic compounds in the laboratory or in urban atmospheres. Thus Arnold et al. considered a system in which $^1\Sigma$ is *deactivated* to the $^1\Delta$ state by quenchers such as hydrocarbons or water, reaction 10, but is simultaneously regenerated by the energy transfer reaction 9.

$$O_2(^1\Sigma) + M \rightarrow O_2(^1\Delta) + M \tag{10}$$

They concluded that when a steady state was established, the rate constant for the chemical reaction with $^1\Sigma$ would have to be ca. 10^8 times *faster* than that for $^1\Delta$ if it were to be competitive (38). In part for this reason and in part because of the lack of information on the liquid or gas phase reactions of $O_2(^1\Sigma)$, we shall focus our primary attention on $O_2(^1\Delta)$ for the remainder of this paper. Two recent reviews by Foote discuss what little is known about the reactivity of $^1\Sigma$ vs. $^1\Delta$ oxygen to organic compounds (34,66) and contain many references to the original literature including the papers of Kearns et al. on the possible involvement of $O_2(^1\Sigma)$ in photosensitized oxidations (67,68).

B. Laboratory Methods for the Generation of Singlet Oxygen

1. *Chemical*

The current surge in laboratory studies of singlet oxygen began in the early 1960's, almost decades following the original proposition by Hans Kautsky that "The extinction of luminescence is the expression of an energy transfer between excited dyestuff molecules and oxygen, in which activated O_2 is produced" (53). Much of the impetus came from studies of the red luminescence observed when chlorine gas was bubbled through an alkaline solution of hydrogen peroxide. Spectroscopic research by Bowen and Lloyd (54), Khan and Kasha (55), and Ogryzlo et al. (56) established that the emission was due to the formation of singlet oxygen in the reaction

$$NaOCl + H_2O_2 \rightarrow NaCl + H_2O + {}^1O_2 \quad (11)$$

Further impetus, particularly among chemists, came from the mechanistic studies of Foote and Wexler who in 1964 first described "a novel and useful synthetic method for the oxidation of olefins and dieneoid compounds to give products identical with those of the well studied dye-photosensitized autooxidations". They went on to state, "The active species appears to be molecular oxygen in an excited state, formed *in situ* by the reaction of sodium hypochlorite and hydrogen peroxide" (57a). In a companion paper, they further proposed that singlet oxygen was "a probable intermediate in photosensitized autooxidations" (57b).

Since then, the solution-phase reaction of alkaline hydrogen peroxide with either sodium hypochlorite or bromine has been employed by an increasing number of researchers investigating oxidations of organic compounds, including many of biochemical interest (34–36,58). It is now generally agreed that $O_2(^1\Delta)$ is the principle reacting species (the $^1\Sigma$ state is rapidly quenched, see above), and that the red chemiluminescence is the "dimol" emission produced by the interaction of two $^1\Delta$ molecules. Laboratory procedures for carrying out the reactions are described in some detail in references 58 and 69.

Chemical methods for generating singlet oxygen in the *gas phase* include decomposing the endoperoxides of polynuclear aromatic hydrocarbons such as 9,10-diphenyl anthracene (60) and the decomposition of the triphenyl phosphite–ozone adduct (61). In both cases the oxygen is evolved in the $^1\Delta$ state but at substantially lower concentrations than observed with the hypochlorite–H_2O_2 system or the microwave discharge.

2. *Physical; Electrical Discharge*

A convenient gas phase method of generating substantial quantities of $^1O_2$ for laboratory studies is to pass normal oxygen at 1–10 torr through an electrodeless microwave discharge (see reference 39 for a review with original references). After the discharge, the gas stream can contain up to 10% of $O_2(^1\Delta)$, depending on the experimental conditions. Oxygen in the $^1\Sigma$ state is also formed, but it is rapidly quenched and, according to Falick and Mahan, amounts to only a few tenths of one per cent

(62). Oxygen atoms are also formed in the discharge but they are removed by saturation of the oxygen gas with mercury vapor prior to its entry into the discharge and by passing the effluent from the discharge over a film of mercuric oxide.

In our laboratory at UCR we have found it convenient to employ a Raytheon microwave discharge unit, Model PGM-10-X1, operating at 2450 MHz for generating $O_2(^1\Delta)$. The microwave energy is coupled from the coaxial cable to the flow system with a tunable wave guide supplied by Ophthos Instrument Company, Rockville, Maryland (63). A diagram of one of our experimental systems developed by Drs. Broadbent, Gleason, and Whittle is shown in Figure 19. It is currently being employed to study the gas-phase chemical reactions of $O_2(^1\Delta)$ with a variety of organic compounds including a number present in polluted atmospheres. For example, we have used tetramethylethylene (**TME**) as the acceptor for the singlet oxygen and monitored the formation of the hydroperoxide by sampling directly into an infrared spectrophotometer fitted with a 10 meter gas cell (the latter was particularly helpful in identifying the ozonide from dimethyl furan, (**III**), which was too unstable to handle in the liquid phase). Alternately, the hydroperoxide product, (**I**), which was the only product, could be condensed out and recovered as indicated in the diagram. Product yields approaching 100% have been obtained with this gas phase system (77); the synthetic possibilities are intriguing.

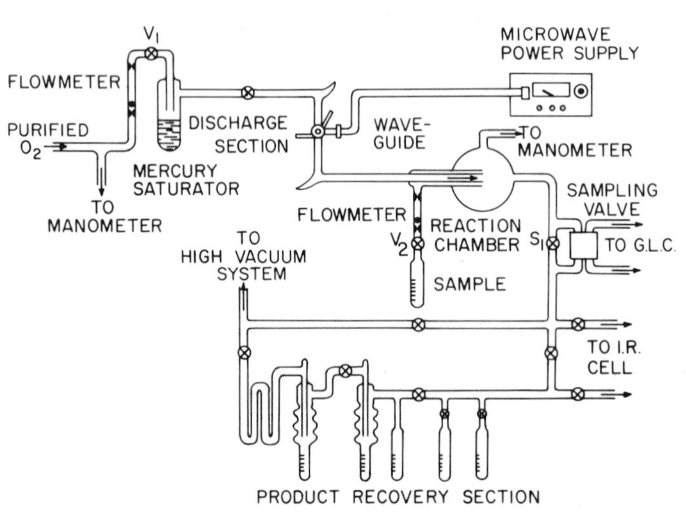

Fig. 19. Diagram of one experimental apparatus used at UCR to study the generation of $O_2(^1\Delta)$ and its gas phase reactions with organic compounds.

3. *Energy Transfer: Photosensitization*

Photooxygenation reactions of organic compounds fall into two general categories: (*a*) those in which the photoexcited sensitizer molecule (*donor*) interacts directly with the substrate to produce free radical species which subsequently react with ground-state molecular oxygen present to give products, and (*b*) those in which the excited sensitizer first transfers its electronic excitation energy to molecular oxygen producing singlet molecular oxygen which subsequently reacts with the substrate to give oxygenated products. Gollnick refers to these as Type **I** and Type **II** photosensitized oxygenation reactions, respectively (36). We shall consider only the second case, Type **II**, as it involves the production of $^1O_2$ molecules. The reviews by Bowen (64), Foote (34,35), Gollnick (36), and Livingston (65) discuss in detail the mechanisms of photosensitized oxidation.

Briefly, the photosensitized oxidation by $^1O_2$ involves the following sequence of reactions; it is now known to be general for both liquid and gas phase systems:

$$\text{Sens}(S_0) + h\nu \xrightarrow{\text{Absorption}} \text{Sens}(S_1) \tag{12}$$

$$\text{Sens}(S_1) \xrightarrow[\text{crossing}]{\text{Intersystem}} \text{Sens}(T_1) \tag{13}$$

$$\text{Sens}(T_1) + {}^3\text{O}_2 \xrightarrow[\text{transfer}]{\text{Energy}} \text{Sens}(S_0) + {}^1\text{O}_2 \tag{14}$$

$${}^1O_2 + \text{Acceptor} \xrightarrow[\text{reaction}]{\text{Chemical}} \text{Product} \tag{15}$$

Processes 12, 13, and 14 are shown schematically on the energy-level diagram, Figure 20. In most cases it is the triplet state of the sensitizer

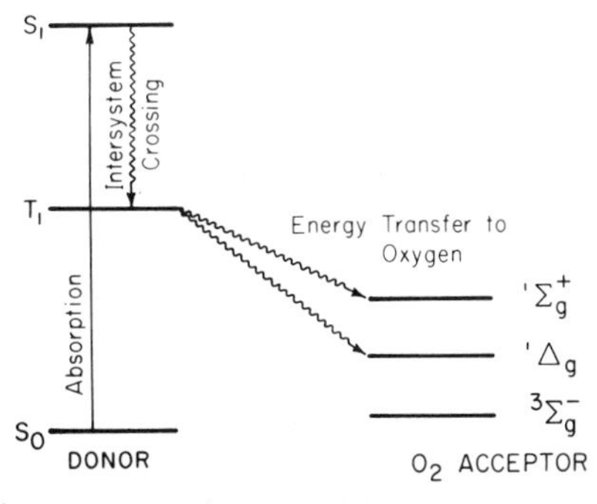

Fig. 20. Schematic representation of triplet energy transfer one from donor organic molecule to molecular oxygen to produce either the $^1\Delta$ or $^1\Sigma$ excited states (processes 12, 13 and 14 in text).

that transfers the electronic excitation energy to oxygen. This is presumably because of the relatively much longer lifetimes of triplet states versus singlet states of organic molecules. Thus, efficient sensitizers have high quantum yields of triplet formation; that is, the radiationless process 13, intersystem crossing, is efficient compared to other competing reactions of the Sens (S_1)species.†

In terms of generation of $^1O_2$ in the laboratory, the procedure is straightforward (see references 37 and 69 for details). The sensitizer may be any one of a number of organic compounds including (*a*) a dye such as rose bengal, (*b*) an aromatic hydrocarbon such as naphthalene which is a (π,π^*) type sensitizer, or (*c*) a ketone such as benzophenone, an (n,π^*) sensitizer.† A list of various sensitizers employed by Gollnick are shown in Table II (36b). One simply bubbles oxygen gas through a solution containing the sensitizer and substrate in a suitable solvent and irradiates the system with a lamp having a high output of radiation at wavelengths absorbed by the sensitizer. The rate of product formation depends upon a variety of factors including the efficiencies of processes 12–15. Thus, while this procedure is excellent for synthetic work, care must be taken in designing experiments upon which detailed kinetic and mechanistic interpretations will be based. A particularly vexing problem is insuring the purity of certain widely used photosensitizers including rose bengal. So-called "pure" samples of the latter may actually be mixtures of a number of compounds; this can play havoc with conclusions from mechanistic studies.

Gas-phase experiments on photosensitized oxygenations analogous to those conducted in solution are currently being conducted in our laboratory. Thus Steer, Sprung, and Pitts have recently observed the formation of oxidation products in an irradiated, gas-phase sensitizer–oxygen–olefin system. In these studies, mixtures of benzene, oxygen, and tetramethylethylene (typically 20, 70, and 2 torr, respectively) were irradiated for one hour at room temperature. The hydroperoxide product 2,3-dimethyl-3-hydroperoxy-1-butene, **I**, was isolated in large yields (70). The significance of this work to photochemical air pollution is discussed in Section IV.

† For a general discussion of processes 12–14, including energy transfer and the nature of singlet and triplet states of organic molecules, see Calvert and Pitts, reference 29. S_0 and S_1 refer to the ground and lowest excited singlet states respectively; T_1 is the lowest triplet state. The wavy arrows indicate radiationless transitions.

TABLE II.

Energy of Triplet State of Some Organic Molecules Used as Photosensitizers to Generate $^1O_2$ Solution

(π,π^*)-Sensitizers	Triplet[a] energy (kcal/mole)	(n,π^*)-Sensitizers	Triplet energy (kcal/mole)
Benzene	84.5	Acetone	—
Fluorene	67.6	Propiophenone	74.6
Triphenylene	66.6	Xanthone	74.2
Biphenyl	65.7	Acetophenone	73.6
Phenanthrene	62.2	Carbazole	70.1
Naphthalene	60.9	Benzophenone	68.5
1-Bromonaphthalene	59.0	Thioxanthone	65.5
1-Iodonaphthalene	58.6	Flavone	62.0
Pyrene	48.7	β-Naphthylphenyl ketone	59.6
1,2-Benzanthracene	47.2	α-Naphthylphenyl ketone	57.5
Anthracene	42.0	Fluorenone	53.3
Acenaphthylene			
Eosin	42.4		
Erythrosin	42.0		
Rose Bengal	39.4		
Azulene	Between 31 and 38		
Tetracene	29.4		

[a] Data supplied by Dr. Klaus Gollnick, reference 36b. For the original source of triplet energies, see page 97 of this reference.

C. Detection of O_2 ($^1\Delta$) and ($^1\Sigma$)

One of the factors that delayed recognition of the role of singlet oxygen, particularly $O_2(^1\Delta)$, in spectroscopy and photochemistry was the lack of convenient, reliable, and sensitive detectors, both physical and chemical in nature. In the last several years, this situation has changed. The sigma state is readily monitored by its emission in the red using highly sensitive, stable photomultiplier tubes and there are now a variety of physical techniques for measuring the $^1\Delta$ state. These include (*a*) direct emission measurements of the red dimol emission using phototubes with red-sensitive cathodes; (*b*) direct emission measurements of the 1.27-μm emission from the transition $O_2(^1\Delta_g) \rightarrow O_2(^3\Sigma_g^-)$ using preferably a liquid-nitrogen cooled germanium diode detector; (*c*) mass spectrometric determination of the appearance potential of the m/e 32 peak; (*d*) calorimetry; (*e*) photoionization techniques; (*f*) electron paramagnetic resonance method (EPR). The pros and cons of each have been reviewed by Wayne (39) and will not be discussed in detail.

The direct detection of emission at 1.27 μm by the cooled germanium detector and the EPR technique are the most useful and unequivocal physical techniques. Thus the first direct observation of the formation of $O_2(^1\Delta)$ in the gas phase was made by Snelling who irradiated benzene–oxygen mixtures and detected the characteristic $^1\Delta$ emission at 1.27 μm (71). Kummler and Bortner have used this technique to monitor $O_2(^1\Delta)$ emission from irradiated gas phase benzaldehyde–oxygen mixtures, a system of great relevance to photochemical air pollution (see below) (72). A germanium diode detector was also employed by Wayne and Pitts in studies of the reaction of O_3 with $O_2(^1\Delta)$ and by Steer, Sprung, and Pitts in observing $^1\Delta$ emission resulting from energy transfer in an irradiated mixture of naphthalene vapor and oxygen (70,73). In the latter studies, the detection system consisted of a chopper, monochromator, and liquid nitrogen cooled detector, the signal from which was fed into a lock-in amplifier and thence to a recorder.*

The EPR technique developed by Mahan et al. is less sensitive but offers the advantage of yielding approximate values of the *absolute* concentrations of $O_2(^1\Delta)$. It is based on the measurement of the characteristic four-line EPR spectrum centered at $|g_J| \backsim 2/3$ (62,74). Recently Kearns et al. and Wasserman et al. independently employed this EPR technique to observe the gas-phase photosensitized generation of $O_2(^1\Delta)$ (75,76).

Chemical detectors involve acceptor molecules which readily react with $O_2(^1\Delta)$ to produce a characteristic product. Particularly useful in this regard is tetramethylethylene (**TME**) which was first used in a gas-phase system by Winer and Bayes who followed reaction 16 by gas chromatographic monitoring of the hydroperoxide product (**I**) (49).

$$(CH_3)_2C{=}C(CH_3)_2 \; (\mathbf{TME}) + O_2(^1\Delta) \longrightarrow CH_3{-}C(CH_3)(OOH){-}C(CH_3){=}CH_2 \; (\mathbf{I}) \quad (16)$$

Recently Broadbent et al. employed the UCR apparatus shown in Figure 19 with both **TME** and dimethyl furan as "traps" for $O_2(^1\Delta)$. They monitored product formation by both gas chromatography and

* We are indebted to Dr. R. P. Wayne for fabricating our first detector from ultra-pure germanium supplied through the courtesy of Mr. Adrian Mears of the Clarendon Laboratory, Oxford.

infrared spectroscopy in which a 10-m long path infrared cell was used. Virtually 100% yields of reaction products were obtained with this system with both acceptors (77).

D. Chemical Reactions of $^1O_2$ with Organic Compounds

There are two general types of reactions of $O_2(^1\Delta)$ with organic molecules: (*a*) the oxygenation of olefins containing allylic hydrogen atoms resulting in the shift of the double bond and formation of an allylic hydroperoxide; this is analogous to the so-called "ene" reaction; (*b*) the oxygenation of polycyclic aromatic hydrocarbons such as cyclopentadienes and heterocycles which give endoperoxides (transannular peroxides); this is analogous to the Diels-Alder reaction.

Reaction 17 is a general example of case (*a*), the "ene" reaction, while reaction 18 shows case (*b*) in which $^1O_2$ acts as a dienophile in a 1,4-cycloaddition reaction (34–37).

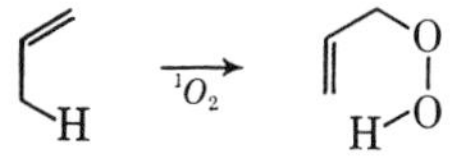

(17)

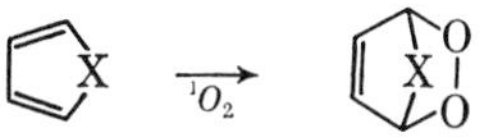

(18)

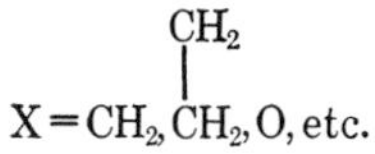

X = CH_2, CH_2–CH_2, O, etc.

Reactions 16 and 19 are specific examples of the "ene" and the Diels Alder types with **TME** and dimethyl furan, (**II**) as acceptor molecules.

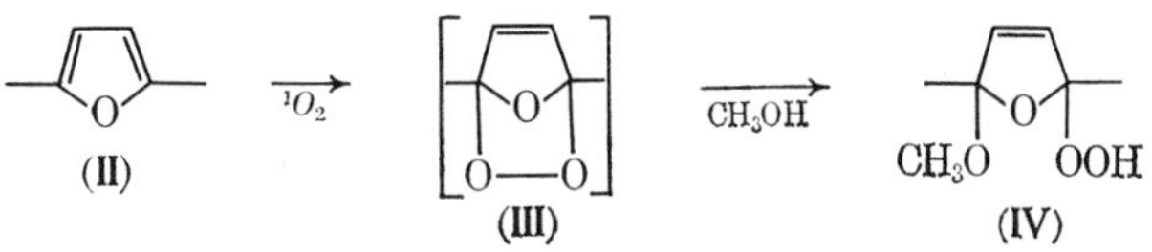

(II) (III) (IV) (19)

The typical ozonide (**III**) formed in reaction 19 is unstable as a pure liquid so in solution phase reactions, methanol is present and compound **IV** is isolated. However, we have found that **III** is readily formed in the gas phase by reaction with $^1O_2$ and can be monitored by use of the 10-m long path infrared cell to monitor its characteristic absorption bands at 1334 and 1131 cm^{-1} (77).

A particularly important reaction of $^1O_2$ is the formation of the endoperoxides, (**VI**), of polynuclear aromatic hydrocarbons such as 9,10-diphenylanthracene, reaction 20.

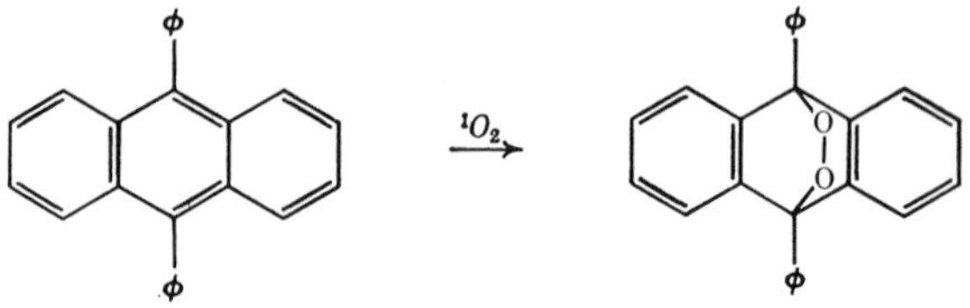

The $^1O_2$ can be generated internally by dye photosensitization or the H_2O_2–NaOCl reaction, or externally by a microwave discharge. The latter technique was used by Corey and Taylor in their definitive paper in 1964 (78).

Table III from Foote shows typical compounds that react with $^1O_2$ and the corresponding oxygenated proucts (35). The reviews by Foote and Gollnick should be consulted for details of the mechanisms of the reactions including their stereospecific aspects (34–37).

E. Reactions of O, O_2, and O_3 Producing $^1O_2$

Singlet molecular oxygen may also be formed in several reactions involving neutral oxygen atoms, molecular oxygen, and ozone, as well as reactions of other simple inorganic molecules. Many of these are important upper atmosphere processes and some may be significant in the lower atmosphere. These have been reviewed by Gill and Laidler (40) and Wayne (39,44), while Thrush has considered the general subject of the "Formation of Electronically Excited Molecules in Simple Gas Reactions" (79). We shall consider the processes most relevant to photochemical air pollution in the next section. Reactions involving singlet molecular oxygen and simple charged species have been reviewed by Kummler and Bortner in terms of their possible involvement in the upper atmosphere (80). We shall not consider charged species in our discussion.

IV. SINGLET MOLECULAR OXYGEN AND THE CHEMISTRY OF THE LOWER ATMOSPHERE

A. Sources of $^1O_2$ in Urban Atmospheres

1. *Direct Absorption of Sunlight*

The emission of radiation from electronically excited singlet oxygen molecules present in the *upper* atmosphere has been known for some time to contribute to the day and night "airglows." Thus the $^1\Sigma_g^+ \rightarrow {}^3\Sigma_g^-$ transition of molecular oxygen gives rise to the "atmospheric

TABLE III

Typical "Acceptor" Molecules and Oxygenated Products for Reactions with Singlet Oxygen[a]

Compound	Product
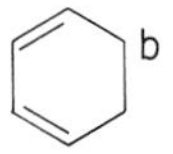 b	 O O c
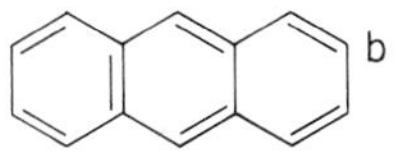 b	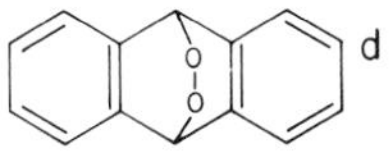 O O d
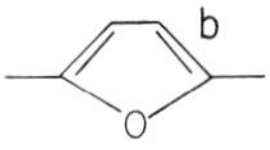 O b	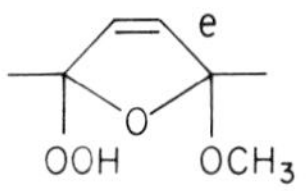 O OOH OCH$_3$ e
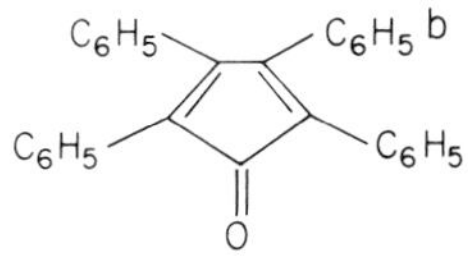 C_6H_5 C_6H_5 C_6H_5 C_6H_5 O b	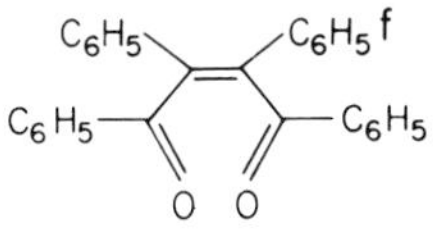 C_6H_5 C_6H_5 C_6H_5 C_6H_5 O O f
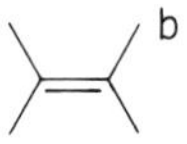 b	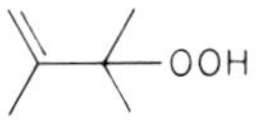 OOH
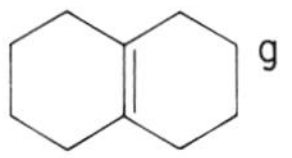 g	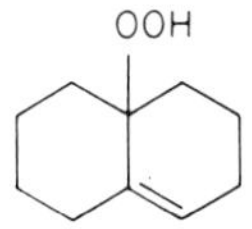 OOH

[a] Table taken from the review by Foote, reference 35. [b] Reference 57. [c] Unstable product. [d] Unreactive acceptor, very low yield. [e] Methanol adduct of unstable endoperoxide. [f] Presumably formed by loss of CO from intermediate peroxide. [g] Reference 69.

band." This emission was first observed in the laboratory in 1947 by Kaplan (81) and observed in the airglow in 1950 by Meinel (82). In 1958 Vallance-Jones and Harrison (83) made the first observation in the airglow of the "infrared atmospheric" band which results from the $^1\Delta_g \rightarrow {}^3\Sigma_g^-$ radiative transition in oxygen (see Sec. III-A-2).

The possibility that singlet molecular oxygen plays a role in the formation of photochemical smog in the lower atmosphere was first considered by Leighton in 1961 (1). He made a critical analysis of the existing spectroscopic and photochemical data on the $^1\Sigma$ and $^1\Delta$ states of molecular oxygen (which at that time dealt almost entirely with the $^1\Sigma$ state) and concluded that excitation by direct absorption of solar radiation could not lead to significant stationary concentrations of the $^1\Sigma_g{}^+$ state in the lower atmosphere. He also raised the possibility that $^1O_2$ could react with hydrocarbons but noted that there was no evidence at that time for such oxidations at room temperature.

In 1964 Bayes demonstrated that $O_2(^1\Delta)$ generated by the H_2O_2–NaOCl reaction reacted rapidly and efficiently with **TME** (reaction 16) and went on to consider the possible role of $^1O_2$ in photochemical smog (49b). In particular, he showed that if the absorption coefficients for the forbidden direct absorption processes forming $^1\Delta$ and $^1\Sigma$ (reactions 1 and 2) were corrected to allow for collisional line broadening (assuming a Lorentzian line shape), the rate of direct absorption of solar radiation by oxygen in the lower atmosphere by the collisionally perturbed process 21a became significant and that it "may be strong enough to contribute significantly to the reactions of photochemical smog".

$$O_2(^3\Sigma_g{}^-) + M + h\nu \underset{b}{\overset{a}{\rightleftarrows}} O_2(^1\Delta_g, {}^1\Sigma_g) + M \tag{21}$$

Furthermore, he noted that collisional deactivation of $O_2(^1\Delta)$ by O_2 or N_2 was highly *inefficient* and concluded that the long lifetime would "permit the average excited oxygen molecule in polluted air to encounter an olefin molecule" (and hence presumably to form a peroxide). Finally, he estimated that in the atmosphere of the Los Angeles Basin the rate of this reaction might compete with the rate of reaction of olefins with oxygen atoms produced by NO_2 photolysis.

Although Bayes' proposal was advanced in 1964, for several years it received relatively little attention partly because direct irradiation of olefin–air mixtures in chambers did not produce smog symptoms. Recently Kummler, Bortner, and Baurer have recalculated the rate of absorption of solar radiation by molecular oxygen using the data of Badger et al. on the efficiency of collisionally induced absorption of oxygen, process 21a (59,85). They conclude that the rate of process 21a at atmospheric pressure is only about twice that of the unperturbed process 1a and 2a. Similar considerations apply to the corresponding rates of emission by processes 21b versus 1b and 2b. Thus the precise role in forming $^1O_2$ player by *direct* absorption of sunlight by oxygen in the lower atmosphere is still not clear and needs further detailed study.

It is clear, however, that $O_2(^1\Delta)$ has a long lifetime at atmospheric pressure of the order of 0.05–0.5 sec (84,85) and that $O_2(^1\Delta)$ exhibits a high degree of reactivity with olefins in the gas phase (see above).

While the matter of direct absorption of sunlight is unresolved, there are several alternate modes of formation of $^1O_2$ which may be relatively efficient mechanisms for its production in urban atmospheres. Some have been discussed earlier in this paper from the viewpoint of laboratory studies; we shall now consider their specific application to photochemical air pollution. They include (*a*) energy transfer, (*b*) direct photolysis of ozone, and (*c*) certain exothermic chemical reactions, some of which are known to occur in smog and in which molecular oxygen is a product.

2. *Production of Singlet Molecular Oxygen by an Energy Transfer Mechanism*

a. Mechanism

The mechanism proposed by Khan, Pitts, and Smith in 1967 (32) and extended by Pitts et al. (33), bypasses the problem of very weak direct absorption of solar radiation by suggesting that significant concentrations of singlet oxygen could be produced *indirectly by triplet electronic energy transfer from electronically excited organic molecules.* Thus a high yield of singlet oxygen can be obtained if solar radiation is absorbed first by an organic molecule and then, on collision, electronic energy is transferred from this donor (in its triplet state) to oxygen to produce electronically excited oxygen, $^1O_2$. Their overall atmospheric mechanism is identical with the Kautsky mechanism for laboratory photosensitized oxygenation reactions discussed in Section III-B-3 and shown in processes 12, 13, and 14 and in Figure 20.

At the time Khan, Pitts, and Smith advanced their proposal, there was substantial laboratory evidence supporting the Kautsky energy transfer mechanism in *liquid* solutions but virtually no evidence for its validity in gas-phase systems. Furthermore, there were conflicting reports as to the rate of the gas phase reactions of $O_2(^1\Delta)$ with olefins such as **TME**. However, during the last year this theory has been greatly strengthened by experimental confirmation of both ideas. Thus, as we have seen, there has been direct spectroscopic confirmation of $O_2(^1\Delta)$ being produced by *gas phase* energy transfer to oxygen from irradiated aromatic hydrocarbons with (π,π^*) triplet states such as benzene and naphthalene (70,71,75,76) and carboyl compounds with (n,π^*) triplets such as benzaldehyde (72) and benzophenone (86). Furthermore, $O_2(^1\Delta)$, generated *extrenally* by a microwave generator, has been found to react rapidly and efficiently in the gas phase with the olefin tetramethylethylene, reaction 16, and the heterocyclic molecule 2,5-dimethyl furan, reaction 19 (77).

Finally, generation of $^1O_2$ by a gas-phase energy transfer process has been confirmed *chemically* by using **TME** as a "trap" for $O_2(^1\Delta)$ and detecting the hydroperoxide product, (**I**), when either benzene (70) or benzaldehyde were also present (87). In the latter case the experiment was carried out in the 40-m path length infrared spectrophotometer by Dr. J. W. Coomber (Sec. I-E, Figs. 14 and 15) using laboratory air to which 650 ppm **TME** and 10 ppm of benzaldehyde were added. Irradiation of this mixture with radiation longer than 3000 Å leads to rapid formation of IR absorption bands at 3594 and 906 cm^{-1}. These are due to the allylic hydroperoxide (**I**) (87).

As Kummler and Bortner note (72), benzaldehyde is a particularly relevant example because it is an important atmospheric contaminant resulting from the oxidation of aromatic compounds in gasoline; because it absorbs relatively strongly in the near ultraviolet, and because its triplet state, T_1, is formed in good yield from its singlet state, S_1 (29). It is interesting that they found no 1.27-μm emission when formaldehyde–oxygen and acetaldehyde–oxygen mixtures were irradiated under conditions similar to the benzaldehyde–O_2 system. This suggests little, if any, energy transfer to O_2 occurs with two simple aldehydes but the results should not be regarded as conclusive evidence until quantitative studies are carried out at a constant rate of absorption UV light by the three different aldehydes. Benzaldehyde is by far the strongest absorber of the three compounds, but the two simple aldehydes photodissociate much more efficiently (29).

It is particularly significant that a wide range of different chemical species of donor molecules may be effective in producing $^1O_2$, and that energy transfer has been shown for heterogeneous gas–solid systems as well as gas–liquid and homogeneous gaseous systems. Thus, in his classic paper, Kautsky adsorbed biologically important dye-stuffs such as chlorophyll and porphyrins on solid supports of silica gel or aluminum oxide and suggested that singlet molecular oxygen was formed on illumination of the surface in the presence of normal oxygen, and that "free" singlet O_2 migrated to the acceptor which was adsorbed on another solid support separated from the donor (53).

Triplet energy transfer at gas–solid interfaces has been demonstrated for the system naphthalene-*cis*-penta-1,3-diene (and in several other systems including those with 9,10-anthraquinone as donor by Daubendiek, Magid, and McMillan). On irradiation of the solid aromatic hydrocarbons in the presence of the diolefin, isomerization of the olefin to the *trans* form is observed (88). It would seem that if gaseous olefins quench triplet aromatics in the *solid phase*, molecular oxygen might also act as a quencher to yield singlet excited oxygen.

The transfer of electronic energy is thus seen to be of considerable generality. Furthermore, the transfer of energy at gas–solid interfaces may be of considerable significance in view of the presence of not only organic but also of inorganic particulate matter in polluted urban air. At present, the possible role of inorganic particulates in such a mechanism is not clear but it merits careful examination.

b. Rate of Formation of $^1O_2$ by Energy Transfer

Formation of $^1O_2$ by direct absorption of solar radiation and by energy transfer from organic molecules are not mutually exclusive processes. Singlet O_2 should be produced by both routes (as well as others, see below) but for several reasons, particularly the low efficiency of the direct absorption of O_2, the energy transfer mechanism should be the more important in urban air and we shall focus our attention on the rate of this process.

The rate of formation of singlet oxygen in the energy transfer mechanism depends on the rate of absorption by the donor molecules. In the lower atmosphere these donors must absorb radiation at wavelengths longer than 2900 Å because the ozonosphere cuts off solar radiation below that wavelength. With this stipulation, the donors may be aromatic hydrocarbons directly emitted to the atmosphere, carbonyl compounds such as aldehydes which may be either primary contaminants or secondary contaminants produced by photochemical reactions of olefins in smoggy atmospheres, or a variety of other absorbing species. The absorbing donor molecule may be in the solid, liquid, or gaseous state, or adsorbed on a solid.

With these considerations in mind, Khan, Pitts et al. calculated an approximate value for the rate of absorption of near ultraviolet solar radiation in a polluted atmosphere by making certain rather drastic simplifying assumptions (32,33). Since it was virtually impossible to estimate accurately the rate of absorption versus scattering of sunlight by particulate matter in smog, they neglected the contributions to the production of singlet oxygen by energy transfer from solids (or aerosol droplets). Even then, it was difficult to estimate the rate of absorption of solar radiation by *gaseous* pollutants alone (89). Thus, to further simplify matters, they noted that the absorption spectrum of a smoggy atmosphere (taken after smog formation had reached a steady state) has two distinct absorption bands (1,90), an absorption band extending from the solar cutoff at 3000 Å up to 3200 Å and a band extending from 3400 to 4000 Å. They confined their attention to the absorption band around 3200 Å because it appears to be a true molecular absorption band due to ozone and organic molecules and not an artifact resulting from molecular and particulate scattering (1,90).* Finally they assumed unit efficiency of the energy transfer process.

On the basis of these severe assumptions, and some admittedly crude estimates of the atmospheric concentrations of polynuclear aromatic hydrocarbons and carbonyl compounds, they calculated that in a column of air extending from the earth's surface to the inversion level, about 1 km, the production rate of excited organic molecules on a smoggy day in the Los Angeles Basin is approximately 4×10^{-12} mole liter $^{-1}$ sec^{-1}. This rate corresponds to the *maximum* rate of singlet oxygen formation by the energy transfer process. In reality the production rate is smaller since alternative deexcitation paths exist for the organic species. This is especially true, of course, in the case of the aliphatic carbonyl compounds which also photodissociate (29).

A comparable calculation of the absorption by NO_2 of solar ultraviolet radiation in the wavelength region 2900–3850 Å (1) indicates that if the drastic assumptions cited above are valid, the absorption rate by NO_2 and by organic compounds will be approximately equal when the nitrogen dioxide concentration is around 10 pphm (33). The calculations therefore suggest that the absorption by polynuclear aromatic hydrocarbon and carbonyl compounds in typical polluted atmospheres may be significant relative to absorption by NO_2 in the photochemical reactive region. Field studies that may yield information on this point are now being conducted (89).

The general implications of this set of calculations, and of the recent laboratory results, to the chemistry of the lower atmosphere are clear. A reasonable mechanism for absorption and energy transfer exists and the rate of absorption of solar UV by organic pollutants is sufficiently large that significant quantities of ${}^1O_2$ could be produced. Confirmation awaits a more accurate evaluation of the *rate* of absorption of light by organic compounds and, most of all, studies in the actual atmosphere. Field experiments using the germanium diode detector for monitoring 1.27-μm emission and tetramethylethylene as a chemical trap for $O_2({}^1\Delta)$ are being initiated in our laboratory and have been proposed by Kummler (105).

* Actually, up to 80% of the solar radiation around 3200 Å may be attenuated on a particularly smoggy day (90).

3. *Singlet Molecular Oxygen from Ozone Photolysis*

If ozone is irradiated with ultraviolet light in the Hartley band (2000–3200 Å), it photodissociates into an electronically excited oxygen atom, $O(^1D)$ and molecular oxygen.

$$O_3 + h\nu(\lambda < 3200 \text{ Å}) \rightarrow O(^1D) + O_2 \text{ (Possible } ^1O_2) \tag{22}$$

Considerable evidence has accumulated that singlet molecular oxygen is produced in the photolysis (91,92) but at the moment it is not clear whether it is formed directly in the spin-allowed primary process 22 or indirectly from the secondary reaction 23, energy transfer from an excited oxygen atom to molecular oxygen (39).*

$$O(^1D) + O_2(^3\Sigma) \rightarrow O_2(^1\Delta \text{ or } ^1\Sigma) + O(^3P) \tag{23}$$

In any case, Wayne has reviewed the implications of several production processes for singlet O_2 in the *upper atmosphere* and concludes that an overall mechanism involving O_3 photolysis is the most probable (44).

With regard to the lower atmosphere, Kummler et al. (85) and Pitts et al. (33) have suggested that O_3 photolysis could also be a source of singlet molecular oxygen in polluted atmospheres. The former authors made detailed calculations involving (*a*) the rate of absorption by O_3 of radiation in the 3000–3200 Å band on a smoggy day in Los Angeles, and (*b*) the quenching rate of $O_2(^1\Delta)$ by $O_2(^3\Sigma)$. Assuming DeMore and Raper's value of 0.4 for the primary quantum yield for reaction 22 (91), they concluded that the first-order rate constant for formation of $O_2(^1\Delta)$ from O_3 photolysis would be approximately 0.01 hr^{-1} and that the stationary state atmospheric concentration of $O_2(^1\Delta)$ would be 3×10^{-14} M/liter (85). While this concentration is very small, nevertheless they calculated the stationary state concentration of atomic oxygen to be two powers of 10 less, 4×10^{-16} M/liter. Since it is well recognized that O atoms play a major role in photochemical air pollution (1,2,4), the small stationary state concentration of $^1O_2$ does *not* mean it is unimportant in the overall mechanism. Not much more can be inferred at this time until accurate values of the gas phase rate constants for $O_2(^1\Delta)$ with various olefins and aromatics become available.

The occurrence of the energy transfer process 23 leads to two further considerations. First, the reaction of $O(^1D)$ with normal O_2 results in the formation of *two oxidizing species*, singlet molecular oxygen and ground-state atomic oxygen: the excess photochemical energy carried

* DeMore and Raper calculated the rate of $O(^1D)$ formation in the lower atmosphere to be 1.2 ± 0.5 pphm/hr for an ozone concentration of 10 pphm and a solar zenith angle of 45° (91).

by $O(^1D)$ has been made available for photooxidation. Secondly, this enhancement of oxidant potential by reaction 23 makes it important to look for other atmospheric sources of $O(^1D)$. The photolysis of NO_2 itself is excluded since the formation of $O(^1D)$ in the reaction

$$NO_2 + h\nu \rightarrow NO + O(^1D) \tag{24}$$

requires more energy than is available in the solar radiation reaching polluted atmospheres (29).

4. *Singlet Oxygen from Exothermic Chemical Reactions*

The possibility that singlet molecular oxygen can be the product of an exothermic chemical reaction in which photochemical excitation is not the direct source of energy (although it may well be the indirect source) is intriguing. In photochemical air pollution, the most obvious reaction is that between nitric oxide and ozone (94).†

$$NO + O_3 \rightarrow NO_2 + O^*_2 \tag{25}$$

The reaction is highly exothermic and leads to electronic excitation of NO_2 (79,94). Furthermore, production of $O_2(^1\Delta_x)$ as well as electronically excited NO_2 seems reasonable because it leads to conservation of spins. Alternatively, in the presence of molecular oxygen an energy transfer process could occur (33).

$$NO^*_2 + O_2 \rightarrow NO_2 + {}^1O_2 \tag{26}$$

While these arguments seem plausible, to date preliminary studies of the emission from the NO–O_3 system give negative results as far as $^1O_2$ production is concerned (95).

Ozone might also react in a manner similar to reaction 25 with a variety of other atmospheric contaminants (such as sulfides) ultimately to yield singlet molecular oxygen as a product (33). Processes of this type are known to lead to $^1O_2$ formation in condensed phases and Murray and Kaplan have also noted the possible significance of such processes to air pollution (61).‡

Whether or not such exothermic excitation mechanisms make a significant contribution in polluted atmospheres depends on both the rate constants for the reactions and on the concentrations of the several

† Asterisks above a formula here represent electronic excitation of unspecified state.

‡ Stephens has commented on "The Formation of Molecular Oxygen by Alkaline Hydrolysis of Peroxyacetyl Nitrate" (107). We are investigating the possibility that at least part of the O_2 produced is in the singlet state and have recently found $O_2(^1\Delta)$ is formed (108).

contaminants. Data are not yet available on these matters, but such possible excitation mechanisms should not be overlooked (33).

Ground state atomic oxygen may participate in several plausible mechanisms for the formation of singlet oxygen

$$O(^3P) + O(^3P) + M \rightarrow {}^1O_2 + M \tag{27}$$

$$O(^3P) + O_3 \rightarrow {}^1O_2 + O_2 \tag{28}$$

$$O(^3P) + NO_2 \rightarrow {}^1O_2 + NO \tag{29}$$

Of these, Young and Black report direct evidence for the production of $O_2(^1\Sigma_g^+)$ in reaction 27 and circumstantial evidence for its formation in reaction 28 (96,97). However, Clyne, Thrush et al. (98) found no evidence for the formation of $O_2(^1\Sigma_g^+)$ in reaction 28 or 29 although Arnold, Brown et al. suggest that some $O_2(^1\Delta_g)$ may be formed (56b). Actually, these processes probably proceed too slowly and the concentrations of oxygen atoms are too low to lead to significant concentrations of $O_2(^1\Sigma_g^+)$ in polluted atmospheres (33).

B. Possible Role of $^1O_2$ in the Photoconversion of NO to NO_2

As we have stated, the photoconversion of NO to NO_2 remains one of the perplexing problems of photochemical air pollution The possible involvement of $^1O_2$ as one part of a very complex mechanism has been proposed by Khan, Pitts et al. (32,33). It is based on the idea that singlet oxygen formed by energy transfer from the donor organic molecules will react with olefinic hydrocarbons to give hydroperoxides. These hydroperoxides, being thermally unstable and sensitive to surface catalysis, then decompose to give free radicals of the same type as those formed in the direct attack of atomic oxygen on olefins. From this point on, the reactions resemble those of the NO_x–olefin–air system and could occur simultaneously. Thus the involvement of $^1O_2$ would be a concurrent process, possibly minor, to the overall reaction. The mechanism includes the following process:

$$O_2(^1\Delta) + \text{—C=C—}\underset{\text{H}}{\underset{|}{\text{C}}}\text{—} \longrightarrow \text{—}\underset{\text{O—O—H}}{\underset{|}{\text{C}}}\text{—C=C—} \tag{30}$$

$$\text{—}\underset{\text{O—O—H}}{\underset{|}{\text{C}}}\text{—C=C—} \xrightarrow[\text{decomposition}]{\text{Thermal}} \text{Radicals (e.g., RCO)} \tag{31}$$

The radicals produced in process 31 can then initiate the sequence

$$RCO + O_2 \text{ (ground state} \rightarrow RCO_3 \quad (32)$$

$$RCO_3 + NO \rightarrow RCO_2 + NO_2 \quad (33)$$

$$NO_2 + h\nu \rightarrow NO + O \quad (34)$$

$$O + O_2 + M \rightarrow O_3 + M \quad (35)$$

$$O_3 + NO \rightarrow NO_2 + O_2 \quad (36)$$

Rections 32–36 are steps in the currently accepted mechanism for the oxidation of NO to NO_2 (1,2,4).

Another possible process for NO → NO_2 conversion, which seems attractive at first glance but in fact is not feasible on energetic grounds, is the spin-allowed reaction 37.

$$NO + {}^1O_2 \rightarrow NO_2 + O \quad (37)$$

Even with $O_2(^1\Sigma)$, the reaction is endothermic when the latter is in its ground vibrational state (33).

Clearly the situation with regard to possible involvement of ${}^1O_2$ in photochemical processes of the lower atmosphere awaits spectroscopic and chemical studies in actual urban atmospheres. Several experimental approaches are possible, the two most promising being a search for the 1.27-μm emission band from $O_2(^1\Delta)$ on smoggy days in the Los Angeles Basin and use of the technique of adding chemical traps such as **TME** to large samples of polluted air and searching for the characteristic hydroperoxide product (**I**).

V. OTHER ENVIROMNENTAL IMPLICATIONS OF SINGLET OXYGEN

Singlet molecular oxygen may not only be an important species in photooxygenation and photochemical air pollution; it may also play a significant role in biochemical processes. Foote discussed this point in some detail in his recent review (34) which opened with a discussion of the phenomenon of "photodynamic action." This term refers to the early observation of Raab that microorganisms are killed when they are irradiated in the presence of oxygen and sensitizing dyes (99). The *in vivo* and *in vitro* damage includes deactivation of enzymes and damage to nucleic acids, polypeptides, and proteins.

In his critical review, Foote discusses, from the viewpoint of possible involvement of singlet oxygen, the existing evidence on the photooxidation of a variety of compounds of biological interest. These include enamines, histidine, methionine, and tryptophan (for reference to original work see reference 34). The conclusion that singlet molecular oxygen is involved in certain of these systems seems reasonable indeed.

Singlet molecular oxygen may also play a significant role in the pathology of man, animals, and plants. In this connection, Foote has mentioned the implications of the production of skin cancer by photosensitizing dyes and polycyclic hydrocarbons (34,35).

Possible effects of singlet oxygen on plants are worth serious consideration. To date there has been no direct evidence that $^1O_2$ is a phytotoxicant but studies with ozone provide a rationale for setting up experiments to find out whether or not this is the case. Thus, specific alteration of plastid structure has been observed with the electron microscope when bean and tobacco tissue is treated with ozone (100). Similarly, ozone is reported to inhibit enzymatic activity (101) and to interfere with photophosphorylation processes within sensitive plants.

A particularly interesting observation has been made recently by Foote and Denny. They found that β-carotene in solution also is an extremely efficient quencher for $^1O_2$ (102); one molecule of β-carotene quenches at least 100 molecules of $^1O_2$. They make the interesting speculation that the well known effect of carotenoids in protecting photosynthetic organisms against the lethal effects of chlorophyll, which is an excellent photosensitizer for oxidation processes, may well be due to the quenching by β-carotene of $^1O_2$ responsible for the damage.

Finally, singlet molecular oxygen may well be involved in the degradation in polluted atmospheres of susceptible natural and synthetic substances with consequent economic loss. Thus, Rawls and Van Santen recently demonstrated the role of singlet oxygen in the autooxidation of fatty acids (103). They passed singlet O_2 from a microwave discharge over a thin film and showed that a conjugated hydroperoxide was formed. They estimated that $^1O_2$ reacted at least 1000 times as fast as the ground-state triplet O_2 molecule. The same oxidation products were formed when biologically significant dyestuffs such as chlorophyll-a were introduced into a fatty acid film illuminated with visible light in the presence of air.

Furthermore, Trozzolo and Winslow have recently produced evidence that singlet oxygen may well be involved in the photooxidation and consequent degradation of polymer film (104).

It is apparent even from these brief considerations above that detailed investigations of the role of singlet oxygen in a variety of chemical, physical and biological environmental processes are timely and challenging.

Acknowledgment

The author acknowledges his indebtedness to Professor Leonard Bowden for furnishing the color infrared photographs and to Mrs. Constance Bennett for assistance in preparation of the manuscript.

The author also acknowledges with appreciation a sabbatical leave from the University of California, Riverside, during which time this manuscript was prepared.

References

1. P. A. Leighton, *Photochemistry of Air Pollution,* Academic Press, New York, 1961.
2. E. A. Schuck and E. R. Stephens, in *Advan. Environ. Sci.,* **1,** 73 (1969).
3. (a) E. R. Stephens, in *Advan. Environ. Sci.,* **1,** 119; (1969); (b) E. R. Stephens, *Intern. J. Air Water Pollution,* **10,** 649 (1966).
4. A. P. Altshuller and J. J. Bufalini, *Photochem. Photobiol.,* **4,** 97 (1965).
5. A. C. Stern, Ed., *Air Pollution,* 2nd ed., Vols. 1, 2 and 3, Academic Press, New York, 1968.
6. E. A. Schuck, J. N. Pitts, Jr., and J. K. S. Wan, *Intern. J. Air Water Pollution* **10,** 689 (1966).
7. J. T. Middleton, "Air Pollution Threat to Flora and Fauna," in *Conservation Catalyst,* Vol. II, No. 2.
8. "The Effects of Air Pollution," U.S. Department of Health, Education and Welfare (1966), U.S.P.H.S. Publication No. 1556.
9. Symposium on Trends in Air Pollution Damage to Plants, *Phytopathology,* **58,** 1075 (1968).
10. "Air Quality Criteria for Sulfur Oxides," U.S. Department of Health, Education and Welfare, Public Health Service, Consumer Protection and Environmental Health Service, National Air Pollution Control Administration, Washington, D.C., January 1969, NAPCA Publication No. AP-50.
11. "Air Quality Criteria for Particulate Matter," ibid, January 1969, NAPCA Publication No. AP-49.
12. (a) Bulletin Los Angeles County Air Pollution Control District, January, 1967; (b) *ibid.,* January 1968.
13. Working Group on Lead Contamination, U.S. Public Health Service, Publication No. 999-AP-12, 1965.
14. J. M. Colucci and C. R. Begeman, *Environ. Sci. Technol.,* **3,** 41 (1969).
15. D. H. Slade, *Science,* **157,** 1304 (1967).
16. H. W. Ford and N. Endow, *J. Chem Phys.,* **27,** 1156, 1277 (1957).
17. H. W. Ford and S. Jaffe, *J. Chem. Phys.,* **38,** 293 (1963).
18. J. Heicklen and N. Cohen, *Advan. Photochem.,* **5,** 157 (1968).
19. J. N. Pitts, Jr., J. H. Sharp, and S. I. Chan, *J. Chem. Phys.,* **40,** 3655 (1964).
20. E. A. Schuck, H. W. Ford, and E. R. Stephens, "Air Pollution Effects of Irradiated Automobile Exhaust as Related to Fuel Composition," Report No. 26, Air Pollution Foundation, San Marino, California, 1958.
21. E. A. Schuck and G. J. Doyle, "Photooxidation of Hydrocarbons in Mixtures containing Oxides of Nitrogen and Sulfur Trioxide," Report No. 29, Air Pollution Foundation, San Marino, California, 1959.
22. R. J. Cvetanović, in *Advan. Photochem.,* **1,** 115 (1963)

23. J. A. Maga, Hydrocarbon Photochemical Reactivity Symposium, American Petroleum Institute, Bartlesville, Oklahoma, September 1966.
24. "Individual Hydrocarbon Reactivity Measurements. State-of-the-Art." Coordinating Research Council, CRC Project CM-4-58, June 1966.
25. A. P. Altshuller, *Intern. J. Air Water Pollution*, **10**, 713 (1966).
26. W. A. Glasson and C. S. Tuesday, General Motors Research Publication 586, August 1966.
27. A. P. Altshuller, S. L. Kopczynski, D. Wilson, W. Lonneman, and F. D. Sutterfield, 61st Annual Meeting, Air Pollution Control Association, Minneapolis, Minnesota, June 23–27, 1968.
28. F. Bonamassa, private communication, 1968.
29. J. G. Calvert and J. N. Pitts, Jr., *Photochemistry*, John Wiley and Sons, Inc., New York, 1966.
30. "The Automobile and Air Pollution," Parts I and II, R. S. Morse, Chairman, U.S. Department of Commerce Document, December 1967.
31. "The Oxides of Nitrogen in Air Pollution," State of California, Department of Public Health, Bureau of Air Sanitation, January 1966.
32. A. U. Khan, J. N. Pitts, Jr., and E. B. Smith, *Environ. Sci. Technol.*, **1**, 656 (1967); J. N. Pitts, Jr., AAAS Divisional Symposium on Air Pollution, University of California, Los Angeles, June 1967.
33. J. N. Pitts, Jr., A. U. Khan, E. B. Smith, and R. P. Wayne, *Environ. Sci. Technol.*, **3**, 241 (1969).
34. C. S. Foote, *Science*, **162**, 963 (1968).
35. C. S. Foote, *Accounts Chem. Res.*, **1**, 104 (1968).
36. (a) K. Gollnick, in *Advan. Photochem.*, **6**, 2 (1968); (b) K. Gollnick, in *Advan. Chem. Ser.*, **77**, 78 (1968).
37. K. Gollnick and G. O. Schenck, in *1,4-Cycloaddition Reactions*, Vol. 8, J. Hamer, Ed., Academic Press, New York, 1967, p. 255
38. S. J. Arnold, M. Kubo, and E. A. Ogryzlo, in *Advan. Chem. Ser*, **77**, 133 (1968).
39. R. P. Wayne, in *Advan. Photochem.*, **7**, 400 (1969).
40. E. K. Gill and K. J. Laidler, *Can. J. Chem.*, **36**, 79 (1958).
41. J. S. Griffith, in *Oxygen in Animal Organisms*, MacMillan, New York, 1964, p. 141
42. G. Herzberg, *Spectra of Diatomic Molecules*, 2nd ed., D. Van Nostrand, Princeton, N. J., 1950.
43. A. Vallance-Jones and A. W. Harrison, *J. Atmospheric Terrest. Phys.*, **13**, 45 (1958).
44. R. P. Wayne, *Quart. J. Roy. Meteorol. Soc.*, **93**, 395 (1967).
45. D. Q. Wark and D. M. Mercer, *Appl. Opt.*, **4**, 839 (1965).
46. L. Wallace and D. M. Hunter, *J. Geophys. Res.*, **73**, 4813 (1968).
47. A. U. Khan and M. Kasha, *J. Am. Chem. Soc.*, **88**, 1574 (1966).
48. S. J. Arnold, R. J. Browne, and E. A. Ogryzlo, *Photochem. Photobiol.*, **4**, 963 (1965).
49. (a) A. M. Winer and K. D. Bayes, *J. Phys. Chem.*, **70**, 302 (1966); (b) K. D. Bayes, Sixth Informal Photochemistry Conference, University of California, Davis, June 1964.
50. T. P. J. Izod and R. P. Wayne, *Proc. Roy. Soc.* (*London*), *Ser. A.*, **308**, 81 (1968).
51. R. A. Young and G. Black, *J. Chem. Phys.*, **42**, 3740 (1965).
52. S. J. Arnold, N. Finlayson, and E. A. Ogryzlo, *J. Chem. Phys.*, **44**, 2529 (1966).

53. H. Kautsky, *Biochem. Z.*, **291,** 271 (1937); *Trans. Faraday Soc.*, **35,** 216 (1939).
54. E. J. Bowen and R. A. Lloyd, *Proc. Roy. Soc. (London), Ser. A.* **275,** 465 (1963); *Proc. Chem. Soc.*, **1963,** 305; E. J. Bowen, *Nature,* **201,** 180 (1964).
55. A. U. Khan and M. Kasha, *J. Chem. Phys.*, **39,** 2105 (1963); **40,** 605 (1964); *Nature,* **204,** 241 (1964).
56. (a) R. J. Browne and E. A. Ogryzlo, *Proc. Chem. Soc.*, **1964,** 117; (b) J. S. Arnold, R. J. Browne, and E. A. Ogryzlo, *Photochem. Photobiol.*, **4,** 963 (1965).
57. (a) C. S. Foote and S. Wexler, *J. Am. Chem. Soc.*, **86,** 3879 (1964); (b) *ibid.*, **86,** 3880 (1964).
58. E. McKeown and W. A. Waters, *J. Chem. Soc. (London), Ser. B,* **1966,** 1040.
59. R. M. Badger, A. C. Wright and R. F. Whitlock, *J. Chem. Phys.*, **43,** 4345 (1965).
60. H. H. Wasserman and J. R. Schaffer, *J. Am. Chem. Soc.*, **89,** 3073 (1967).
61. R. W. Murray and M. L. Kaplan, *J. Am. Chem. Soc.*, **90,** 4161 (1968).
62. A. M. Falick and B. H. Mahan, *J. Chem. Phys.*, **47,** 4778 (1967).
63. W. S. Gleason and J. N. Pitts, Jr., unpublished results.
64. E. J. Bowen, in *Advan. Photochem,* **1,** 23 (1963).
65. R. Livingston, in *Autooxidation and Antioxidants,* Vol. 1, W. O. Lundberg, Ed., Interscience, New York, 1961, p. 249.
66. R. Higgins, C. S. Foote and H. Cheng, in *Advances in Chemistry Series,* Vol. 77, American Chemical Society, Washington, D.C., 1968, p. 102.
67. D. R. Kearns, R. A. Hollins, A. U. Khan, and P. Radlick, *J. Am. Chem. Soc.*, **89,** 5455 (1967); **89,** 5456 (1967).
68. A. U. Khan and D. R. Kearns, in *Advan. Chem. Ser.*, **77,** 143 (1968).
69. C. S. Foote, S. Wexler, W. Ando, and R. Higgins, *J. Am. Chem. Soc.*, **90,** 975 (1968).
70. R. P. Steer, J. L. Sprung, and J. N. Pitts, Jr., *Environ. Sci. Technol.*, In press.
71. D. R. Snelling, *Chem. Phys. Letters,* **2,** 346 (1968).
72. R. H. Kummler and M. H. Bortner, private communication, 1968; submitted *Environ. Sci. Technol.*, March 1969.
73. R. P. Wayne and J. N. Pitts, Jr., *J. Chem. Phys.*, **50,** 3644 (1969).
74. A. M. Falick, B. H. Mahan, and R. J. Myers, *J. Chem. Phys.*, **42,** 1837 (1965).
75. D. R. Kearns, A. U. Khan, C. K. Duncan, and A. H. Maki, *J. Am. Chem. Soc.*, **91,** 1039 (1969).
76. E. Wasserman, V. J. Kuck, W. M. Delevan, and W. A. Yager, *J. Am. Chem. Soc.*, **91,** 1040 (1969).
77. A. D. Broadbent, W. S. Gleason, J. N. Pitts, Jr., and E. Whittle, *Chem. Commun.*, 1315 (1968).
78. E. J. Corey and W. C. Taylor, *J. Am. Chem. Soc.*, **86,** 3881 (1964).
79. B. A. Thrush, *Chem. Britain,* **1966,** 287.
80. R. H. Kummler and M. H. Bortner, G. E. Report R67SD20 (1967).
81. J. Kaplan, *Nature,* **159,** 673 (1947).
82. A. B. Meinel, *Astrophys. J.*, **112,** 464 (1950).
83. A. Vallance-Jones and A. W. Harrison, *J. Atmos. Terrest. Phys.*, **13,** 45 (1958).
84. I. D. Clark and R. P. Wayne, unpublished data.
85. R. H. Kummler, M. H. Bortner, and T. Baurer, *Environ. Sci. Technol.*, **3,** 248 (1969).
86. A. M. Trozzolo and S. R. Farenholz, Abstracts, 155th Meeting, American Chemical Society, San Francisco, March 31–April 5, 1968, p. 138.

87. J. W. Coomber and J. N. Pitts, Jr., preliminary results.
88. R. L. Daubendiek, H. Magid, and G. R. McMillan, *Chem. Commun.*, **1968,** 218.
89. J. S. Nader, "Pilot Study of Ultraviolet Radiation in Los Angeles," October 1965, Public Health Service Document No. 999-AP-38 (1967).
90. R. Stair, Proceedings of Third National Air Pollution Symposium, 1955, p. 48.
91. W. B. DeMore and O. F. Raper, *J. Chem. Phys.*, **44,** 1780 (1966).
92. R. G. W. Norrish and R. P. Wayne, *Proc. Roy. Soc.* (*London*), **A288,** 200 (1965).
93. E. Schuck, private communication.
94. M. A. A. Clyne, B. A. Thrush, and R. P. Wayne, *Trans. Faraday Soc.*, **60,** 359 (1964).
95. R. P. Wayne, R. P. Steer, and J. N. Pitts, Jr., unpublished data.
96. R. A. Young and G. Black, *J. Chem. Phys.*, **42,** 3740 (1965).
97. R. A. Young and G. Black, *J. Chem. Phys.*, **44,** 3741 (1966).
98. M. A. A. Clyne, B. A. Thrush, and R. P. Wayne, *Photochem. Photobiol.*, **4,** 957 (1965).
99. O. Raab, *Z. Biol.*, **39,** 524 (1900).
100. W. W. Thomson, W. M. Dugger, Jr., and R. L. Palmer, *Can. J. Botany*, **44,** 1677 (1966).
101. L. Ordin and M. A. Hall. *Plant Phys.*, **42,** 2, 205 (1967).
102. C. S. Foote and R. Denny, *J. Am. Chem. Soc.*, **90,** 6233 (1968).
103. H. R. Rawls and P. J. Van Santen, *Tetrahedron Letters*, 1675 (1968).
104. A. M. Trozzolo and F. H. Winslow, *Macromol.*, **1,** 98 (1968).
105. R. H. Kummler, private communication.
106. E. R. Stephens, *Weatherwise*, **18,** 172 (1965).
107. E. R. Stephens, *Atmos. Environ*, **1,** 19 (1967).
108. R. P. Steer, K. R. Darnell, and J. N. Pitts, Jr., *Tetrahedron Letters*, In press.
109. E. A. Ogryzlo, In *Photophysiology*, Vol. 5, Ed. A. C. Geise, In press.

Author Index

Numbers in parentheses are reference numbers and show that an author's work is referred to although his name is not mentioned in the text. Numbers in *italics* indicate the pages on which the full references appear.

Abram, L. E., 208(77), *232*
Adams, C. R., 245(54), *283*
Adams, K. F., 223(147), *234*
Agnew, J. T., 256(61), *283*
Agnew, W. G., 239(1), 241(1), 257, 261 (61), 268(1), *280*, *283*
Air Products Co., 278
Albert, M. M., 225(161), *234*
Allard, H. A., 207(73), *232*
Altschuller, A., 82, *117*, 267(122), 268 (106,107), *286*, *287*, 291(4), 293(4), 300, 302(4), 303(4), 305, 307, 308, 329 (4), 332(4), *334*, *335*
American Cyanamid Co., 276, 279
American Public Health Association, 43, 45, *71*, 155(13,14), *194*
American Water Works Association, 154 (10), 155, *193*
Andersen, A. A., 226, *235*
Anderson, R. B., 246(37), *282*
Anderson, R. G., 246(38), *282*
Ando, W., 318(69), 323(69), *336*
Andreatch, A. J., 268(40), *282*
Andreoli, A., 154(12), *194*
Applied Science and Technology Index, 243(2), *280*
Arguembourt, P., 228(185), *235*
Arnold, S. J., 309(38), 311(38), 312–314, 315(56), 331, *335*, *336*
Arthus, M., 199(5), *230*
Ascher, M. S., 221(126), *233*
Association of American Soap and Glycerine Producers, 156, *194*
Aucoin, E., 228(184), *235*
Austen, K. F., 199(7), 200(12), *230*
Avery, S. B., 223(145), *234*
Axworthy, A. E., Jr., 83(38,42), 98(42), *118*
Ayen, R. J., 250(30), *281*

Bach, F., 223(146), *234*
Badger, R. M., 311(59), 315(59), 324, *336*
Bagnold, R. A., 213, *232*
Baker, R. A., 246(49), 250(31), *281*, *282*
Bambrick, W. E., 246(60), 248(60), 249 (60), 254(60), *283*
Barbee, R. A., 227(178), 228(178,181, 188), *235*
Barkin, G. D., 205(57), *231*
Barlow, P. P., 205(62), 206(62), 207(68), *231*, *232*
Barnard, J. H., 203(41), *231*
Barnett, H. L., 223(152), *234*
Barth, E. F., 189, *196*
Bauer, D. G., 245, *282*
Bauer, E. R., 246(32), *281*
Baurer, T., 309(85), 312(85), 324, 325 (85), 329(85), *336*
Baxter, W. P., 91, *118*
Bayes, K. D., 309, 312, 313(49b), 320, *335*
Beard, L. L., 268(9), *280*
Beck, J. A., 241(72), *284*
Becker, E. L., 201(22), *230*
Becker, R. J., 203(39), *231*
Beckman, E. W., 243(62), 256(62), *283*
Begeman, C. R., 301, 302, *334*
Behrens, M. D., 245(45), 246(45), 278, *282*
Bellas, T. A., 267(122), *287*
Belman, S., 202(26), *231*

Benson, J. D., 260(63), *283*
Benson, S. W., 83(38,42), 98(42), *118*
Berl, E., 82, *117*
Bernstein, I. L., 221(129), *233*
Berrens, L., 217(113), 218(115), *233*
Bienstock, D., 246(32), *281*
Bier, K. C., 241(64), *283*
Birmingham, D. J., 156, 190(68), *194*, *196*
Blacet, F. E., 74(1), 91, 95(1), 99(1), *117*, *118*, 290
Black, G., 314, 331, *335*, *337*
Black, J. H., 205, *231*
Blemel, K. G., 267(119), *287*
Bloch, K. J., 200(20), *230*, 277, 278
Blumenthal, J. L., 250(56), *283*
Bodenstein, M., 79(12), 81, 82, *117*
Bold, H. C., 212(92), 221(92), *232*
Bolt, J. A., 239(3), *280*
Bolze, C., 241(111), *286*
Bonamassa, F., 307, *335*
Bortner, M. H., 309(85), 312(85), 320, 322, 324, 325(72,85), 326, 329(85), *336*
Bouhuys, A., 229(210), *236*
Bowden, L., 293
Bowen, 278
Bowen, E. J., 315, 317, *336*
Bowen, R., 205(54), *231*
Boyd, W. C., 198(3), *230*
Braithwaite, D., 202(30), *231*
Brandenburg, J. T., 245(45), 246(45), *282*
Braverman, M. M., 228(190), *235*
Brenner, T. E., 147
Briggs, 276, 278
Briner, E., 82, *117*
Bringhurst, L. S., 229(191), *235*
Britt, 279
Broadbent, A. D., 316, 320, 321(77), 326 (77), *336*
Brocklehurst, W. E., 200(12), *230*
Broder, I., 205, 206(62), 207(68), *231*, *232*
Brooks, C., 139, *146*
Brown, F. B., 82, *117*
Brown, R. M., 212(92), 221(92), *232*
Browne, R. J., 312(48), 315(56), 331, *335*, *336*
Brownson, D. A., 262(65), *283*
Brubacher, M. L., 238(108), 240(108), *286*
Bruce, R. A., 221(127), *233*
Bruckner, H. C., 223(145), *234*
Brugsch, H. G., 223(144), *234*
Buchanan, L. M., 223(155), *234*
Buechner, H. A., 228(184,185), *235*
Bufalini, J. J., 82, 102, *117*, 291(4), 293 (4), 300(4), 302(4), 303)4(, 305 (4), 307(4), 308, 329(4), 332(4), *334*
Buffam, W. P., 205(56), *231*
Bunch, R. L., 167, *195*
Burleson, F. R., 125(16), 126(16), 129 (16), *146*
Burrage, W. S., 229(194), *235*
Buttgereit, W., 256(66), *283*
Buttivant, J., 256(81), *284*
Byerly, T. C., 18(1), *23*
Byrne, R. N., 229(191), *235*

California Motor Vehicle Pollution Control Board, 242(4), *280*
California Research Corp., 149(4), *193*
California State Department of Public Health, 240(6), *280*, 308(31), *335*
Callies, Q. C., 227(175), *235*
Calvert, J. G., 94, 110, *118*, 277, 309(29), 311(29), 318, 326(29), 328(29), 330 (29), *335*
Campbell, D. H., 217(110), *233*
Cannon, W. A., 246(33), *282*
Caplan, J. D., 242(67), 243(67), *284*
Cardiff, E. A., 125(16), 126(16), 129(16), 138(25), *146*
Cavagna, C., 217(112), *233*
Cavanaugh, J. J. A., 203(34), *231*
Cazort, A. G., 222(140), *234*
Chamberlain, A. C., 19(2), *23*, 216(106), *233*
Chambers, C. W., 167, *195*
Chambers, V. B., 204(43), *231*
Chan, S. I., 90(45), 91(45), *118*, 302(19), *334*
Chandler, J. M., 243(68), 256(68), *284*
Chang, L., 157(19), *194*
Chao, J., 268(109), *286*
Chase, J. O., 260(115), *286*
Chemical & Engineering News, 5(3), *23*, 161(33), 181(59), 182(59), *194*, *195*

Cheng, H., 314(66), *336*
Chipman, J. C., 268(109), *286*
Chow, T. J., *23*
Christensen, C. M., 212, *232*
Clark, I. D., 312(84), 325(84), *336*
Clarke, J. A., 207(71), *232*
Clarke, N. A., 157(19), *194*
Cleary, E. J., 62, *71*
Clewell, D. H., 243(8), *280*
Clifford, H. T., 219(118), *233*
Cluff, L. E., 203(42), *231*
Clyne, M. A. A., 83 *117*, 330(94), 331, *337*
Cobb, R. M., 267(110), *286*
Coca, A. F., 202, *231*
Cochrane, C. G., 199(6), *230*
Cohen, I. R., 82(16), *117*
Cohen, J. M., 157(21,22), *194*
Cohen, M. B., 217(109), *233*
Cohen, N., 302(18), 308(18), *334*
Cohen, S., 200(10), *230*
Cole, A. E., 45(6), *71*
Colucci, J. M., 301, 302, *334*
Connell, J. T., 201(21), *230*
Conte, J. F., 261(105), *286*
Cooke, R. A., 202, 203(36,40), 204, *231*
Coomber, J. W., 326, *337*
Coordinating Research Council, 268 (116), *286*, 305(24), *335*
Corcoran, W. H., 82(28), *117*
Corey, E. J., 322, *336*
Corn, M., 212(91), *232*
Coughlin, F. J., 148, 157(18,20,23,24), 158(26), *193*, *194*
Cowan, D. W., 222(137), *234*
Crawford, N. J., 205(60), *231*
Crist, R. H., 82, *117*
Crocker, W., 219(119), *233*
Crooks, R. N., 19(15), *23*
Cross, H. D., 111, 160(32), *194*
Cryst, S., 225(163), 227(173), *234*, *235*
Culp, R. L., 157(19), *194*
Cummings, M. M., 229(203), *236*
Curran, W. S., 204, *231*
Cvetanovic, R. J., 104–107, *118*, 305

Dahlgren, C. M., 223(155), *234*
Daigh, H. D., 260(71), 264(71), *284*
Daniel, W. A., 260(69,70), 261(69), *284*
Daniels, F., 79(11), 82, 91, 95(11), 95 (52), 101, *117*, *118*
Darley, E. F., 74(5), 111(61), *117*, *118*, 119(1–3), 121(3,12), 123(3), 131(3), 137(20), 141(3), *145*, *146*, 294
Darnell, K. R., 330(108), *337*
Daubendiek, R. L., 326, *337*
Davidson, N., 83(41), *118*
Davis, T. C., 206(100), *285*
Daws, L. F., 212(89), *232*
Dean, P. M., 217(108), *233*
Decker, H. M., 223(155), *234*
Deeter, W. F., 260(71), 264(71), *284*
DeGara, P. F., 204, 206(48), *231*
Delevan, W. M., 320(76), 325(76), *336*
DeMore, W. B., 329, *337*
Dennis, E. G., 201(23), *230*
Denny, R., 333, *337*
Derbes, V. J., 206(66), 229(198), *232*, *235*
DeRossit, 278
De Rycke, D., 237(34), 250(34), *282*
Dickerson, R. C., 229(198), *235*
Dickie, H. A., 227(175,176,178), 228 (178,181), *235*
Dickinson, R. G., 91, *118*
Dingle, A. N., 224(159), *234*
Dingwall-Fordyce, I., 229(192), *235*
Dixon, F. J., 199(6), *230*
Dixon, F. R., 53
Dodson, V. N., 233(145), *234*
Doerr, R. C., 120(7–9,11), 121(7), 122 (7), 123(8), 125(9), 138(9), *145*, 250 (31), *281*
Dondes, S., 83, *118*
Doyle, G. J., 305, 306, *334*
Dozois, C. L., 260(115), *286*
Duckstein, L., 268(9), *280*
Duckworth, F. S., 214(98), *232*
Duffy, R., 246(41), 247(41), 249(41), 250(41), *282*
Dugger, W. M., 119(2), *145*, 333(100), *337*
Duncan, A. B. F., 74(2), 95(2), *117*
Duncan, C. K., 320(75), 325(75), *336*
Dunstan, H. J., 19(2), *23*
Du Pont de Nemours, E. I., & Co., 276, 278, 280
Durham, O. C., 223(148), 224(157), *234*
Dybas, B., 217(108), *233*

Eastwood, 279
Ebel, R. H., 237
Eberhardt, J. E., 241(72), *284*
Eisen, H. N., 202(26), *231*
Elias, L., 106, *118*
Elkins, H. B., 223(144), *234*
Ellis, C. F., 260(115), *286*
Ellis, M. M., 45(5), *71*
El-Messiri, I. A., 260(85), *285*
Elrod, R. J., 211(87), *232*
Elston, J. C., 240(127), *287*
Eltinge, L., 256(73), 262(74), *284*
Emanuel, D., 228(187), *235*
Emery, E. M., 160(32), *194*
Emik, L. O., 120(5), *145*
Endow, N., 78, 79, 84, 85, 88–91, 92(7), 93(7), 97, 98(7), 101, 104, 105, 110, *117*, *118*, 302(16), *334*
Engelhardt, H. T., 206(66), *232*
Environmental Science and Technology, 13(5), 14(5), *23*, 240(11), 243(10), *280*
Epstein, S., 202(27), *231*
Erdtman, G., 214(102), *233*
Esso-American Petroleum Institute, 264(96), *285*
Esso Co., 276, 277, 280
Ethyl Corp., 276, 277
Ettinger, M. B., 189, *196*
Evans, R. L., 190(67), *196*
Eyzat, P., 259(75), *284*

Fagley, W. S., 243(62), 256(62), *283*
Fair, G., 46, *71*, 174(53), *195*
Faith, W. L., 238(36), 241(111,112), 242(35), *282*, *286*
Falick, A. M., 315, 320(62,74), *336*
Farenholz, S. R., 325(86), *336*
Farr, R. S., 199, *230*
Feenan, J. F., 246(37), *282*
Fegraus, C., 240(127), *287*
Feinberg, B., 205(56), *231*
Feinberg, S. M., 198(1), 203(39), 222 (141), *230*, *231*, *234*
Feldberg, W., 200(8), *230*
Ferrin, P. E., 260(115), *286*
Field, T. H., 246(32), *281*
Figley, K. D., 211(87), 221(134), *232*, *233*
Fine, A. J., 208(77), *232*
Finlayson, N., 314(52), *335*
Finulli, M., 217(112), *233*
Firestone Tire and Rubber Co., *279*
Fleming, D. M., 19(13), *23*
Flynn, J. M., 154(12), *194*
Foote, C. S., 309, 310, 314, 315, 317, 318 (69), 321(34,35), 322, 323, 332, 333, *335–337*
Ford, H. W., 78, 79, 84, 85, 88–92, 93(7), 97, 98(7), 101, 104, 105, 110, *117*, *118*, 302(16,17), 303(20), *334*
Forman, J. A. S., 228(181), *235*
Franck-Phillipson, A., *275*
Frankowski, J. J., 241(64), *283*
Frazer, J. C. W., 275
Fries, J. H., 204(49), *231*
Fromer, J. L., 229(194), *235*
Funder, S., 223(151), *234*

Gagliardi, J. C., 254(78), 256(78), 261 (97,98), 266(78), *284*, *285*
Gandhi, H., 250(55), *283*
Gelfand, H. H., 223(143), *234*
Gerhold, R. M., 171(49), *195*, *277*
Gershon-Cohn, J., 229(191), *235*
Geyer, J., 174(53), *195*
Ghannam, F. E., 261(98), *285*
Gilbert, L. F., 252(114), *286*
Gill, E. K., 309(40), 322, *335*
Gill, G. C., 225(164), *234*
Gilman, J. C., 223(153), *234*
Glaser, J., 204(43), 205(60), *231*
Glass, A. C. 152(8), *193*
Glasson, W. A., 82, 83, 96, *117*, 119(24), 120(24), 134, 135, 139, 145, *146*, 305, *335*
Gleason, W. S., 316, 321(77), 326(77), *336*
Goldfarb, A. R., 220(124), *233*
Goldman, G., 204, *231*
Gollnick, K., 309, 315(36), 317–319, 321 (36,37), 322, *335*
Gonyon, D. K., 241(64), *283*
Good, R. A., 217(111), *233*
Goodwin, J. E., 227(174), *235*
Goodwin, J. T., Jr., 241(111), *286*
Grace, W. R., Co., 276, 278, 279
Graiff, L. B., 259(99), *285*
Grant, E. P., 238(108), 240(108), *286*

Green, H. L., 210, *232*
Green, J. B., 148(1), 157(18,20,23,24), 158(26), *193*, *194*
Gregory, P. H., 210(82,84), 213(82), 214 (82), 215, 223, 225(166), 228(180), *232*, *234*, *235*
Greig, J. D., 82, *117*
Griffin, B. I., 19(13), *23*
Griffith, J. S., 309(41), *335*
Gross, P., 210(86), 230(204), *232*, *236*, *280*
Groulx, A. D., 220(121), *233*
Guerrera, A. A., 154(12), *194*
Guibet, J. C., 259(75), *284*
Gurney, C. W., 225(163), 227(173), *234*, *235*
Gussey, P., 246(38), *282*
Gyenes, L., 200(19), *230*

Haagen-Smit, A. J., 237(13), 268(113), *281*, *286*, 290
Hall, L. B., 223(155), *234*
Hall, M. A., 333(101), *337*
Hall, P. G., 82, *117*
Hall, T. C., 74(1), 91(48), 95(1), 99(1), *117*, *118*
Hamblin, 279
Hamkins, L. R., 256(76), 261(76), 262, *284*
Hamming, 305
Hanna, G. P., Jr., 171(49), 187(60), *195*
Hansen, E., 225(163), *234*
Hanson, W. C., 19(13), *23*
Hanst, P. L., 120(7–9,11), 121(7), 122(7), 123(8), 125(9), 138(9), *145*
Harger, J., 275
Hargreave, F. E., 228(189), *235*
Harkins, J., 130, 136, 140, 141, 143, 144(17), *146*
Harrington, J. B., 225(164), *234*
Harris, M. M., 223(155), *234*
Harrison, A. W., 311(43), 323, *335*, *336*
Harteck, P., 83, *118*
Hasche, R. L., 82, *117*
Hass, N. de, 83, *117*
Hatch, T. F., 210(86), *232*
Hawkins, D. F., 200(11), *230*
Hayes, W. J., Jr., 22(6), *23*
Haywood, T. J., 221(130), *233*
Hegsted, D. M., 57(9), *71*
Heicklen, J., 302(18), 308(18), *334*
Heinen, C. M., 237(14), 242(14), *281*
Hemphill, F. M., 227(174), *235*
Henderson, 278
Herron, J. T., 83, 84(44), 97(44), *117*, *118*
Hersh, C. K., 159(28,29), 160(31), *194*
Herxheimer, H., 200(11,14,15), 221(128), 222, *230*, *233*, *234*
Herzberg, G., 309–311(42), *335*
Hewson, E. W., 219(120), *233*
Higgins, R., 314(66), 318(69), 323(69), *336*
Hill, A. C., 139, *146*
Hill, D. M., 240(124), *287*
Hirschler, D. A., 252(114), *286*
Hirschorn, K., 223(146), *234*
Hirst, J. M., 214(103), 216, 225, *233*, *234*
Hittler, D. L., 256(76), 261(76), *262*, *284*
Hocker, A. J., 268(109), *286*
Hoekstra, 276
Hofer, L. J. E., 246(37,38), *282*
Hollins, R. A., 314(67), *336*
Holmes, H. H., 79(11), 91, 95(11), 101, *117*
Hornbrook, M., 201(23), *230*
Horstmann, 279
Horton, R. J. M., 205(62), 206(62), 207 (68), 229(198), *231*, *232*, *235*
Houdry, 277, 278
Howell, J. B., 229(196), *235*
Houk, 276, 280
Huess, J. M., 119(24), 120(24), 134, 135, 139, 145, *146*
Hughes, R. F., 208(78), 209, *232*
Huls, T. A., 256(77), 259(77), *284*
Hume, N. B., 182(63), *196*
Hunt, E., 14(7), *23*
Hunter, D. M., 311(46), *335*
Hurn, R. W., 238(15), 260(100,115), *281*, *285*, *286*
Huron, W. H., 228(186), *235*
Hurst, G. W., 214(103), *233*
Husmann, W., 187, *195*
Hyde, H. A., 221(128), 223(147), *233*, *234*

Imhoff, K., 46, *71*
Ingold, C. T., 213, 216(93), *232*

Innes, W. B., 237(39), 238(39), 246(41), 247(39,41), 248(39), 249(41), 250(39, 41), 253(39), 268(39,40), 276, 279, *282*
Ishizaka, K., 201(23), *230*
Itkin, I. H., 218(117), *233*
Izod, T. P. J., 313, *335*

Jaeschke, W. H., 227(175), *235*
Jaffe, S., 78(9), 90(9), 91, 92, *117*, 277, 302(17), *334*
Jenkins, P. A., 227(177), 228(177), 229 (193), *235*
Jensen, D. A., 242(5), *280*
Johnson, J. E., 203(42), *231*
Johnson, L. L., 246(42), *282*
Johnson, W. C., 246(42), *282*
Johnston, H. S., 82, *117*
Johnston, T. G., 222(140), *234*
Johnstone, M. S., *23*
Jones, J. H., 155, *193*, 254(78), 256(78), 266(78), *284*
Jones, W. M., 83(41), *118*
Jordan, W. S., 205(59), *231*
Joyce, R. S., 264(43), *282*
Justice, D., 149(5), 151, 159(27), 160, 161, 162(35), *193*, *194*

Kailin, E. W., 203(38), *231*
Kali Chemie, 277
Kallio, 163
Kaplan, J., 323, *336*
Kaplan, M. L., 315(61), 330, *336*
Kasha, M., 312(47), 315, *335*, *336*
Katz, G., 200(10), *230*
Kaufman, F., 83, *118*
Kautsky, H., 314, 326, *336*
Kayman, H., 203(35), *231*
Kearby, 276, 277
Kearns, D. R., 314, 320, 325(75), *336*
Kelly, C. D., 214(101), *233*
Kelso, J. R., 83(39), *118*
Kern, R. A., 222(139), *234*
Kettner, K. A., 111(61), *118*, 137(20), *146*
Khan, A. U., 115(65), *118*, 309(32,33), 312(47), 314(67,68), 315, 320(75), 325, 327, 328–330–33), 331, 332(33), *335*, *336*
Kiefer, D., 3(8), *23*
King, J., 91(49), *118*
Kinosian, J. R., 267(123), *287*
Kistiakowsky, G. B., 83, *117*
Klein, F. S., 83, 84(44), 97(44), *117*, *118*
Klein, S. A., 158(25), 172, 175(51), 176 (51), 187(66), 177(66), 189, 190, *194–196*
Kline, W. A., 171(47), *195*
Klingler, E., 82, *117*
Klinksiek, K. E., 267(117), *286*
Knapp, J. W., 169(46), *195*
Knopp, P. V., 171(48), *195*
Kobayashi, M., 227(176,178), 288(178, 181), *235*
Kobayashi, Y., 229(202), *236*
Koepernik, 277
Kollar, K. L., 152(7), *193*
Kolodny, R. L., 223(146), *234*
Kontsoukos, E., 246(44), *282*
Kopa, R. D. 260(79), *284*
Kopczynski, S. L., 82(16), *117*, 268(106, 107), *286*, 307(27), 308(27), *335*
Kornfeld, G., 82, *117*
Korth, M. V., 239(118), *287*
Kramer, C. L., 221(131), *233*
Krusé, C. W., 41
Kruse, R. E., 240(124), *287*
Kubo, M., 309(38), 311(38), 312, 313 (38), 314(38), *335*
Kuck, V. J., 320(76), 325(76), *336*
Kuhns, W. J., 203(37), *231*
Kumke, G., 172(50), *195*
Kummler, R. H., 309, 312(85), 320, 322, 324, 325(72,85), 326, 328, 329, *336*, *337*
Kurtzrock, R. C., 246(32), *281*

Lacey, M. E., 228(180), *235*
Lachmann, P. J., 200(17), 229(193), *230*, *235*
Laidler, K. J., 309(40), 322, *335*
Lair, J. C., 237(51), *283*
Lamb, F. W., 252(114), *286*
Lamberti, V., 149(5), 151, 159(27), 160, 161, 162(35), *193*, *194*
Landahl, H. D., 210(85), *232*
Lane, W. R., 210, *232*
Langston, P. D., 264(43), *282*
Lanoff, G., 205(60), *231*
Larborn, A. O. S., 256(80), 258(80), *284*

Larsen, R. I., 264(16), 267(119), *281*, *287*
Larson, D. A., 212(92), 221(92), *232*
Lawrence, G., 256(81), *284*
Lawson, S. D., 249(101), 264(102), *286*
Lawton, B., 228(187), *235*
Lawton, G. W., 191, *196*
Lararowitz, L. C., 208(76), *232*
Lerk, R. J., 245(45), 246(45), 276, *282*
Leary, R. D., 183(64), 188(64), *196*
Lecomte, J., 200(9), *230*
Lee, D. H. K., 7(9), *23*
Lee, R. C., 264(103), *286*
Leigh, D., 202(30), *231*
Leighton, F., 83(40), *118*
Leighton, P. A., 8, *23*, 76, 78(10), 79, 83, 84, 89(10), 90(10), 91(48), 94(50), 95 (10), 96, 99(10), 112, 114, *117*, *118*, 241(120), 268(120), *287*,, 290, 291(1), 293(1), 294(1), 300–303(1), 305(1), 307(1), 308, 323, 328(1), 329(1), 332 (1), *334*
Leopold, H. C., 207(71), *232*
Leskowitz, S., 203(32–35), *231*
Lestz, S. S., 256(94), *285*
Leupen, M. J., 217(114), *233*
Levy, R. M., 245, *282*
Lewis, A. H., 149(4), *193*
Lewis, J. P., 199(7), *230*
Lichtenstein, E. P., *23*
Lidwell, O. M., 226(172), *235*
Lienesch, J. H., 239(121), *287*
Lindell, S. E., 229(201), *236*
Livingston, R., 317, *336*
Lloyd, R. A., 315, *336*
London, A. L., 237(82), *284*
Longbottom, J. L., 228(189), *235*
Lonneman, W., 267(122), 268(106,107), *286*, *287*, 307(27), 308(27), *335*
Los Angeles County Air Pollution Control District, 299, *334*
Loveless, M. H., 201(25), 203(40), *230*, *231*
Lovell, R. G., 229(195), *235*
Lowell, F. C., 203(32,34), *231*
Lowry, T., 228(183), *235*
Ludwig, J. H., 239(17), 268(17,18), *281*
Lunge, G., 82, *117*
Lyklema, A. W., 217(114), *233*
Lynn, D. A., 267(119), *287*
McCabe, L. S., 53
McDonald, J. E., 216, *233*
McElhenney, T. R., 221(130), *233*
McGauhey, P. H., 158(25), 172, 174, 175 (51), 176(51), 187(66), 188(66), 189, 190, *194–196*
McGovern, J. P., 205(57), 221(130), *231*, *233*
McKeown, E., 315(58), *336*
McLean, J. A., 227(174), *235*
McMichael, W. R., 240(124), *287*
McMillan, G. R., 326, *337*
Maga, J. A., 267(123), *287*, 305, 306, *335*
Magid, H., 326, *337*
Mahan, B. H., 315, 320, *336*
Mahon, W. E., 229(193), *235*
Maillie, 279
Maki, A. H., 320(75), 325(75), *336*
Malet, G., 82(26), *117*
Marley, E., 202(30), *231*
Marsee, F. J., 262(74), *284*
Marvin, H. N., 222(140), *234*
Masciello, F., 228(190), *235*
Massoud, A., 229(200), *236*
Maternowski, C. J., 206(63), 208, *231*
Mathews, K. P., 206(63), 207(72), 208, 218(72), 222(135,142), 229(195), *231–235*
Matsumura, T., 222(136), *234*
Maunsell, K., 210(83), *232*
May, K. R., 226(171), *235*
Mayrsohn, H., 139, *146*
Meinel, A. B., 323, *336*
Menzel, A., 203(36), *231*
Mercer, D. M., 311(45), *335*
Metcalf, R. L., 1
Metzler, D. F., 157(19), *194*
Meyer, G. H., 226(168), *235*
Meyer, W. E., 256(94), *285*
Meyers, P., 203(36), *231*
Meyers, R. L., 220(125), *233*
Michalko, 277, 278
Mick, S. H., 243(87), 256(87), *285*
Middleton, E., 200(18), *230*
Middleton, F. M., 13(13), *23*, 153(9), 157(9), *193*, *194*
Middleton, J. T., 74(4,5), *117*, 119(1,4), 120(5,6), 121(12), *145*, 269(19), *281*, 290, 294–296, 297(7), *334*

Mielke, P. W., Jr., 222(137), *234*
Mills, 278
Milner, F. H., 217(108), *233*
Mine Safety Appliances, 275
Mobil Oil Co., 279
Moltulsky, A. G., 204(51), 206, *231*
Mongar, J. L., 200(11), *230*
Monsanto Co., 277
Moor, J. F., 249(101), *286*
Morgan, J. M., Jr., 169(46), *195*
Moriss, F. V., 241(111), *286*
Morris, J. H., 217(113), *233*
Morrison, M. E., 82, *117*
Morrow, M. B., 226(168), *235*
Moss, H. V., 148(1), 157(18,20,23,24), 158(26), *194*
Mudd, J. B., 119(2), 142, 144, 145, *145*, *146*
Mueller, P. K., 130, 136, 140, 141, 143, 144(17), *146*
Murray, R. W., 315(61), 330, *336*
Myers, R. J., 320(74), *336*

Nader, J. S., 327(89), 328(89), *337*
Nakajima, K., 256(92), 257(92), *285*
National Sanitation Foundation, 173 (52), *195*
Nelson, E. E., 264(104), *286*
Nelson, T., 217(109), *233*
Neuberger, D., 74(2), 94(2), *117*
Neville, H. A., 275
Newhall, H. K., 259(84), 260(83,85), *284*, *285*
Newmark, F. M., 218(117), *233*
Ng, Y. S., 250(30), *281*
Nichol, H. A., 256(77), 259(77), *284*
Nicholls, J. E., 206(85), *285*
Nichols, C. W., 74(5), *117*, 119(1), *145*
Nichols, M. S., 171(48), *195*
Nicksic, S. W., 130, 136, 140, 141, 143, 144, *146*
Niebylski, L. M., 252(114), *286*
Niepoth, G. W., 243(87), 256(87), *285*
Nissen, W. E., 267(125), *287*
Nobe, K., 246(44), *282*
Noble, W. C., 226(172), *235*
Norrish, R. G. W., 74(3), 91, 92, 94, 95, 110(46), *117*, *118*, 329(92), *337*
Noyes, W. A., Jr., 94(50), *118*

O'Brien, D. L., 246(42), *282*
Ogden, E. C., 224(160), 225(165), *234*
Ogryzlo, E. A., 83, *118*, 309, 310, 311(38), 312, 313(38), 314(38,52), 315, 331 (56b), *335*, *336*
Onashi, S., 229(202), *236*
O'Neill, C. G., 256(81), *284*
Ordin, L., 119(2), *145*, 333(101), *337*
Ordman, D., 211(88), *232*
Orgel, G., 171(47), *195*
Orris, L., 202(26), *231*
Osgood, H., 221(133), *233*
Ostertag, M., 205(55), *231*
Otto, K., 250(55), *283*
Oxy-Catalyst Co., 277, 278

Pady, S. M., 221(131), *233*
Pahnke, A. J., 261(105), *286*
Palmer, C. M., 157(19), *194*
Palmer, H. E., 19(13), *23*
Palmer, R. L., 333(100), *337*
Papa, L. J., 260(126), *287*
Pappenheimer, A. M., Jr., 203(37), *231*
Parkhurst, J. D., 184–186(65), 188(65), *196*
Patrick, W. A., 82, *117*
Patterson, S. J., 167(44a,44b), *195*
Pattison, J. N., 240(127), *287*
Paulus, H. J., 222(137), *234*
Pearson, R. S. B., 204, *231*
Pepys, J., 227(177), 228(177,189), 229 (193), *235*
Perlman, F., 221(132), 222(132), *233*
Perna, F., Jr., 267(110), *286*
Pernis, B., 217(112), *233*
Peshkin, M. M., 207(69), *232*
Peterson, R. D. A., 217(111), *233*
Pfeiffer, W., 82(26), *117*
Phelphs, H. W., 229(199), *236*
Phillips, E. W., 209, *232*
Phillips, L. F., 84(43), 97(43), 98(43), *118*
Phillips, M&C, 278
Pierce, J. V., 200(16), *230*
Pitts, J. N., Jr., 1, 90(45), 91, 94, 98(53), 100(53), 110, 115(65), *118*, 289, 293, 302(19), 308(19), 309(29,32,33), 311 (29), 316(63,77), 318, 320, 321(77), 325, 326(29,70,77,87), 327, 328(29,33),

329, 330(29,33,95,108), 331, 332(33), *334–337*
Polunin, N., 214(101), *233*
President's Science Advisory Committee, 5(14), 6(14), 16(14), 17(14), *23*
Prettre, M., 246(48), *282*
Price, M. A., 111(63), *118*, 139(21), *146*
Prince, H. E., 226(168), *235*
Pringle, R. B., 222(140), *234*
Product Engineering, 263(50), *282*

Quintero, J. M., 208(75), *232*

Raab, O., 332, *337*
Rackemann, F. M., 218(116), *233*
Radlick, P., 314(67), *336*
Ramirez, M. A., 201(24), *230*
Rankin, J., 227(175,176,178), 228(178, 181), *235*
Rapaport, H. G., 207(70), *232*
Raper, O. F., 329, *337*
Rather, J. B., Jr., 249(101), *286*
Ratner, B., 204, *231*
Rawls, H. R., 333, *337*
Raynor, G. S., 224(160), 225(165), *234*
Reckner, L. R., 267(128), 268(128), *287*
Reddi, M. M., 246(49), *282*
Reed, C. E., 228(188), *235*
Reed, P. W., 152(6), *193*
Reinarz, B. H., 217(109), *233*
Renn, C. E., 166, 171(47), 172(50), *195*
Renshaw, E. F., 69, *71*
Rice, O. K., 79(12), *117*
Ridgway, S. J., 237(51), *283*
Rinker, R. G., 82(28), *117*
Robinson, 278
Rohlich, G. A., 171(48), *195*
Ronayne, J. J., 223(145), *234*
Rooks, R., 214(100), 226(169), *233*, *235*
Rose, A. H., Jr., 239(118), *287*
Ross, H. R., 239(21), *281*
Rossbach, E. A., 203(38), *231*
Roth, J. F., 245, *282*
Rubinfeld, J., 160(32), *194*

Safferman, R. S., 221(129), *233*
Saltzman, B. E., 101(54), *118*
Salvaggio, J. E., 203(33–35), 228(185), *231*, *235*
Samter, M., 201(22), *230*
Sandberg, J. S., 214(98), *232*
Sarto, J. O., 243(62), 256(62), *283*
Sato, S., 105(57), *118*
Sawyer, R. F., 259(129), *287*
Schachner, H., 246(52), *283*
Schaffer, J. R., 315(60), *336*
Schaldenbrand, H., 263(53), *283*
Schenck, G. O., 309, 318(37), 321(37), 322(37), *335*
Schepegrell, W., 206(64), *231*
Schiff, H. I., 83, 84(43), 97(43), 98(43), 106, *118*
Schild, H. O., 200(11,13), *230*
Schilter, C., 256(66), *283*
Schlaffer, W. G., 245(54), *283*
Schnyder, U. W., 202(29), 205(29), *231*
Schrock, R. R., 78(8), 87–93(8), 96–99 (8), 101(8), 192(8), 105(8), 106(8), 109 (8), *117*
Schuck, E. A., 73, 74(4), 78(8), 87(8), 88(8), 89–91, 92(8), 93(8), 96, 97–99 (8), 101(8), 102(8), 105(8), 106(8), 109(8), 111(62,63), *117*, *118*, 119(4), 120(10), *145*, 291, 293, 299(2), 302(2), 303, 305, 306, 307(2), 308(2), 329(2), 332(2), *334*, *337*
Schuman, L. M., 228(183), *235*
Schwartz, M., 204(47), *231*
Scivally, 277
Scott, C. C., 167(44a,44b), *195*
Scott, W. E., 119(3), 120(7–9,11), 121 (3,7,13), 122(7,13), 123(3,8), 125(9), 131(3), 138(9), 141(3), *145*, 267(128), 268(128), *287*
Seabury, J. H., 228(185), *235*
Sehon, A. H., 200(19), *230*
Sergeys, F. J., 276
Settipane, G. A., 201(21), *230*
Shapiro, R. S., 214(100), 226(169), *233*, *235*
Sharp, J. H., 90(45), 91(45), *118*, 302 (19), 308(19), *334*
Sheets, W. D., 171(49), *195*
Sheldon, J. M., 207(72), 218(72), 219 (120), 222(140), 227(174), 229(195), *232–235*
Shelef, M., 250(55), *283*
Sherman, W. B., 200(18), 201(21), *230*

Shilkret, H. H., 208(76), *232*
Siegel, B. B., 220(122), *233*
Silberman, D. E., 204(52), *231*
Silverman, I. J., 204(46), *231*
Simon, F. A., 220(123), *233*
Simon, G., 205(59), *231*
Skaggs, J., 203(36), *231*
Skalig, P., 223(155), *234*
Slade, D. H., 302, *334*
Slentz, L. W., 82, *117*
Sleva, S. F., 82(16), *117*
Smith, C., 228(190), *235*
Smith, E. B., 115(65), *118*, 309(32,33), 325, 327(32,33), 328–330(33), 331 (32,33), 332(33), *335*
Smith, G., 223(150), *234*
Smith, J., 203(42), *231*
Smith, J. H., 82, 109(29), *117*
Smith, R. S., 213(94), *232*
Snelling, D. R., 320, 325(71), *336*
Soap and Detergent Association, 149(3), 150(2), 155, 192, *193*
Šolc, M., 82, *117*
Solomon, W. R., 197, 207(72), 218(72), 223(148,154), 226(167), *232*, *234*, *235*
Sosman, A., 228(188), *235*
Sourirajan, S., 250(56), *283*
Spaich, D., 205(55), *231*
Spain, W. C., 204, *231*
Sparks, D. B., 203(39), *231*
Spieksma, F. T. M., 217(114), *233*
Sponitz, M., 229(197), *235*
Sprung, J. L., 318(70), 320, 325(70), 326 (70), *336*
Stahmann, M. A., 227(176), *235*
Staines, F. H., 228(181), *235*
Stair, R., 328(90), *337*
Stanford Research Institute, 238(22), 246(22), *281*
Starkman, E. S., 259(84,129), 260(86), 284, *285*, *287*
Stebar, R. F., 262(65), *283*
Steer, R. P., 318(70), 320, 325(70), 326 (70), 330(95,108), *336*, *337*
Steigerwald, B. J., 268(18), *281*
Stein, F., 212(91), *232*
Steinhagen, W. K., 243(87), 256(87), *285*
Stephan, D. G., 153(9), *193*
Stephens, E. R., 73, 74(4), 77(6), 78(8), 82, 87–90(8), 91, 92(8), 93(8), 96–99 (8), 101(8), 102, 105(8), 106(8), 109(8), 111(61,63), *117*, *118*, 119, 120(7–11), 121(3,7,13), 122(7,13,14), 123(3,8), 125(9,16), 126(16), 129(16), 131(3,18, 19), 137(20), 138(9,25), 139(21), 141 (3), 143(23), *145*, *146*, 276, 277, 291, 293(2,3), 295, 299(2), 302(2), 303(2, 20), 305, 307(2), 308(2), 329(2), 330, 322(2), *334*, *337*
Stern, A. C., 237(88), 267(130), *285*, *287*, 291(5), 293(5), 300(5), 302(5), 303(5), 305(5), 307(5), 308(5), *334*
Stetson, D. M., 223(145), *234*
Stevenson, D. D., 222(135), *233*
Stewart, H. E., 260(86), *285*
Stewart, N. O., 19(15), *23*
Stiles, 276, 278
Stivender, D. L., 259(89), 262(89), 264 (89), *285*
Stokinger, H. D., 156, 190(68), *194*, *196*
Stoltenberg, H. A., 157(19), *194*
Stoneburner, G. R., 264(43), *282*
Stover, 276
Stressmann, E., 200(14,15), *230*
Struck, J. H., 243(68), 256(68), 263(53), *283*, *284*
Stull, A., 203(40), *231*
Stunkard, C. B., 264(43), *282*
Sullivan, W. T., 190(67), *196*
Surgeon General (U.S.), 267(20), *281*
Sutterfield, F. D., 268(107), *286*, 307(27), 308(27,) *335*
Swan, H. W., 246(37), *282*
Sweany, H. C., 228(186), *235*
Sweeney, M. P., 253(57), *283*
Sweet, A. H., *281*
Swisher, R. D., 162, 163, *194*

Tateno, K., 222(136), *234*
Taylor, A. W., 16(16), 18(16), *23*
Taylor, C. F., 254(90), 259(90), *285*
Taylor, E. S., 254(90), 259(90), *285*
Taylor, G., 229(200), *236*
Taylor, O. C., 116(66), *118*, 119(2,3), 120 (5), 123(3), 131(3), 138(25), 141(3), *145*, *146*

Taylor, W. C., 322, *336*
Tees, E. C., 217(108), *233*
Terry, L. L., 174(54), *195*
Texaco Co., 276, 278
Theophile, C., 228(190), *235*
Thomas, J. M., 247(58), 248(58), 252 (58), *283*
Thomas, W. J., 247(58), 248(58), 252 (58), *283*
Thommen, A. A., 206, 208(67), *232*
Thompson, C. E., 275
Thompson, H. J., 222(137), *234*
Thompson, W. B., 261(91), *285*
Thomson, W. W., 333(100), *337*
Thrush, B. A., 83, *117*, 322, 330(79,94), 331, *336*, *337*
Tingey, D. T., 139, *146*
Tips, R. L., 206, *231*
Tobias, G. S., 264(43), *282*
Toda, T., 256(92), 257(92), *285*
Tom, M., 268(9), *280*
Toshisasa, K., 222(136), *234*
Towey, J. W., 228(186), *235*
Toyoda, S., 256(92), 257(92), *285*
Trautz, M., 82, *117*
Treacy, J. C., 82, *117*
Tribble, P. E., 260(100), *285*
Trozzolo, A. M., 325(86), 333, *336*, *337*
Tsu, K., 239(39), 238(39), 247(39), 248 (39), 250(39), 253(39), 268(39), *282*
Tucker, K. B. E., 167(44a,44b), *195*
Tuesday, C. S., 82, 83, 96, *117*, 124, *146*, 305, *335*
Tufnell, P. G., 229(192), *235*
Tunney, J., 25
Tyldesley, J. B., 213(97), *232*

Uhren, L. J., 171(48), *195*
United States Congress, 67(12), *71*, 238 (29), *281*
United States Department of Commerce, 238(24), 239(24), 242(24), 253 (24), 261(24), 266(24), 267(24), *281*, 308(30), *335*
United States Department of Health, Education and Welfare, 12, *23*, 56, *71*, 238(27,28), 240(27), 243(25), 268(25, 26), 270(28), 271(27), *281*, 295(8), 297(10,11), *334*
United States Department of the Interior, 179(58), *195*
United States Public Health Service, 44, *70*, 267(7), 268(12), *280*, 300(13), *334*
Universal Oil Products, 276–279
Urbach, H. B., 83(40), *118*

Vallance-Jones, A., 311(43), 323, *335*, *336*
Van Arsdel, P. P., Jr., 204(51), 206, *231*
Vander Veer, A., 202, 204(28), *231*
Vannier, W. E., 217(110), *233*
Van Santen, P. J., 333, *337*
Varekamp, H., 217(114), *233*
Vaughan, W. T., 205, *231*
Verhoek, F. H., 95(52), *118*
Versie, R., 217(113), *233*
Vigliani, E. G., 217(112), *233*
Vignes, A. J., 228(184), *235*
Voges, C. H., 256(66), *283*
Volonte, A. F., 152(7), *193*
Volpi, G. G., 83, *117*
Von Pirquet, C., 198, *230*
Voorheis, W. J., 243(68), 256(68), *284*
Voorhorst, R., 217(114), *233*

Wade, R. W., 239(121), *287*
Wagner, H. C., 218(116), *233*
Waldbott, G. L., 221(126), *233*
Wallace, L., 311(46), *335*
Wallin, O. W., Jr., 260(71), 264(71), *284*
Walton, G., 157(19), *194*
Walzer, M., 203(38), 220(122), *231*, *233*
Wan, J. K. S., 293, *334*
Wark, D. Q., 311(45), *335*
Warr, B. R., 225(164), *234*
Warren, A. J., 262(74), *284*
Wasserman, E., 320, 325(76), *336*
Wasserman, H. H., 315(60), *336*
Waters, W. A., 315(58), *336*
Watrous, R. M., 222(141), *234*
Wayne, R. P., 309, 311(44), 312(84), 313, 314(44), 320, 322, 325, 327(33), 328 (33), 329, 330(33,94,95), 331(33,98), 332(33), *335–337*
Weaver, E. E., 246(59), 248(59), 249(59), 254(59), 268(59), *283*
Weaver, P. J., 148(1), 157(18,20,23,24), 158(26), 164, 171(49), *193–195*

Webster, M. E., 200(16), *230*
Weible, S. R., 53
Weidner, R. B., 53
Weigle, W. O., 199(6), *230*
Weill, H., 228(184), 229(198), *235*
Weinberger, L. W., 153(9), *193*
Weiner, A. S., 204(46,49), *231*
Weinhold, A. R., 214(99), *232*
Welling, C. E., 246(33), *282*
Wentworth, J. T., 260(70,93), 261(93), *284*, *285*
Wenzel, F., 228(187), *235*
Westenberg, A. A., 83, *117*
Westveer, J. A., 267(131), *287*
Wexler, S., 315, 318(69), 323(57,69), *336*
Whitlock, R. F., 311(59), 315(59), 324 (59), *336*
Whittle, E., 316, 321(77), 326(77), *336*
Wicklund, P. E., 217(111), *233*
Williams, D. A., 208, 221(128), *232*, *233*
Williams, J. V., 227(179), 228(179), *235*
Wilson, D., 268(106), *286*, 307(27), 308 (27), *335*
Wilson, J. N., 245(54), *283*
Wimmer, D. G., 264(103), *286*
Windle Taylor, E., 193(69), *196*
Winer, A. M., 309(49), 312, 320, *335*
Winkler, G. L., 256(94), *285*
Winneberger, J. H., 174, *195*
Winslow, F. H., 333, *337*
Witting, H. J., 205(60), *231*
Wnuk, R. J., 83(40), *118*
Wodehouse, R. P., 223(149), *234*
Wojtowicz, J. A., 83(40), *118*
Wolf, H. W., 223(155), *234*
Woodward, R. L., 156, 157(19), 190(68), *193*, *194*, *196*
World Health Organization, 7(18), *23*
Wraith, D. G., 228(189), *235*
Wright, A. C., 311(59), 315(59), 324(59), *336*
Wright, G. L. T., 219(118), *233*

Yager, W. A., 320(76), 325(76), *336*
Yamaguchi, I., 229(202), *236*
Yamamoto, I., 229(202), *236*
Yarrington, R. M., 246(60), 248(60), 249(60), 254(60), *283*
Young, R. A., 314, 331, *335*, *337*
Yugami, S., 222(136), *234*

Zackrisson, F. E. S., 256(80), 258(80), *284*
Zaslowsky, J. A., 83(40), *118*
Zeman, H., 203(36), *231*
Zieve, I., 204(49), *231*
Zimmer, E. E., 267(119), 279, *287*
Zimmerschied, W. J., 159(30), *194*
Ziskind, M. M., 229(198), *235*
Zoberi, M. H., 213(95), 216(95), *232*
Zvonow, V. A., 260(86), *285*

Subject Index

Abatement, 31
ABS biodegradability, 158
ABS levels in river and streams, 156
ABS toxicity, 156
Accidental contamination, 19
Acid mine drainage, 17
Actinometry, 98
Activated sludge process, 48, 153
Aeroallergens, 198, 210
 in atopic disease, 216
 in nonatopic diseases, 227
 as particulates, 210
Air, 7
 contaminants, 11
 fuel (A/F) mixture, 254
 pollutants, 237
 pollution, 21, 25
Air Quality Act of 1967, 32
Air resource of the earth, 7
Air sampling methods, 223
Algal growth, 16, 43
Alkaline complex phosphates, 147
Alkaline hydrolysis of peroxyacetyl nitrate, 143
Alkyl benzene sulfonate (ABS), 149, 154
Allergenic exposure, 207
Allergens, 199
Allergic contact dermatitis, 229
Allergic mechanisms, 198
Allergic populations, 202
Allergy, in nonatopic populations, 209
 a unified concept, 198
American Water Works Assoc., 56, 155
Analyses of crude house dust fractions, 217
Analysis of the PANs, 137
Animal emanations, 221
Animal wastes, 18
Annual soap and synthetic detergent sales, 150
Antibodies in hay fever and asthma, 200
Antibody-dependent mechanisms, 199
Antigens, 199
Aromatic oxidation, 163
Atmospheric contaminants, 11
Atmospheric dilution,
Atmospheric ozone levels, 113
Atmospheric particulate matter, 11
Atomic Energy Commission, 34
Atop, 202
Atopic group, 202
Auto exhaust, 237
 emission standards, 265

Bacterial contamination, 12
Basic formulation, 147
Behavior of NO_2 during photolysis, 113
Beta-oxidation, 162
Biochemical oxygen demand, 13, 153
Biodegradability, 14, 22, 166
 testing, 163
Biodegradable, detergents, 147
 insecticides, 16
 products, 63
 surfactants, 159, 166
Biofiltration, 50
BOD bond, 153
Books and documents, 62

California requirements, 269
Carburetor appurtenances, 256
Catalyst granules, 244
Catalysts, 249
Catalyst technology, 244
Catalytic devices, 239
Catalytic device technology, 250
Catalytic oxidation of HC, CO and H_2, 246, 247
Catalytic removal, 237
CCE concentration, 63
Cell-dependent mechanisms, 201
Changes to the carburetor, 258

Chemical carcinogens, 14
Chemical contamination, 5
Chlorination, 20
Chloroform-soluble carbon filter extract method, 63
Clear Rivers Restoration Act of 1966, 61
Coal production, 5
Coliform bacterial density, 43
Coliform organisms, 66
Components, 260
Concentrations, 139
Construction requirements for water utilities, 58
Continuous activated sludge, 165
Control, of automotive air pollution, 238
 of urban typhoid fever, 42
Controlled fluoridation, 57
Costs of extending water and sewerage services, 54
Council of Ecological Advisers, 39
Cracked wax alpha-olefin, 161
Cylinder flame-front temperature, 260

Damage, 10
 to agricultural crops, 119
Deactivation of exhaust catalysts, 238
Degraded, 162
Degree of sewage treatment, 43
Depletion of dissolved oxygen, 45
Department of Agriculture, 34
Department of Commerce, 33
Department of Defense, 34
Department of Health, Education and Welfare, 33
Department of Housing and Urban Development, 34
Department of the Interior, 34
Detergent, 147, 149
Determination of CO and HC Levels in Engine Exhaust, 271
Development and demonstration, 28
Dilution ratio, 43
Discard costs, 26
Diseases of unknown cause, 229
Disinfection, 48
Dissolved oxygen, 46, 66
Distribution of nutrient sources, 66
Dollar value of water, 69
Domestic heating and air conditioning arrangements, 212
Drinking water, 20
 standards, 44

Early findings, 155
Early history, 154
Ecological Advisers Act of 1967, 35
Ecology, 19
Economics of environmental quality, 30
Effect of detergent residue on water-softening, ion-exchange resins and waste treatment processes, 157
Effect of rain upon particulate concentrations, 216
Elemental analysis, 140
Emission levels of CO, HC and NO_x, 240
Emission standards for specific pollutant sources, 267
Engine-device systems, 261
Enteric viruses, 65
Environmental contamination, 2
Environmental disease, 42
Environmental sciences, 1
Estimated emissions of future control systems, 263
Estimated total U.S. emissions of various pollutants in millions of tons per year, 266
European situation, 192
Eutrophication, 12, 16, 43, 66
Evaporative emissions, 264
Exhaust control devices, 269
Exhaust hydrocarbons, 260
Explosions, 141
Eye, irritants, 74
 irritation, 119

Familial aspect of atopy, 204
Farmer's lung, 227
Farmer's lung group, 227
Fate of bacteria in polluted waters, 46
Fatty acids, 162
Federal, state and local roles, 27
Federal laws, 152
Federal participation in water supply construction, 58
Federal role, 25

Federal Water Pollution Control Act, 54
Federal Water Pollution Control Administration, 60, 151
Fertilizers, 16
Field studies, 167
  at sewage treatment plants, 168
Filter design and operation, 57
Filtration, 63
Fish kill, 45, 157
Fish toxicity, 157
Flame front, 259
Flat catalyst bed, 250
Fluoridation, 20
Foam, 154
40-m Folded path cell of an infrared spectrophotometer, 100
Food, 21
  quality, 21
Formation and destruction of NO_3, 114
Formation of Nitrate ion, 144
Formation of peroxyacyl nitrates, 121
Free chlorine in heavily polluted water, 65
Free chlorine residual, 64
Free residual chlorination, 57
Fuel technology related to emission control, 263
Funding, 27
Fungi, 220

Gas chromatography, 137
Gaseous, 11
Gas flow through the bed, 250
Gasoline engine technology related to emission control, 254
Gastroenteritis and diarrhea, 64
General environment, 2
Global pollution, 22
Grain dusts, 222
Grass pollens, 219
Gravity slide or Durham sampler, 224
Growth, 149

Half-lives, 18
Health, effects, 63
  -related sections, 44
  relationships, 30
Herbaceous weeds, 219
Histamine, 200
History of catalytic devices, 241
House dust, 217
Household wells, 52
Housing boom, 51
Human population, 2
Hydrocarbons, 267
Hydrolysis, 139, 142

Immediate environment, 2
Importance of NO_2, 74
Individual hydrocarbons, 268
Industrial effluents, 20
Industrialization, 4
Infrared, 137
Infrared absorptivities (a) of the peroxyacyl nitrates, 133
Infrared spectra, 130
Infrared spectrum of peroxypropionyl nitrate, 131, 132
Iodine, 65
Insect emanations, 221
Instruments and techniques, 77
Interaction with hydrocarbons, 106
Internal environment, 2
Irrigation, 59

Kinetics, 122, 247

LAS, in sewage treatment, performance, 168
  removal, 168, 187
Lead, 11
Lead deactivation of exhaust catalysts, 248
Lead traps, 252
Level of NO_x in exhaust, 260
Life of catalyst in exhaust devices, 246
Light absorbed by N_2O_4, 95
Lightoff temperature for particular catalyst, 247
Linear alkylate sulfonate (LAS), 160
Located stations, 223
London type smog, 10
Long-distance transport of potentially allergenic particles, 214
Long-path infrared spectrophotometer, 101

Los Angeles type smog, 10
Low molecule weight combustion products, 260
Low quality uses, 46

Management, 41
Manufacture ABS, 149
Mass spectra of PAN homologs, 135
Mechanism, 114, 121
Mechanisms of biodegradation, 162
Mediators of tissue reactions, 200
Methemoglobinemia, 16
Methods of preparation, 123
Methyl oxidation, 163
Methylene blue technique, 155
Microchemical pollutants, 12, 13
Miscellaneous allergens, 221
Molecular sieves, 159
Muffler acid, 251
Multiple use of our water resources, 47

Nature and sources, 210
Nitrates, 16
Nitrogen dioxide, 98
  photolysis, 77
    in the presence of other species, 102
Nitrogen and phosphorus in natural water, 66
NO, 116
  level characteristic of cruising auto exhaust, 75
NO_2, as a gas phase actinometer, 98
  /NO ratio in the atmosphere, 74
  photolysis, 114
  toxicity, 116
N_2O_4 absorption, 95
Non-health related aspects of water quality, 56
Nonmonochromatic light source, 109
Nonuniform light intensities, 109
n-paraffins, 159
Nuclear magnetic resonance, 136
  proton lines, 136
Nutrient-removal processes, 67

Otto cycle, 258
Oxidation of CO and HC, 246
Oxides of nitrogen, 73
Oxygen atom concentration, 88
  during photolysis of NO_2, 80
Oxygen atoms, with hydrocarbons, 104
  with nitric oxide, 83
  with nitrogen dioxide, 84
  with oxygen, 83
Oxygen content of peroxyacetyl nitrate, 141
Outline of environmental sciences, 1
Overall quantum yields, 89
Over fertilization, 66

PAN concentrations, 138
Particulate components, 11
Peroxyacetyl nitrate, 111, 120
Peroxyacyl nitrate, 111, 119
Peroxybenzoyl nitrates, 120
Peroxybutyryl nitrate, 120
Peroxyisobutyryl nitrate, 120
Peroxypropionyl nitrate, 120
Persistence, of PAN, 138
  of pesticides, 15
  of radionuclides, 19
Persistent chemicals, 62
Pesticides, 15
Petroleum production, 5
Phenylmercuric propionate, 222
Phosphate fertilizers, 16
Phosphorus, 16
Photochemical air pollution, 8
Photochemical reactivity, 268
Photochemical smog, 10, 25, 241
  peaks, 10
Phytotoxicants, 74
Plasma kinins, 200
Pollens, 218
Pollinosis and pollen counts, 226
Polluted air masses, 74
Pollution, 26
  control, 25, 41
  -related research and development, 27
  research, 28
  and self-purification of streams, 46
Population, 3
  of the continental United States, 4
  living in urban areas, 5
Porosity of catalysts, 244
Powdered activated arbon, 63
Preparation of peroxyacetyl nitrate, 125

Present, projected and target engineer exhaust emissions, 274
Pressure drop through the catalyst, 251
Prevalence, of atopic allergy, 205
of symptoms, 207
Production, of solid wastes, 17
of synthetic organic chemicals, 6, 7
Progress, 48
Promoter, 244
Properties, 198
Public health, 20, 25
Public Health Service, 33
Public sewerage, 53
Public water supply, 55
Purification, 127

Quality standards of water for shellfish growing, 44
Quantitative limitations of the gravity slide sampler, 224
Quantum yield of NO_2 photolysis, 91

Radial exhaust flow device, 250
Radionuclides, 18, 19
Ragweeds, 219
Ragweed pollen, 207, 208
Rate, of atmospheric reaeration, 46
of biodegradability, 22
of deoxygenation, 46
of NO_2 formation during atmospheric dilution, 76
Rate constant ratios, 84, 88, 89, 96
Rate constants, 96
Reaction of nitric oxide with oxygen, 81
Reactions in the fractional pphm range, 77
Reactions of the PANs, 141
Recycle costs, 26
Redesign of combustion chamber, 261
Reduced the amount of hydrocarbons emitted by the average, 238
Reduction of NO emission levels, 260
Relatively short range transport, 215
Removal of NO_x, 250
Resource conservation, 26
River die-away, 164

Salinity of soils, 17
Sand filter plant, 48
Sanitary reform movement, 41
Selectivity of catalysts, 249
Semicontinuous activated,
Sensitive lymphoid cells, 201
Separation of *n*-paraffins, 159, 160
Septic tanks, cesspools, 168
Septic tank systems, 52
Sequence of steps in a catalyzed reaction, 247
Sewage, 14
Sewage-borne detergent residues, 157
Shake flask method, 164
Simple chemical substances, 222
Singlet oxygen, 115
Sludge, 164
Sludge disposal, 62
Smog, 10
SO_2 oxidation, 249
Soap, 147
Soap and Detergent Assoc., 154
Soil, 14
Soil persistence of pesticides, 15
Sources of aeroallergens, 210
Spectra, 130
Spectrum of acetyl nitrate, 130
SRS-A, 200
Stability, 141
Stratosphere, 7
Stratospheric pollution, 8
Stream pollution, 151
Strip mining, 17
Strontium 90, 18
Studies on individual household disposal units, 173
Subcommittee on Science, Research and Development, to the Committee on Science and Astronautics, 38
Summary of principal field test results, 170
Supersonic transport aircraft, 8
Surface-active agents, 147
Swimming, in polluted waters, 45
Symptoms of nasal allergy, 201
Synthesis and purification of the PANs, 123
Synthetic detergent, 13
Synthetic organic chemicals, 13

Taste and odors, 157
Temperature inversions, 8

Temperature potential, 253
Terminal disinfection of sewage effluents, 65
Terminal setting velocities, 214
Thermal degradation of typical catalyst supports, 245
Thermal pollution, 14, 59
Third surge of effort on catalytic devices, 243
Toluene diisocyanate, 223
Total U.S. investment in water and sewerage facilities, 54
Toxicity to fish, 63
Toxicological aspects, 156
Transmission of virus, 64
Transport of aeroallergens, 212
  processes in closed spaces, 212
  processes out-of-doors, 212
Tree pollens, 218
Tropopause, 7
Troposphere, 7
Two-step procedure, 165
Types of allergic response, 199
  of smog, 8
Typical chromatogram, 129

Ultimate economic development of water supply, 67
Ultraviolet spectrum, 134
  of peroxyacetyl nitrate, 134
Unit processes employed in water treatment, 57
Urban epidemics, 41
Urbanization, 3
U.S. Public Health Service Drinking Water standards, 156

Viral inactivation, 65
Viral infectious hepatitis, 64
Virus, inactivation, 65
Virus inactivity, 65

Waste, 17
  oxidation ponds, 67
  water treatment, 59
Water, 12, 41
  conversion, 69
  pollution, 21, 147, 151
    control, 62
  quality, 12
    standards, 66
  sanitation practice, 44
  shortages, 56
  supplies, 25
  treatment practice, 63
Water-borne disease, 52, 64
  infectious hepatitis, 65
  viral tansmission, 65
Water Pollution Control Act of 1948, 52
Water Quality Act, 31
  of 1965, 60
Water resource management, 47
Water Resources Planning Act of 1965, 70
Water and waste water treatments, 48
Wind, as a factor in human exposure to allergenic particles, 215
World population, 3
World production of fossil fuels, 6

Ziegler polymerization of ethylene, 161